TANKER OPERATIONS

The fourth "State Class," double-hull tanker built at the NASSCO yard undergoing sea trials. Courtesy Ken Wright, General Dynamics NASSCO staff photographer.

TANKER OPERATIONS

*A Handbook
for the
Person-In-Charge (PIC)*

FIFTH EDITION

Mark E. Huber

Based on earlier editions of
*Tanker Operations:
A Handbook for the Ship's Officer*
by G. S. Marton

CORNELL MARITIME PRESS
A Division of Schiffer Publishing, Ltd.

With love to my wife Jody

Type set in New Baskerville BT/Aldine 721 BT

ISBN: 978-0-87033-620-1
Printed in China

Published by Schiffer Publishing, Ltd.
4880 Lower Valley Road
Atglen, PA 19310
Phone: (610) 593-1777; Fax: (610) 593-2002
E-mail: Info@schifferbooks.com
Web: www.schifferbooks.com

For our complete selection of fine books on this and related subjects, please visit our website at www.schifferbooks.com. You may also write for a free catalog.

Schiffer Publishing's titles are available at special discounts for bulk purchases for sales promotions or premiums. Special editions, including personalized covers, corporate imprints, and excerpts, can be created in large quantities for special needs. For more information, contact the publisher.

We are always looking for people to write books on new and related subjects. If you have an idea for a book, please contact us at proposals@schifferbooks.com.

Contents

vi

Preface to the Fifth Edition

The fifth edition contains a wealth of new and updated material as well as many new photos and graphics to visually support the text. I am also happy to announce that several new contributors have joined the ranks of the talented group of individuals from the fourth edition to lend us their expertise in this subject. It is particularly gratifying to be able to call on these professionals who willingly give of their time to bring you the very latest information in the industry. Readers of TANKER OPERATIONS will note the addition of new chapters on Chemical Tankers, by Margaret Kaigh Doyle, and Cargo Calculation, by Kelly Curtin and Kevin Duschenchuk, each a recognized expert in the field. They are joining the contributors from the fourth edition---Richard Beadon, Scott Bergeron and John O'Connor, each of whom is professionally in a class of their own.

The chapter review questions have been enhanced with the addition of challenge questions which are meant to provoke thought and additional research on the part of the reader. As noted in the preface of the fourth edition it has never been more apparent to me that TANKER OPERATIONS is—and will always be—a work in progress. The tanker industry continues to evolve with each passing day in vessel design, equipment, regulation and operational procedures. The public continues to press tanker operators to improve performance in the area of protecting the marine environment while at the same time customers expect competitive rates and efficiencies in the transportation of these cargoes.

As previously stated, this text is primarily directed at individuals entering the workforce in the tanker industry. However, seasoned tanker mates, barge tankermen, and many of the shoreside staff may find the information in this edition of practical value.

The rules governing the qualifications of personnel serving on tank vessels continue to change particularly on the international level through the IMO. In addition to obtaining practical sea experience, individuals serving on tank vessels must complete an approved training program in cargo handling and fire fighting. As of this writing, the IMO is moving forward with defining in a more detailed way the topics that should be addressed and experience necessary in the qualification process toward receiving an endorsement on the STCW certificate to serve on oil and chemical tankers. With these requirements in mind, it seemed appropriate for the fifth edition to be revised with the goal of becoming the standard reference for this specialized cargo training. Successful completion of the aforementioned requirements qualifies a person to receive an endorsement on the following documents:

In the United States, the individual receives an endorsement on the Merchant Mariner Credential (MMC) as a Tankerman Person-in-Charge (PIC) Dangerous Liquids (DL) or Liquefied Gas (LG). Under the International Convention on Standards of Training, Certification, and Watchkeeping for Seafarers, 1978 (STCW), as amended in 1995, an individual receives an endorsement on the STCW certificate. This endorsement states that the person is qualified for service on tankships carrying dangerous oils, chemicals, or gas in bulk. The following references should be consulted for details concerning each of these endorsements: the United States *Code of Federal Regulations*, Title 46 CFR Part 13, Certification of Tankermen; and the International Maritime Organization's STCW Convention, 1978, as amended in 1995, Annex 2, Chapter V, "Special Training Requirements for Personnel on Tankers" (Regulation V/1).

There are many individuals I would like to acknowledge for their assistance with this revision. First and foremost, I am eternally grateful to my wife Jody and family for their patience and support in this project and the endless task of trying to stay abreast of this constantly changing industry. This text would not be possible without the support of the following individuals that have exhibited patience above and beyond the call of duty with my constant requests for photos, drawings and information. In particular, I would like to thank Capt. George McShea, Karen Davis, and Andrew Smith of Polar Tankers, John O'Connor of International Marine Consultants, Knut Kaupang with AIR PRODUCTS AS, Mike Newton of Herbert Engineering Software Solutions, Chris Deschenes and Mike Blunt of OSG, Don Sherwood, John Quagliano, Mark Homeyer, Capt. Vic Goldberg, Stacy DeLoach, Kevin Schroeder and Larry Miles of Crowley Petroleum Services. I also wish to thank my colleagues at the U.S. Merchant Marine Academy: Captain George Edenfield, Captain John Hanus, Captain Douglas A. Hard, Captain Tim Tisch, Cdr. Rob Smith (USCG), Paul Zerafa, and Brian Holden. Finally, I would like to thank the following individuals and organizations for providing information and many of the illustrations:

Air Products AS; Alaska Tanker Company; American Petroleum Institute; American Waterway Operators; Atlantic Richfield Company; Avondale Shipyard; Mary Jen Beach; Ian-Conrad Bergan, Inc.; Bethlehem Steel Corporation; BP Pipelines (Alaska) Inc.; British Petroleum Company, Ltd.; Butterworth Systems, Inc; Calhoon MEBA Engineering School; California Maritime Academy; Chevron Shipping Company; Clement Engineering Services; College of Nautical Studies; Coppus Engineering Corporation; Crowley Petroleum Services; Design Assistance CorporationDAC; Dixon Valve and Coupling Co.; Environmental Protection Agency; Dresser Inc; Exxon/Mobil Corporation; Thomas J. Felleisen; Bill Finhandler; Foster Wheeler Boiler Corporation; Gamlen Chemical Company; General Dynamics Corporation; Keith Gill; Global Maritime and Transportation School; Gulf Oil Corporation; Eric Halbeck; Brian Hall; Hamworthy Moss AS; Haywood Manufacturing Company; Herbert Engineering Software; Howden Engineering; Lynn Huber; Hudson Engineering Company; IMO Industries, Inc., Gems Sensors Division; Ingersoll Dresser Pump; International Association of Classification Societies; International Association of Independent Tanker Owners (INTERTANKO); International Association of Ports and Harbors (IAPH); International Chamber of Shipping (ICS); International Marine Consultants; International Maritime Organization (IMO); International Paint; International Tanker Owners Pollution Federation; Keystone Shipping Company; Keystone Valve Division of Keystone International, Inc.; Lee Kincaid; Kockumation AB; Laurin Maritime; Brian Law; Library of Congress Photo Duplication Service; Kimberly Lorenzo; Louisiana Offshore Oil Port (LOOP); Eric Ma; Maine Maritime Academy; *Marine Log;* Maritime Institute of Technology and Graduate Studies; Captain John Mazza; MEDAL/Air Liquide Advanced Technologies; Metritape, Inc.; Steven Miller; Mine Safety Appliances Company; MMC International Corporation; Frank Mohn AS (FRAMO); Ron Monel; National Academy Press and the National Academy of Sciences; National Audubon Society; National Fire Protection Association; National Geographic Society; National Maritime Union of America; National Research Council; National Steel and Shipbuilding Company; National Transportation Safety Board; Nautical Institute; Newport News Shipbuilding; John O'Connor; Oil Companies International Marine Forum (OCIMF); OSG America; Penn-Attransco Corporation; Permea Maritime Protection; Phillips Petroleum Company; Polar Tankers; George Rozanovich; Saab Electronics; Saab-Scania, Aerospace Division; Sailors Union of the Pacific; Salen & Wicander AB; Salwico, Inc.; San Francisco Maritime Museum; E.W. Saybolt & Company, Inc.; Ed Schultz; Seafarers International Union; Seamen's Church Institute; SeaRiver Maritime; Servomex (U.K.), Ltd.; Shell International Petroleum and Shell Oil Company (U.S.A.); Shipbuilders Council of America; Skarpenord Data Systems AS; Southern Oregon State College; Sperry Marine Systems; Stacey Valve Co., Inc.; Star Enterprise; State University of New York Maritime College; Stolt Nielsen Transportation Company; Sun Shipbuilding and Dry Dock Company; Texaco, Inc.; Tosco; Transamerica Delaval, Inc.; TS Tanksystem SA; Underwriters Laboratories, Inc.; U.S. Coast Guard; U.S. Department of Transportation; U.S. Hose Corporation; U.S. Maritime Administration; U.S. Salvage Association; Valve Manufacturers Association; Viatran Corporation; Vitronics, Inc.; Rosalie Vitale; West Coast Ship Chandlers, Inc.; Terra White; Jeff Williams; William E. Williams Valve Corp.; Wilson Walton International; Worthington Pumps.

An earlier edition of *Tanker Operations* offered these words of advice to the reader: "You can't learn tankers from a book; don't try to do so. Ships are designed and equipped differently, and no two are exactly alike. In the end, there is no substitute for seeing the actual equipment and operating it yourself." As a follow-up to that thought, I would add that no text on this subject can adequately address every vessel design, piece of equipment, or procedure. Ultimately, a thorough working knowledge of the particular cargo and ballast system on your vessel is the best defense against potential mishaps. Remember, the specialized training and practical experience gained as an apprentice on tankers is just the beginning of a lifetime of learning.

M. E. Huber

Preface to the First Edition

A number of years ago, when I was beginning my career on oil tankers, I often felt the lack of a simple, straightforward handbook on the basic problems of tanker operations. Hence, this book. Tanker Operations: A Handbook for the Ship's Officer is directed primarily toward the newcomer to tankers; specifically, the new officer. Generally speaking, it is not a step-by-step manual covering every possible situation. Instead, it is intended as:

1. An introductory guide designed to make the new officer's adjustment to tanker life smoother, less perilous.

2. A source of useful information for the more experienced officer.

3. A reference book for other individuals interested in the operation of oil tankers, particularly those aspiring to the rating of tankerman.

I should point out, however, that tankers cannot be learned entirely from a book. The tankerman's job is too complex and, in many ways, intuitive. Moreover, each tanker is unique and must be learned individually.

Fortunately, the learning process is not an entirely lonely task. Shipmates—pumpmen, fellow officers, sailors—have knowledge to share, and some make excellent teachers. In the end, however, the way to learn a tanker is to put on a boiler suit and, flashlight in hand, explore every corner of the vessel, learning pumproom, piping systems, valves. This is a tedious, sometimes exhausting process, but it must be done. An officer unwilling to make this effort should forget about a career, even a brief one, on tankers.

Some tankers, old and rusty, are relics of a bygone era. Others are so futuristic, so thoroughly automated, that their crewmembers feel more like astronauts than tankermen. And, in all likelihood, the future tankerman will need the training and temperament of an astronaut.

Regardless of age or equipment, however, all tankers perform the same basic task—they carry oil. Their voyages span the globe, from the blazing deserts of Saudi Arabia to the frozen shores of the Arctic. Through it all, tankermen are accompanied by the pungent smells of crude oil and gasoline, by loneliness, tension, exhaustion . . . and the satisfaction of doing a job well. No individual can adequately describe this unique way of life. It must be experienced firsthand.

I would like to take this opportunity to thank the many individuals and organizations who were kind enough to help me in this effort. Some showed remarkable patience with my repeated requests for information, research materials, and illustrations.

Special thanks to: The American Bureau of Shipping; American Cast Iron Pipe Company; American Institute of Marine Underwriters; American Institute of Merchant Shipping; American Petroleum Institute; the Ansul Company; Apex Marine Corporation; Atlantic Richfield Company; the Scott Aviation Division of ATO, Inc.; Mrs. Gerry Bayless; Bethlehem Steel Corporation; Bingham-Willamette Company; British Petroleum Company, Ltd.; Henry Browne & Son, Ltd.; Butterworth Systems, Inc.; Chevron Shipping Company; Coppus Engineering Corporation; Exxon Corporation and Exxon Company (U.S.A.); FMC Corporation; Mr. Steve Faulkner; Mr. Bill Finhandler; Gamlen Chemical Company; General Dynamics Corporation; General Fire Extinguisher Corporation; Mr. R.W. Gorman; Gulf Oil Corporation; Mr. Arthur Handt; Hendy International Company; the Penco Division of the Hudson Engineering Company; Mr. John Hunter; Huntington Alloys, Inc.; the Keystone Valve Division of Keystone International, Inc.; Kockums Automation AB; Mr. Gene D. Legler; the Harry Lundeberg School; Mine Safety Appliances Company; Mr. C. Bradford Mitchell; National Audubon Society; National Foam System, Inc.; National Maritime Union of America; National Steel and Shipbuilding Company; Miss Maureen Ott; the Ralph M. Parsons Company; Paul-Munroe Hydraulics, Inc.; Mrs. Pia Philipp; Phillips Petroleum Company; Sailors' Union of the Pacific; Salen & Wicander AB; San Francisco Maritime Museum; E.W. Saybolt & Company, Inc.; Mr. W.F. Schill;

Seafarers International Union; Shell International Petroleum and Shell Oil Company (U.S.A.); Shipbuilders Council of America; Sperry Marine Systems; Sun Shipbuilding and Dry Dock Company; Mr. Bob Sutherland; Underwriters Laboratories, Inc.; United States Coast Guard; United States Maritime Administration; U.S. Salvage Association; Valve Manufacturers Association; West Coast Ship Chandlers, Inc.; Worthington Pump Corporation.

G. S. MARTON

G. S. Marton graduated from the California Maritime Academy in 1969. During his seagoing career, he served on all types of merchant ships, including tankers of all types and sizes.

TANKER OPERATIONS

CHAPTER 1

Tank Vessel Design and Classification

The first tanker appeared over a century ago, and since that time the movement of liquid cargoes by tank vessel has evolved into one of the most efficient global modes of transportation. Seagoing tankers represent some of the largest and most technologically advanced man made vehicles that ply the oceans of the world. Modern refinements in the design of these vessels have resulted in the development of a versatile carrier capable of transporting a wide array of bulk liquid cargoes. Today, tank vessels (both ships and barges) are responsible for the movement of tremendous volumes of liquid cargoes. This chapter focuses primarily on vessels that are designed to carry cargoes classified as "dangerous liquids," which encompasses both oils and chemicals in bulk.

The following definitions are provided to eliminate confusion about the types of vessels described in the text. The United States Coast Guard (USCG) defines a tank vessel as "a vessel that is constructed or adapted primarily to carry, or that carries, oil or hazardous material in bulk as cargo or cargo residue." The USCG further categorizes a tank vessel as a tank ship (if it is self-propelled) or a tank barge (if it has no means of propulsion). Throughout the text, efforts have been made to use the term "tank vessel" if the topic applies to both ships and barges.

OIL TANKER

As we move into a new era of tanker design a retrospective look at how we got where we are today is in order. The earliest design of tank vessel involved construction with a single hull. Figure 1-1 shows a cross section of a traditional single-hull design.

In the early part of the twentieth century, the shift toward longitudinal construction resulted in a unique subdivision of the cargo tank area. As seen in Figure 1-1, the use of twin longitudinal bulkheads created a three-tank configuration athwartships in the vessel: a center tank flanked by a set of wing tanks. A series of oil-tight, transverse bulkheads completed the subdivision of the cargo area, as required, creating the total number of tanks necessary for the particular trade of the vessel. This method of construction was well suited for the transportation of bulk liquid cargoes; resulting in a structure that was inherently stable. This design was credited with virtually eliminating the free surface problems experienced in earlier tanker designs.

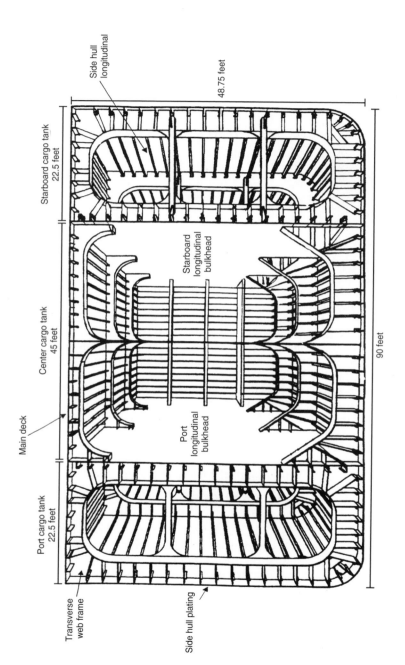

Figure 1-1. Cross-sectional view of a single-hull tanker. The complex internal structure of the tanks made cargo stripping and tank cleaning operations more difficult. Courtesy National Transportation Safety Board.

Figure 1-1a. Courtesy Kevin Duschenchuk

Free surface is an effect created when liquids move about in an unrestricted fashion within a compartment such as a cargo or ballast tank. The resultant shift of weight has an adverse impact on the stability of the vessel, so every effort is made to minimize shifting. Typical methods of reducing the free surface effect include keeping the number of slack cargo and ballast tanks to a minimum, constructing smaller compartments (subdivisions), and utilizing partial bulkheads (swash plates or swash bulkheads). The success of the single-hull design is evidenced by the fact that it had withstood the test of time and deadweight (dwt) tonnage. Single-hull construction predominated until the late 1960s when political and environmental pressures drove the tanker industry to seek other methods of construction. By the 1970s a number of owners had shifted to double-bottom construction (Figure 1-2) to meet the new segregated-ballast requirements.

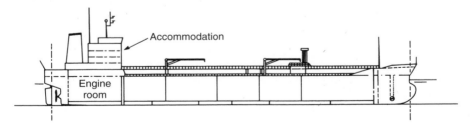

Figure 1-2. Profile view of a double-bottom tanker. The double-bottom space serves as the segregated-ballast capacity for the vessel. Copyright ©International Maritime Organization (IMO), London.

The grounding of the single hull *Exxon Valdez* in 1989 prompted domestic and international requirements calling for newly constructed oil tankers to be fitted with a double hull. Double-hull tankers had been successfully operated for a number of years, hence this design took center stage as the most likely response to the public's outcry for heightened protection of the marine environment. The use of two pieces of steel (inner and outer hulls) to separate the cargo area from the sea is expected to minimize oil outflow from the majority of tanker casualties—grounding, collision, or minor shell damage—that involve a breach of the hull. The construction scenes of the Pelican State at the NASSCO yard (Figure 1-3) clearly illustrate the protection afforded the cargo tanks within the double hull.

The Oil Pollution Act of 1990 called for new tankers contracted after June 30, 1990, to be constructed with a double hull. The U.S. construction requirements are contained in Title 33 CFR Part 157.10d, which specifies minimum spacing between the hulls as follows:

For vessels of 5,000 dwt and above—
Double sides (W)
W = 0.5 + dwt/20,000 or 2 meters the lesser and in no case less than 1 meter

Double bottom (H)
H = Breadth/15 or 2 meters the lesser and in no case less than 1 meter

Figure 1-3. Double Hull tanker under construction.
Courtesy George Edenfield.

For vessels of less than 5,000 dwt—
Double sides (W)
W = 0.4 + (2.4) (dwt/20,000) in meters, but in no case less than 0.76 meter

Double bottom (H)
H = Breadth/15 in meters, but in no case less than 0.76 meter

Figure 1-4 shows the double-hull tanker *American Progress*.

DOUBLE HULLS

The requirement for new oil tankers to be constructed with a double hull grew out of a provision in the Oil Pollution Act of 1990 (OPA '90) however the International Maritime Organization (IMO) adopted similar mandates in 1992. Internationally, the double hull requirement appeared as an amendment to Annex I of MARPOL 73/78 contained in regulation 13F (new regulation 19 in the latest revision that entered into force on 1 January 2007). Further to the change in design for new tankers came the mandatory phase out/conversion of existing single hull tank vessels both under OPA '90 and Annex I MARPOL 73/78 regulation 13G (new regulation 20 under the latest revision that entered into force on 1 January 2007). The mandatory phase out of the existing single hull fleet continues until 2010 and for existing double bottom tankers the phase out runs to 2015. Since the inception of these design changes in the early 1990's a number of marine casualties involving single hull tankers led to calls for the acceleration of the phase out of these vessels. Incidents such as the break up of the ERIKA off the coast of France in 1999, the PRESTIGE off the coast of Spain in 2002 and most recently the holing of the HEBEI SPIRIT off South Korea exemplify the influence such events have in spurring legislative bodies to action.

Industry experts have long debated the effectiveness of the double hull design in high energy groundings and collisions with the potential for significant loss of cargo and resulting environmental impact (Figure 1-4a).

Consequently, the United States Congress and International Maritime Organization (IMO) left the door open to alternative designs and control technologies that offered equivalent or better protection of the marine environment than the double hull. The stakeholders in the transportation industry engaged in a study of these alternatives and in 1993 the United States Coast Guard reaffirmed the double hull design as the only method of construction for new tankers that should be permitted in U.S. waters. This conclusion created international controversy as the IMO embraced two alternative concepts to the double hull namely the mid-deck design and the Coulombi egg design.

Figure 1-4a. Collision and resulting oil spill involving a double hull tanker. Courtesy USCG.

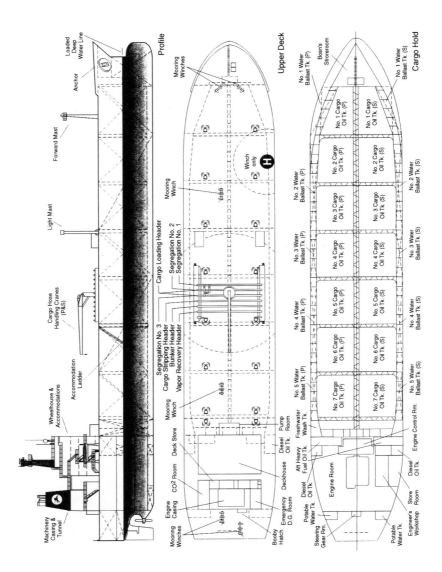

Figure 1-4. The general arrangement drawing for the double-hull tanker *American Progress*. Courtesy Mobil Shipping and Transportation Company.

EXPERIENCE TO DATE WITH DOUBLE HULLS

The most common cargo and ballast tank arrangements in the double hull tankers seen to date are illustrated in Figures 1-4b, 1-4c, and 1-4d. As of this writing, the world tanker fleet is comprised of approximately 3600 tankers of which roughly 2600 are double-hulled.

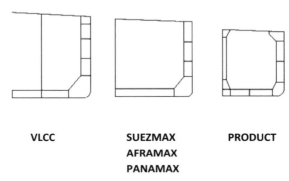

VLCC   SUEZMAX   PRODUCT
AFRAMAX
PANAMAX

Figure 1-4b. Double hull arrangements. Courtesy International Association of Classification Societies (IACS).

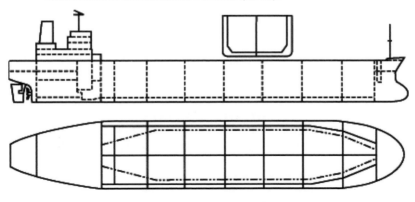

Figure 1-4c. Double hull tanker 150,000 DWT or less.

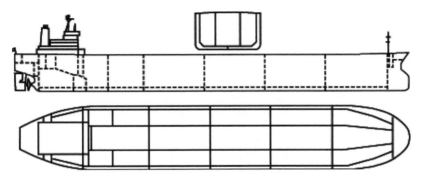

Figure 1-4d. Double hull tanker 150,000 DWT and greater.

As the percentage of double hull vessels operating worldwide has increased, operators have gained greater insight into the various concerns associated with this method of construction such as:

1. Overall quality of construction
2. Accelerated corrosion rates
3. High stress areas leading to fatigue cracks particularly in the inner bottom
4. Proper application and longevity of ballast tank coatings
5. The need for ongoing inspection and maintenance of ballast tank coatings
6. The need to coat designated "storm" ballast (cargo) tanks
7. The need for (hydrocarbon) gas detection in the ballast, pumproom and void spaces
8. Safety concerns associated with ventilation and access for inspection of the double hull
9. The ability to inert the empty ballast space between the hulls
10. Fire safety risks related to empty ballast spaces (air) surrounding cargo tanks
11. Increased construction and maintenance costs
12. Loss of intact stability due to free surface problems on certain double hull designs
13. Damaged stability concerns related to raking damage during a grounding incident

Without a doubt, the structural complexity of the space between the hulls is expected to pose significant challenges to operators over the lifetime of these vessels.

MID-DECK DESIGN

The mid-deck design is an alternative approach to the double hull in which an intermediate oil tight deck essentially creates an upper and lower cargo tank (Figure 1-5).

The basis of this design is a simple concept that involves placing an oiltight 'tween deck at about mid-height in the vessel, which results in the lower tank being located substantially below the waterline of the vessel when fully loaded. In the event of a breach of the bottom tank the outflow of cargo is minimal based on the fact that the external seawater pressure on the hull exceeds the head pressure of the cargo in the lower tank. Therefore, water would enter (press up) the lower cargo tank as opposed to cargo gravitating out the bottom of the vessel to the sea. Additionally, the wider double sides in this design resemble traditional wing tanks potentially providing better protection of the marine environment from side damage to the vessel. Model testing performed by a number of engineering groups worldwide confirmed the viability of this design in minimizing oil outflow from major damage to the hull. These findings ultimately led to the acceptance of the mid- deck design by the international maritime community. In the United States, the Coast Guard cited inexperience with the mid-deck concept as one of the reasons for not recommending it as an alternative to the double hull.

Another design closely related to the mid-deck is the Coulombi Egg, shown in Figure 1-6. After several years of evaluation, IMO has also accepted this design as

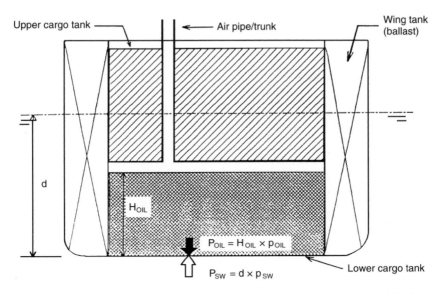

Figure 1-5. The mid-deck design shown here uses hydrostatic pressure to minimize oil outflow in the event that the cargo tanks are breached. Courtesy *Marine Log*.

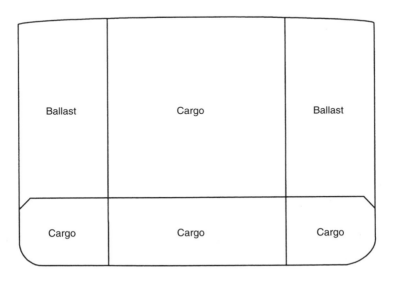

Figure 1-6. The Coulombi Egg design.

affording a measure of protection of the marine environment equivalent to that of the double hull. As in the case of the mid-deck design, however, the United States opposes the idea of equivalence and will not permit either design into U.S. waters. At this point, the controversy appears to be over as evidenced by the worldwide acceptance of double hull construction as the standard for the next generation of tanker.

PARCEL TANKER

As consumer demand for chemicals and other specialty products increased world-wide, the need for vessels designed specifically to transport these cargoes also expanded. As a quick fix, some owners modified existing product carriers into what were termed "drugstore" vessels, carrying limited quantities of many different products. Ultimately, these vessels paved the way for parcel carriers, vessels specially designed and constructed from the keel up to accommodate the growing market. Figure 1-7 shows one such vessel, the *Stolt Innovation*, built and operated by Stolt-Nielsen S.A.

Figure 1-8 shows the deck of an externally framed vessel. This method of construction allows for smooth internal surfaces within the tanks. Due to the nature of the cargoes transported, parcel tankers are designed to maintain a high degree of segregation between cargoes. Figure 1-9 shows the complexity of deck piping on one coastal chemical carrier.

Toward the end of the twentieth century, the demand for parcel tankers increased as the transport of these cargoes by such vessels proved to be safe and cost-effective while maintaining the highest standards of quality assurance. The list of different cargoes carried by parcel tankers is exhaustive; however, the rules governing the safe transport of these cargoes are well defined in the international bulk chemical codes. The construction and survivability requirements for chemical vessels can be found in U.S. regulation Title 46 CFR Part 151 (barges) and Part 153 (ships) as well as in the bulk chemical codes (IBC/BCH) from the International Maritime Organization (IMO). (IBC is the International Code for the Construction and Equipment of Ships Carrying Dangerous Chemicals in Bulk. BCH is the Code for the Construction and Equipment of Ships Carrying Dangerous Chemicals.)

The marine environment is afforded three levels of protection against an uncontrolled release of the cargo resulting from a breach of the cargo tank. Figure 1-10 illustrates the spacing requirements for Types 1, 2, and 3 containment in the cargo area, as specified in the bulk chemical codes.

1. Type 1 containment provides the maximum level of protection possible when transporting substances that pose the greatest environmental risk if an uncontrolled release from the vessel should occur. In addition to the spacing requirements between the side and bottom shown in Figure 1-10, vessels constructed in accordance with these rules must also be capable of surviving a certain prescribed level of damage to the hull.
2. Type 2 containment is required when transporting substances that pose a significant hazard to the environment. The spacing requirements and the survivability requirements of the vessel are less than those for Type I containment.
3. Type 3 containment affords a moderate level of protection. No special spacing requirements are necessary and the survivability criteria in the event of vessel damage are not as stringent as those for Type 1 or 2.

Figure 1-7. The *Stolt Innovation* serves in the parcel trade worldwide. Courtesy Stolt-Nielsen Transportation Group Ltd.

Figure 1-8. External framing on deck. Courtesy Maximillian Paul

To further limit the environmental impact from hazardous cargoes classified as Type 1 and Type 2, parcel tankers are governed by a limit on the quantity of cargo that can be carried in any one tank as follows:

1. Type 1—maximum allowable cargo quantity transported in any one cargo tank shall not exceed 1250 m3 (7.862 bbls).
2. Type 2—maximum allowable cargo quantity transported in any one cargo tank shall not exceed 3000 m3 (18,869 bbls).
3. Type 3—no limit on the quantity of cargo transported in any one tank.

The chemical codes further classify cargo tanks according to their construction.

Independent tanks are cargo tanks not designed as a part of the hull structure. An example of an independent tank would be a cylindrical cargo tank installed above the deck. An independent tank is used to eliminate or at least minimize the forces or stresses that may be working on the adjacent hull structure. An independent tank is installed in such a manner that it can be moved relative to the vessel.

Integral tanks are cargo tanks that form an essential part of the hull structure and contribute to the strength of the vessel. Integral tanks are subject to the forces and stresses experienced by the hull structure as a result of cargo operations and motion of the vessel. Figures 1-11 and 1-11a illustrate several cargo tank configurations on parcel tankers.

Figure 1-9. Complexity of deck piping on a chemical tanker. Courtesy Christopher Adams.

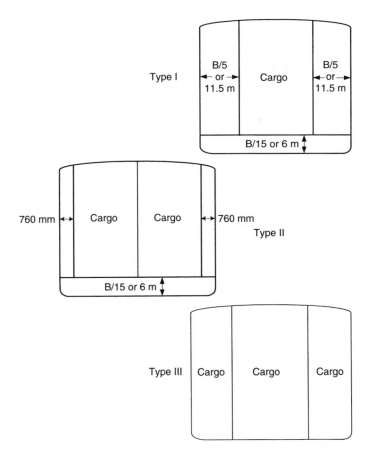

Figure 1-10. The spacing requirements for Types 1, 2, and 3 containment under the bulk chemical codes. Copyright © International Maritime Organization (IMO), London.

Gravity tanks are those tanks having a design pressure not greater than 10 pounds per square inch gauge and of prismatic or other geometric shape where stress analysis is neither readily or completely determinate.

Pressure tanks are independent tanks, whose pressure is above 10 pounds per square inch gauge and fabricated in accordance with domestic rules.

COMBINATION CARRIER

The USCG defines a combination carrier as any vessel designed to carry oil or solid cargoes in bulk. They are specially built vessels often referred to as ore/bulk/

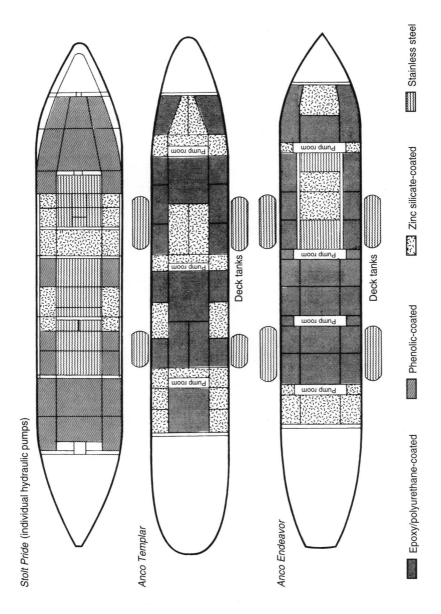

Figure 1-11. Cargo tank layouts of parcel tankers. Copyright © International Maritime Organization (IMO), London.

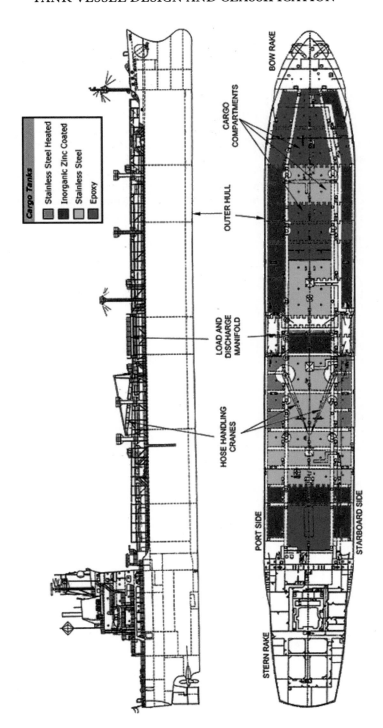

Figure 1-11a. General arrangement of a parcel tanker.

oil carriers (OBOs) capable of alternating between carrying oil cargoes and bulk commodities such as iron ore or coal. Combination carriers can also be adapted to meet the specialized requirements of a customer such as alternating between the transport of caustic and alumina for the aluminum industry. The advantages of this design include the ability to carry cargo in both directions during a voyage and to shift trades as market conditions and freight rates change.

Figure 1-12 illustrates a typical cross section of an OBO. The design is characterized by large raised hatch openings as well as a double-bottom and topside ballast tanks for trimming of solid cargoes. Some of the concerns expressed with this design include damage to the tank coatings and high stresses from the loading of dry cargoes. Problems also arise in situations where major components of the cargo system (such as pumps, valves, inert gas systems, and so forth) experience extended periods of inactivity. To combat these problems, combination carriers require frequent inspection and ongoing preventive maintenance to ensure the

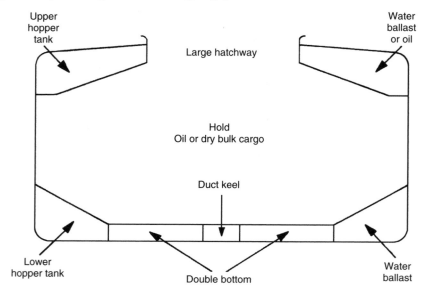

Figure 1-12. OBO: typical section. Reprinted with permission from the International Safety Guide for Oil Tankers and Terminals (ISGOTT), 4th Edition. Courtesy OCIMF, ICS, and IAPH.

continued reliability of cargo system equipment.

BARGES

The tank barge industry has undergone a renaissance in recent years with the wave of new double hull vessels entering service. The modern tank barge fleets represent a safe, cost effective method of transporting vast quantities of bulk liquid cargoes. In the United States the tank barge industry consists of approximately 3,700 barges that account for the transport of millions of tons of cargo annually. Tank barges deliver products throughout the inland waterway

Petroleum & Petroleum Products
Moved by Tank Barge

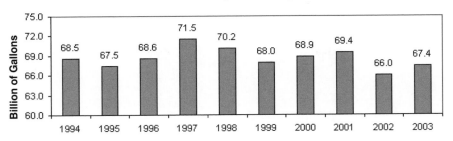

Figure 1-12a. Barge transportation statistics. Courtesy American Waterway Operators (AWO).

system (rivers, lakes, bays, and sounds) of the U.S. as well as in the coastwise trade. To fully appreciate the role played by the barge industry in the transportation of bulk liquid cargoes consider the most recent data available from the American Waterway Operators. (Figure 1-12a)

These versatile vessels transport the full range of cargoes carried by tank ships. Figure 1-13 shows a typical plan view of a double hull barge. Most barges are constructed with a centerline bulkhead and a series of transverse bulkheads that result in a port-and-starboard cargo tank configuration as seen in Figure 1-13. The number of cargo compartments found on a barge is generally dictated by the trade of the vessel. Under the Oil Pollution Act of 1990, the barge industry was also confronted with the mandatory replacement of the existing single hull fleet with double hull vessels. The Crowley Petroleum Service ATB shown in Figure 1-13a is another addition to their expanding coastwise fleet, equipped with the latest advances in cargo system design. An articulated tug and barge unit such as the Crowley vessel employ a deep notch and intercon connection unit to lock the tug and barge together providing a safe and efficient method for operating offshore without the need for a lengthy tow wire (Figures 1-13b and 1-13c).

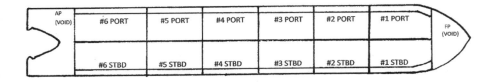

Figure 1-13. Double hull arrangements. Courtesy OSG America.

Figure 1-13a. Crowley ATB. Courtesy Stacy DeLoach.

In addition to this advanced connection system between the tug and barge, the cargo systems on the newest barges incorporate many state-of-the-art features, such as:

-inert gas generator
-nitrogen generator
-vapor recovery system
-closed gauging system
-segregated ballast system
-flexible and highly efficient cargo system
-fixed tank washing system
-thermal oil heating systems
-automated cargo control system
-increased number of cargo segregations

It is apparent that versatility is a key element in the design of the next generation of tank barges giving owners the ability to adjust to changing market conditions and customer demands. Today's operators realize they must compete for work in an increasingly demanding and competitive oil and chemical transportation market.

Barges transporting cargoes other than oil must meet the construction requirements outlined in Title 46 CFR Part 151, which call for heightened protection of the cargo area from side or bottom damage to the barge. Barge hulls are categorized according to structural strength, collision and grounding requirements, and surviv-

Figure 1-13b. Deep notch on barge and Intercon connection Courtesy Stacy DeLoach.

Figure 1-13c. Intercon Connector. Courtesy Stacy DeLoach.

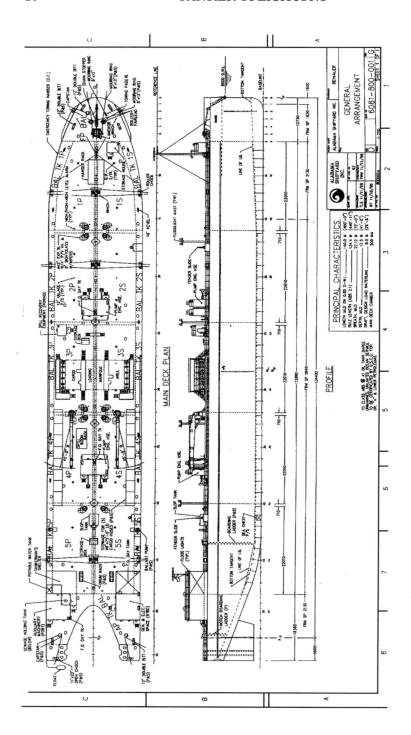

Figure 1-14. General arrangement drawing of 135,000-barrel, double-hull barge built at Alabama Shipyard. Courtesy Alabama Shipyard Inc.

ability in the event of flooding from specified damage to the hull. The hull types are categorized in three ways as follows:

1. Type 1 barge hulls are designed to carry products which require the maximum preventive measures to preclude the uncontrolled release of the cargo.
2. Type 2 barge hulls are those designed to carry products which require significant preventive measures to preclude the uncontrolled release of the cargo.
3. Type 3 barge hulls are those designed to carry products of sufficient hazard to require a moderate degree of control.

SPECIAL PURPOSE TANKERS

There are a number of tank vessels that do not fall within the traditional classification of a commercial tanker. Such vessels are purpose built in a class of their own based on the needs of the particular operation they serve. Examples of these special purpose tank vessels include:

UNDERWAY REPLENISHMENT TANKERS

The underway replenishment tanker is specifically designed to serve the needs of the military by providing fuel for the operation of vessels and aircraft. The vessel seen in Figure 1-14a has a conventional cargo and ballast tank arrangement below deck in addition to the necessary hoses, rigging and winches on deck to conduct the refueling of military vessels while underway at sea. The refueling operation is an "all hands" operation that requires unique planning and coordination by everyone involved to ensure a safe and efficient operation.

Figure 1-14a. Underway Replenishment with Aircraft Carrier. Courtesy Brian Roscovius.

Figure 1-14aa. Underway replenishment operation with naval vessel.
Courtesy Brian Roscovius.

FLOATING PRODUCTION, STORAGE AND OFFLOADING (FPSO) AND FLOATING STORAGE AND OFFLOADING (FSO)

Offshore oil production platforms normally transfer the oil to the mainland via subsea pipelines or by tanker. When shuttle tankers are employed to move the oil to the mainland, the oil is typically stored in an FPSO or FSO that is either permanently moored or able to disconnect.

An FSO is often a converted oil tanker that is used to temporarily store the oil prior to transfer to a shuttle tanker.

An FPSO is a floating tank system used to receive oil from nearby platforms or wells, process the oil onboard and store it until it can be transferred to a shuttle tanker. (Figure 1-14b)

BARRIERS

In the construction of a tank vessel, a physical barrier is generally required to separate the cargo and non-cargo areas of the vessel. Several approaches to meet this requirement are outlined in the construction regulations. The most common method is the use of a void—dead air space, known as a cofferdam—that places two bulkheads between the cargo and non-cargo areas as seen in Figure 1-15.

Alternative methods of separation include the use of a cargo or ballast pump room, an empty cargo tank, or a tank carrying a grade E cargo (flashpoint of 150°F and above). This barrier extends the breadth and depth of the vessel creating the transition between the gas-safe areas of the vessel (the superstructure and engine

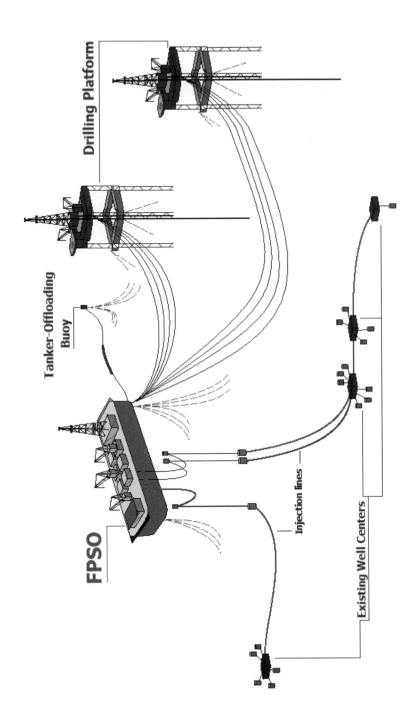

Figure 1-14b. FPSO diagram. Courtesy Gunnernett.

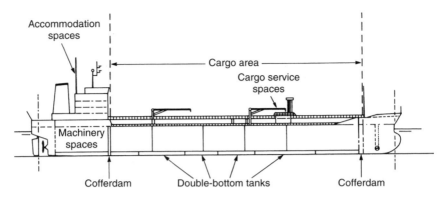

Figure 1-15. The physical separation between the cargo and noncargo areas of a tanker can be seen in this drawing. Copyright © International Maritime Organization (IMO), London.

spaces) and the potentially hazardous cargo tank area. Several other design features contribute to this barrier mentality, including the fact that the forward side of the after house facing the cargo tank area is sealed, and access to the house is limited to doors located at the side of the superstructure. These changes have improved the safety of the vessel over earlier designs by enhancing two basic principles of construction: (1) minimizing the accumulation of flammable cargo vapors in and around the superstructure and (2) separating the cargo area from potential sources of ignition.

CLASSIFICATION

Tank vessels are usually classified by the trade in which they are engaged and according to deadweight tonnage.

The trade of a vessel is defined by the type of cargoes routinely carried over a number of voyages. In the tanker industry three broad categories predominate:

1. Crude-oil carriers
2. Product carriers
   Clean (gasoline, jet, diesel, etc.)
   Dirty (black oils—residual fuel oils, vacuum gas oils, asphalt, etc.)
3. Parcel carriers (chemical/specialty cargoes, etc.)

Tankers tend to remain in one trade. However, as market conditions and customer requirements change, a vessel may move back and forth between trades during the lifetime of the vessel. To change the trade of a vessel is a substantial commitment on the part of an owner as extensive cleaning and even modification of the vessel may be necessary.

Tanker personnel often refer to the vessel according to its deadweight tonnage (dwt). The deadweight tonnage is used as a rough measure of the cargo carrying capacity of the vessel and is usually expressed in long tons (1 long ton = 2,240 pounds) or metric tons (1 metric ton = 2,204.6 pounds). The deadweight tonnage of a vessel is defined as the amount of cargo, fuel, water, and stores a vessel can carry when fully loaded. Tankers are typically divided into four broad categories as seen in table 1-1 and Figure 1-16.

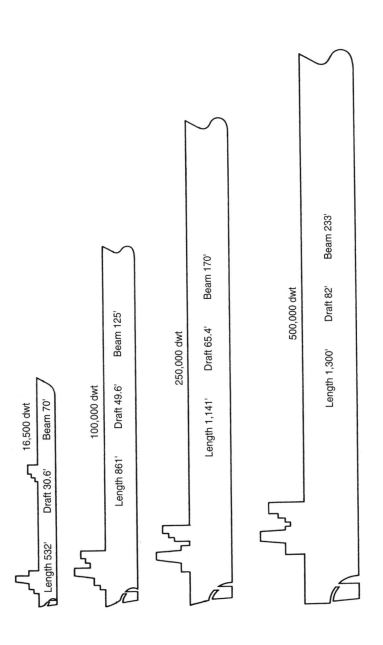

16,500 dwt

Length 532' Draft 30.6' Beam 70'

100,000 dwt

Length 861' Draft 49.6' Beam 125'

250,000 dwt

Length 1,141' Draft 65.4' Beam 170'

500,000 dwt

Length 1,300' Draft 82' Beam 233'

Figure 1-16. Relative sizes of tankers. Tanker size has increased dramatically since WWII. The top figure represents a T-2 tanker. Courtesy Exxon.

Table 1-1
Classification of Tankers
According to Deadweight Tonnage

Category	Tonnage Range	Trade
Handy/Coastal/Parcel/Barge	5,000 to 35,000 dwt	Product/parcel
Medium	35,000 to 160,000 dwt	Product/crude oil
VLCCs (very-large crude carrier)	160,000 to 300,000 dwt	Crude oil
ULCCs (ultra-large crude carrier)	300,000 dwt and above	Crude oil

DEVELOPMENT OF THE SUPERTANKER

During the post–World War II era, the tanker industry experienced dramatic changes in both the dimensions and the trade routes of these vessels. The ever popular T-2 tanker of the war years gave way to modern construction (Figure 1-17) in order to create more economical ways of transporting oil to meet the growing demands of the industrialized world.

Figure 1-17. "State Class" double hull tanker under construction at the NASSCO yard. Courtesy Ken Wright, General Dynamics NASSCO staff photographer.

Figure 1-18. Lightering operation. Courtesy Abigail Robson.

A number of factors contributed to the rapid increase in tanker size, including the hostilities in the Middle East that resulted in the closure of the Suez Canal, a choke point for tanker traffic to and from the oil fields of the Persian Gulf. Nationalization of the oil refineries in the Middle East and fierce competition among international ship owners all played a role in accelerating the development of the modern-day supertanker. VLCCs and ULCCs ply the most solitary trade routes of the oceans, typically loading at offshore platforms or single-point moorings and discharging at SPM's, deepwater terminals or by lightering offshore (Figure 1-18).

These vessels can enter only a limited number of ports in the world when fully loaded and therefore remain at sea for extended periods of time, a typical voyage often taking seventy to seventy-five days.

REVIEW

1. Define the term "tank vessel."
2. What is the effect of free surface on a vessel?
3. How can the effects of free surface be reduced or eliminated?
4. Describe the method of construction of single-hull tank vessels.
5. The Oil Pollution Act of 1990 mandates double hulls for new construction. What are the minimum spacing requirements between the hulls? List some of the concerns associated with the double hull design.
6. Draw a cross section of a mid-deck tanker and explain the method employed to reduce oil outflow in the event of a casualty (grounding/collision).

7. In the transport of hazardous chemicals, explain the requirements for Types I, II, and III containment.
8. In the construction of a modern tanker, the cargo and non-cargo areas of the vessel must be physically separated through what means?
9. List three factors that contributed to the development of the modern supertanker.
10. List the various trades in which a tank vessel is typically engaged.

CHALLENGE QUESTIONS

11. The absence of a centerline bulkhead on certain double hull tankers resulted in the loss of intact stability during cargo and ballast operations. Why?
12. List three concerns associated with the double hull method of construction.
13. The _____ design attracted the most interest as a likely alternative to the double hull method of construction mandated by the Oil Pollution Act of 1990.
14. Describe the various ways one might detect that cargo has migrated into the empty ballast space of a double hull tanker.
15. In the event of a collision in which the cargo tank boundary is breached (inner hull), describe your actions to limit the spill and safeguard the vessel.

CHAPTER 2

Cargo Characteristics

RICHARD BEADON AND MARK HUBER

Numerous potential hazards are associated with the seagoing transport of bulk liquid cargoes. To minimize those risks it is imperative for the person-in-charge (PIC) to have a keen understanding of the physical properties of the cargo being transported. Experience has shown that a thorough working knowledge of the cargo is vital to intelligent decision-making with respect to safe carriage as well as to efforts to maintain quality assurance. Improper transfer procedures, stowage, and care of the cargo have all factored into incidents that resulted in harm to personnel and damage to vessel, cargo, and the environment. This chapter seeks to address the main characteristics and hazards presented by the cargo as it relates to the role of the vessel PIC. Many of the properties and hazards discussed in this chapter apply to all bulk liquids. However, due to their special nature, liquid chemicals may present significantly different characteristics and hazards.

BULK LIQUID CARGOES

Tank vessels transport a wide variety of liquids in bulk (unpackaged). These fall under three broad classifications: petroleum liquids, chemical liquids, and special liquids.

Petroleum Liquids

Petroleum liquids consist of naturally occurring crude oil and the various products derived (refined) from this raw material, including the following:

Gasoline	Kerosene	Residual fuel oil
Fuel oil	Jet fuel	Asphalt
Diesel	Lubricants	Coke
Residual fuel oil	Asphalt	

Chemical Liquids

A liquid chemical is any substance used in, or obtained by, a chemical process. There are literally hundreds of different chemicals transported by tank vessels. These substances are derived from many sources and have diverse characteristics. They may be categorized as organic or inorganic chemicals. Table 2-1 shows a sampling of each.

Table 2-1
Chemical Liquids

Organic Chemicals	*Inorganic Chemicals*
Aromatic hydrocarbons	Boric acid
Vinyl chloride	Sulfuric acid
Acetone	Phosphoric acid
Acetic acid	Caustic soda
Styrene monomer	Hydrochloric acid
Acrylonitrile	Molten sulfur

Special Liquids

Liquid substances other than those classified as petroleum or chemical are described as special liquids. Table 2-2 shows some examples.

Table 2-2
Special Liquids

Animal/Vegetable Oils	*Miscellaneous Liquids*
Palm oil	Freshwater
Soybean oil	Beer
Sunflower oil	Wine
Other vegetable oils	
Animal oils	
Tallow and greases	
Molasses	

PROPERTIES OF PETROLEUM

Crude oil and the products derived from the raw material are considered petroleum liquids. Crude oil is a mixture of a wide range of long-chain *hydrocarbon* molecules. A hydrocarbon molecule is essentially one or more hydrogen atoms linked with one or more carbon atoms, hence the term hydrocarbon. The composition of crude oil varies widely (paraffins, naphthenes, or aromatics) depending on its geographic source. Crude oil can be described as either "heavy" or "light" based upon its specific gravity. The number of carbon atoms in the hydrocarbon molecule influences the specific gravity of a crude oil. The greater the number of carbon atoms in a molecule, the heavier the molecule will be.

A *compound* is a chemical substance made up of two or more elements bonded together and not separable by physical means. Crude oil is a mixture of hydrocarbon compounds ranging from those that are partly gaseous under normal atmospheric conditions to those that are liquid or solid. Also present are traces of nitrogen, sulfur, oxygen, and metals. Crude oils containing sulfur compounds such as hydrogen sulfide are known as sour crudes and are characterized by a vile and nauseating rotten-egg odor.

The refining process involves separating the various hydrocarbon compounds in crude oil into groups or *fractions* of compounds having similar boiling point ranges. A number of methods are used in the refining process including the following:

Distillation, or physical separation, consists of boiling off the crude oil and splitting it into a number of fractions.

Cracking is a chemical conversion that results in splitting the heavier fractions into lighter fractions. Each fraction has its own boiling point and a unique set of physical properties.

Purification is the process of removing certain impurities (such as sulfur) from the petroleum products during the refining process.

FLAMMABILITY CHARACTERISTICS OF BULK LIQUID CARGOES

There are serious fire risks associated with the transport of petroleum and certain chemical cargoes; hence a discussion of these characteristics is in order. To enable the PIC to judge the degree of risk, most cargoes are categorized according to their flammability. Following is a review of some basic terms used in the classification of cargoes:

Volatility: In a fire involving a flammable liquid, the vapor that is given off by the liquid burns, not the liquid itself. Therefore, the flammability of a liquid cargo will depend primarily on the ability of the liquid to produce flammable vapor. Volatility is a term used to describe the tendency of oils or chemical products to produce flammable vapor. To assist the PIC, there are a number of ways of expressing the volatility of a liquid.

True vapor pressure (TVP): Vapor pressure indicates the volatility of a liquid. For example, when a petroleum (liquid) cargo is loaded into a tank, it will begin to vaporize into the space above. When the vapor and liquid in the space reach equilibrium, the pressure exerted on the liquid is its true vapor pressure. The true vapor pressure of a petroleum liquid will vary due to differences in composition and temperature; consequently another method of expressing the vapor pressure is employed—the Reid vapor pressure (RVP).

Reid vapor pressure (RVP): Reid vapor pressure is the measured vapor pressure that results when a sample of liquid in a closed container is heated to a standard temperature of 100°F (37.8°C). It is determined in a standard laboratory experiment using Reid testing apparatus. This test is of practical value to the PIC as it replicates the conditions that may exist during transport of a cargo at sea. It does so by providing an indication of the behavior of a particular cargo in the sealed tanks of a vessel when subjected to changing ambient (sea and air) temperatures. Reid vapor pressure is used in the classification of flammable liquids, as shown in table 2-4.

Flash point: Another term frequently encountered in the classification of liquids is flash point. The flash point of a flammable liquid is the lowest temperature at which the liquid gives off sufficient vapor to form an ignitable mixture near its surface. This mixture of vapor and air is ignitable by an external source of ignition, but the rate of vaporization is usually insufficient to sustain combustion.

Fire point: The fire point of a flammable liquid is the lowest temperature at which the liquid will produce sufficient vapor to ignite and continue to burn. This temperature is higher than the flash point of a liquid. The principal use of the terms flash point and fire point is to indicate the relative fire hazard associated with different products.

Autoignition temperature: The autoignition point of a liquid is the lowest temperature at which sustained combustion will occur in a liquid without the application of a spark or flame (external source of ignition). This temperature is above the fire point of a liquid.

Flammable limits: A liquid cannot burn unless it emits flammable vapors. In order to burn, the correct proportions of oxygen, vapor, and heat must be present. The flammable vapor of a liquid must therefore mix with air in the proper proportions to form an ignitable mixture.

Lower explosive limit (LEL) or lower flammable limit (LFL): The lower explosive limit is the smallest percentage of vapor in air that will form an ignitable mixture (point C in Figure 2-1). If the concentration of vapor is below the LEL, the mixture is considered "lean" and will not support combustion.

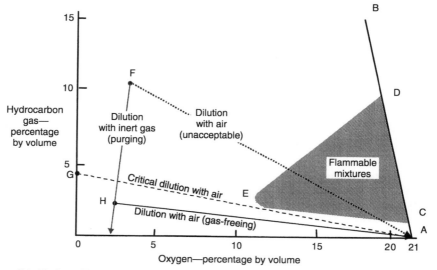

Note: The flammable range diagram shows characteristics of a typical crude oil and does *not* reflect the range of flammability of all substances. The range of flammability can differ substantially between oils and chemicals; therefore, this diagram should be used for general informational purposes only.

Figure 2-1. Flammable range diagram. Reprinted with permission from the *International Safety Guide for Oil Tankers and Terminals (ISGOTT)*, 5th edition. Courtesy OCIMF, ICS and IAPH

Upper explosive limit (UEL) or upper flammable limit (UFL): The upper explosive limit is the greatest percentage of vapor in air that will form an ignitable mixture (point D in Figure 2-1). If the percentage of vapor present exceeds the UEL, the mixture is considered "rich" and will not support combustion.

Flammable range or explosive range: The flammable or explosive range diagram in Figure 2-1 illustrates all the possible combinations of vapor in air (between the upper and lower flammable limits) that form an ignitable mixture. Mixtures of hydrocarbon vapor and air that lie outside the flammable range (shaded area in the curve) will not support combustion. In the case of oil cargoes, if the hydrocarbon concentration is below the lower explosive limit, there is insufficient vapor to support combustion. Conversely, if the hydrocarbon concentration is above the upper explosive limit, there is insufficient air to support combustion.

Table 2-3
Typical Flammable Limits of Sample Cargoes

Product	LEL	UEL	Range
Crude (average)	1.0%	10.0%	9.0%
Gasoline	1.3%	7.6%	6.3%
Kerosene	0.7%	6.0%	5.3%
Benzene	1.4%	8.0%	6.6%
Ethylene oxide	2.0%	100.0%	98.0%
Ammonia	15.5%	27.0%	11.5%
Naphtha	0.9%	6.7%	5.8%

Source: United States Coast Guard

The flammable limits can vary substantially for different petroleum and chemical cargoes. Table 2-3 lists the typical limits of flammability for several cargoes. To determine the flammable limits of a particular cargo, consult the data sheets or the laboratory analysis of the cargo.

Vapor density: Vapor density is the ratio of the weight of a vapor or gas with no air present compared to an equal volume of air at the same temperature and pressure. The vapor density of a liquid can only be accurately determined in a standard laboratory experiment. A vapor density of 1 indicates that the gas weighs the same as that of an equal volume of air. Values less than 1 indicate that the gas is lighter than air and will tend to rise. Values greater than 1 would indicate that the gas is heavier than air and would tend to settle.

An understanding of vapor density is important because most petroleum cargo vapors are heavier than air and will settle in lower regions of a tank or pump room. This is an important consideration when determining the method and adequacy of testing an atmosphere for the presence of cargo vapors.

Vapor density is also an important element that contributes to the accumulation of flammable vapors on deck and around the superstructure while tanks are venting during a loading operation.

AVOIDANCE OF THE FLAMMABLE RANGE

To enhance the overall safety of transporting flammable cargoes, many tank vessels are equipped with inert gas systems (see chapter 16 for a detailed discussion). The purpose of this system is to maintain the atmosphere of the cargo tanks in a nonflammable condition throughout the voyage cycle (operating life) of the vessel. This is achieved through the use of a gas or a mixture of gases that is deficient in oxygen and therefore incapable of supporting combustion. Although the atmosphere of a tank may contain flammable vapors in varying concentration, there can be no fire or explosion if the tank is starved of oxygen.

A properly inerted cargo space is any compartment with an atmosphere containing 8 percent or less oxygen by volume and maintained under positive pressure. The application of an inert gas system during a typical voyage can best be illustrated relative to a flammable range diagram (Figure 2-1). The goal of this system is to prevent the atmosphere of the cargo tanks from ever entering the flammable range.

Voyage Cycle (Inert Gas Cycle)

A convenient starting point for this discussion is a vessel in the shipyard with the entire cargo system clean and gas free (see chapter 15 for a discussion of the gas-free state). As shown in the flammable range diagram (Figure 2-1), the atmosphere of the cargo tanks would likely be found near position A with an oxygen content of 21 percent by volume and a reading of less than 1 percent of the LEL on a combustible-gas indicator. Prior to departing the yard or while en route to the first loading port, the inert gas system is operated to carry out the *primary inerting* of the cargo tanks. With vents open, the fresh air is driven out of the cargo tanks and replaced with good quality inert gas. The net effect of this operation is to lower the oxygen content of the atmosphere in the cargo tanks. Primary inerting is illustrated in Figure 2-1 by moving to the left along the horizontal axis from point A until the atmosphere reaches 8 percent or less oxygen by volume. It should be noted that as the oxygen content of a space is lowered, the range of flammability for most petroleum products decreases progressively until it terminates at about 11 percent oxygen by volume.

At the loading port, cargo entering the empty tanks will start to displace the inert atmosphere. Due to the turbulence of the loading operation, flammable cargo vapors are generated, resulting in the atmosphere moving up the vertical (hydrocarbon) axis of the flammable range diagram toward point F. There is no material change in the oxygen content provided a positive pressure is maintained within the space, thereby preventing the ingress of air. At the completion of the loading operation, the atmosphere above the cargo in the topped-off tank(s) is likely to be a rich mixture, yet still in an inert state.

During the sea passage to the discharge port, the oxygen content and tank (deck) pressure should be monitored. Some operators have reported a significant increase in the oxygen content of the atmosphere of inerted cargo tanks during the loaded passage. The only cause that could be attributed to the oxygenation of the atmosphere involved a characteristic of the cargo. Upon leaving the dock at the loading terminal, the vessel cargo tanks were reported to have a measured oxygen content of 5% by volume, and upon arrival at the discharge terminal the oxygen content was found to be between 14% and 15% by volume. Also, due to the fact that liquid cargoes expand and contract with changes in sea and air temperature, significant fluctuations in the tank (deck) pressure occur during the voyage. If, for example, the deck pressure rises as a result of increasing ambient temperatures, it may be necessary to vent off the excess pressure. Conversely, when colder temperatures and a corresponding drop in tank (deck) pressure are encountered, it may be necessary to start the inert gas system and *top up* the pressure in the tanks. Topping-up is defined as the introduction of inert gas into a tank already in the inert condition with the object of increasing the tank pressure to prevent any ingress of air.

Upon arrival at the discharge port, the inert gas system is started and operated for the duration of the cargo discharge. Inert gas is delivered to the tanks to replace the cargo being discharged. To ensure that positive pressure is maintained, the inert gas supply must exceed the cargo discharge rate. During the discharge operation (Figure 2-1) the hydrocarbon concentration of the atmosphere will drop as the cargo vapors are diluted with inert gas. Thereafter, during each successive load and discharge, the tank atmosphere moves up and down the vertical (hydrocarbon) axis. This vertical change is acceptable provided oxygen (air) is not introduced into the space, possibly compromising the inert status of the tank or vessel.

If a problem should develop at the discharge port, this may necessitate a return to the shipyard. During the ballast trip it may become necessary to prepare the cargo tank(s) for entry by personnel. The tank or tanks should be water-washed in accordance with recommended guidelines while maintaining an inert condition. Following the wash, the tanks are *purged* with inert gas prior to ventilation with air.

PURGING

Purging is the introduction of inert gas into a tank that is already in an inert condition with the object of reducing the hydrocarbon concentration to a point where subsequent ventilation with fresh air will not result in the creation of a flammable atmosphere. The purging process is illustrated in Figure 2-1 by moving from point F to point H.

Safe industry practice dictates that purging of a tank should continue until the hydrocarbon concentration of the space is 2 percent or less by volume as determined by using a suitable hydrocarbon analyzer.

Upon completion of purging, the space is ventilated with air using portable fans or the inert gas system in the gas-free mode. Ventilating with air at this point further reduces the hydrocarbon concentration while increasing the oxygen content of the space. The ventilation process continues until the atmospheric tests reveal a return to safe readings (21 percent oxygen by volume and less that 1 percent LEL on a combustible-gas indicator). The process of ventilating the tank with air is shown in Figure 2-1 by moving from point H to point A.

It is important to realize that avoidance of the flammable range in this way is the expectation of the transportation industry and is only possible if the operator thoroughly understands the use of the inert gas system.

CLASSIFICATION OF PETROLEUM

Petroleum liquids are classified in many ways throughout the world. The following information addresses two common approaches.

International Classification

In many safety-related rules and regulations, petroleum cargoes are broadly classified as volatile liquids and nonvolatile liquids.

Volatile liquids: Petroleum liquids that have closed-cup flash points below 140°F (60°C) are considered volatile. Over the normal range of ambient temperatures encountered during transport, cargoes in this category are capable of producing gas/air mixtures within and above the flammable range. For this reason, volatile cargoes are frequently transported in a tank with a controlled (inerted) atmosphere.

Nonvolatile liquids: These are petroleum products that have closed-cup flash points of 140° F (60° C) and above. Over the normal range of ambient temperatures encountered during transport the atmosphere above these cargoes (headspace) typically contains gas concentrations below the lower flammable limit. Cargoes in this category include residual fuel oils and diesel oils. Due to other properties associated with these cargoes, however, the application of heat is often necessary during the voyage. Caution must be exercised with heated cargo, as the creation of a flammable atmosphere is possible if it is heated to or near the flash point.

United States Coast Guard Classification

The United States Coast Guard (USCG) separates petroleum liquids into two categories: flammable and combustible.

Flammable liquids: Liquids that have an *open-cup* flash point at or below 80°F (26.7°C) are classified as flammable liquids.

Combustible liquids: Liquids that have an *open-cup* flash point above 80°F (26.7°C) are classified as combustible liquids.

The flammable and combustible liquids are subdivided into grades based on their flash points and Reid vapor pressures. Tables 2-4 and 2-5 show the USCG classification system contained in Title 46 CFR Parts 30.10.15 and 30.10.22.

Table 2-4
USCG Classification of Flammable Liquids
Flash Point at or below 80°F (26.7°C)

Grade	Flash Point	Reid Vapor Pressure	Examples
A	80°F or below	14 psi and above	Natural gasoline, naphtha
B	80°F or below	More than 8.5 but less than 14 psi	Most commercial gasoline
C	80°F or below	8.5 psi and below	Most crude oils Aviation gasoline

Table 2-5
USCG Classification of Combustible Liquids
Flash Point above 80°F (26.7°C)

Grade	Flash Point	Reid Vapor Pressure	Examples
D	Above 80°F but below 150°F	N/A	Kerosene Commercial jet fuels
E	150°F and above	N/A	Heavy fuel Lube oils Asphalt

CARGO WEIGHT, CAPACITY, AND FLOW

To plan the loading of a liquid cargo into a tank vessel, it is necessary to know the unit weight of the cargo and the amount of space it will occupy. Once a cargo has been loaded into a tank, it is then necessary to determine the quantity in the tank. At the discharge port, it is necessary to again determine the quantity of cargo onboard prior to the discharge.

The safe, efficient, and accurate determination of the quantity of cargo in the vessel's tanks is a key responsibility of the PIC. The quantity of cargo is an important factor for proper accounting (billing); for cargo calculation (determining draft, trim, and stress in order to ensure the vessel is not overloaded); and for calculation of transfer rates. The PIC should therefore be familiar with the following terms relating to volume and weight, as they are used in conjunction with the transport of liquid cargoes in bulk.

Density

The density of a substance is the weight per unit volume at a standard temperature of 60°F (15.6°C). The density of a liquid is expressed in ounces per cubic foot. For example, at a standard temperature of 60°F (15.6°C), the density of freshwater is 1,000 ounces (62.5 lbs) per cubic foot and the density of salt water is 1,025 ounces (64.06 lbs) per cubic foot.

If the density and volume of a liquid are known, the weight of the volume occupied by the liquid can be found by using the following formula:

$$\text{Weight} = \text{Volume} \times \text{Density}$$

If the weight and volume of the liquid are known, then the density can be found by transposing the formula:

$$\text{Density} = \text{Weight} / \text{Volume}$$

Specific Gravity (SG)

The specific gravity (SG) of a substance is the ratio of a given volume of a substance at a standard temperature of 60°F (15.6°C) to the weight of an equal volume of freshwater at the same temperature. For example:

If the density of a liquid is known, it can be converted into specific gravity by dividing its density by the density of freshwater.

API Gravity

API gravity is an arbitrary scale developed by the American Petroleum Institute and used in the transportation industry as an alternative means of expressing the weight of a measured volume of a liquid.

The API gravity of a liquid is expressed in a scale of degrees API at a standard temperature of 60°F (15.6°C). Freshwater has an arbitrary gravity of 10 degrees. Liquids lighter than freshwater have an API gravity greater than 10 and liquids heavier than freshwater have an API gravity less than 10.

For information purposes, the API gravity is derived using the following formula:

$$\text{API gravity in degrees} = (141.5/\text{Specific gravity @ } 60°F) - 131.5$$

Determination of Density, Specific Gravity, and API Gravity

A *hydrometer* is one of the instruments commonly used to measure density. Hydrometers are calibrated to measure density in ounces; however, those that measure API gravity are marked in degrees API. To obtain the specific gravity of a liquid, a density hydrometer is used and the reading is divided by 1,000 (the density of freshwater).

To obtain the density/API gravity of cargo in a tank, a sample is drawn and the appropriate hydrometer used. Due to the fact that liquid cargoes expand or contract with changes in temperatures, the reading obtained is the density/API gravity of the liquid at which the sample was tested. Therefore, it is essential to take a temperature reading of the sample to accurately calculate the density or API gravity.

Units of Measure

Table 2-6 shows the typical units of measure used in the transportation industry.

Table 2–6
Units of Measure

Unit	Measure
1 barrel	42 gallons (US)
1 cubic meter	6.2898 barrels
1 ton metric (tonne)	1,000 kilograms (2,204.6226 pounds)
1 ton (long)	2,240 pounds (1,016.0469 kilograms)
1 gross barrel	42 gallons at actual temperature in the tank
1 net barrel	42 gallons adjusted to standard temperature of 60°F

Viscosity

Viscosity is a measure of the internal friction of a liquid or its resistance to flow. It is an important consideration when determining the pumpability of a liquid cargo. The viscosity of a liquid changes with different temperatures. For example, as the temperature of a liquid increases, the viscosity decreases. For efficient loading and discharging of the vessel, the PIC should be aware of the optimum viscosity of the cargo. This value is useful in determining the heating requirements of a cargo and the proper temperatures to be maintained during cargo transfer and in transit .

There are many standards for expressing viscosity. Controlled laboratory experiments are used to determine the viscosity of a liquid. In one method, Saybolt Seconds Universal (SSU), viscosity is measured by the time in seconds that it takes for a liquid at a prescribed temperature to drain from a standard viscosimeter. This information is typically derived from a laboratory analysis report of the cargo.

Pour Point

The pour point of a liquid is the lowest temperature at which the liquid will remain fluid. It is expressed as a temperature either in degrees Fahrenheit or Celsius. The PIC should be mindful of this temperature when transporting cargoes with elevated pour points. Examples of such cargoes include residual fuel oils, vacuum gas oils, wax, and asphalt. During transport, the cargo temperature in the tanks should be closely monitored and the heating system adjusted to maintain the recommended temperature. To avoid possible solidification of the cargo, the temperature should never be allowed to approach the pour point of the substance.

Cloud Point

The temperature below which sludge deposition can be expected in the cargo tanks. In the case of certain crude oils that require the application of heat it represents the temperature at which the paraffin wax in the crude separates from the liquid phase of the cargo.

TOXICITY—MEASUREMENT AND REGULATIONS

Toxicity refers to the poisonous nature and potential health risks associated with a particular substance. The toxicity of a substance is difficult to measure and is subject to revision as more detailed information about the ramifications of exposure become available. The threshold limit value–time weighted average provides a convenient indicator of the relative toxicity of gases and assists individuals in reducing health risks. Studies performed on animals and extrapolated for the human body form the basis of rating toxicity levels.

Threshold Limit Value–Time Weighted Average (TLV-TWA)

The threshold limit value–time weighted average (TLV-TWA) is a designation established by the American Conference of Governmental and Industrial Hygienists (ACGIH) for various substances. These designations are used as recommended guidelines in the workplace; they are subject to review and may be updated annually, in which case the results will be published in ACGIH publications. The term threshold limit value–time weighted average (TLV-TWA) is used in the transportation industry to express the toxicity of vapors from a substance. The TLV-TWA of a substance is usually expressed as the number of parts per million (ppm) by volume of vapor in air.

According to ACGIH, "Threshold Limit Values refer to airborne concentrations of substances and represent conditions under which it is believed that nearly all workers may be repeatedly exposed day after day without adverse health effects." When expressed as a time weighted average, the concentration is considered over a normal eight-hour workday and a forty-hour workweek.

Permissible Exposure Limit–
Time Weighted Average (PEL-TWA)

The permissible exposure limit (PEL) of a substance is a designation used by the Occupational Safety and Health Administration (OSHA) and the United States Coast Guard (USCG). The PEL represents a regulatory value (as opposed to a recommended guideline) that must not be exceeded in the workplace. For example, the PEL-TWA for cargoes covered by the benzene regulation is 1 ppm.

Threshold Limit Value–
Short-Term Exposure Limit (TLV-STEL)

The threshold limit value–short-term exposure limit (TLV-STEL) defines the concentration of a substance to which workers can be exposed continuously for a short period of time, provided that the daily TLV is not also exceeded.

The STEL is a fifteen-minute time weighted average exposure that should not be exceeded at any time during the workday, even if the eight-hour time weighted average is within the TLV.

Exposures at the STEL may not be longer than fifteen minutes and cannot be repeated more than four times per workday. There must also be at least sixty minutes between successive exposures at the STEL.

Threshold Limit Value–Ceiling (TLV-C)

The threshold limit value–ceiling (TLV-C) is the maximum concentration of vapor in air, expressed as either a TLV or PEL, that must not be exceeded even for an instant. In situations when there is no established ceiling limit, the TLV-STEL is used.

Immediately Dangerous to Life or Health (IDLH)

The designation IDLH (immediately dangerous to life or health) was established by the National Institute for Occupational Safety and Health (NIOSH), an agency of the Public Health Service.

IDLH is defined by NIOSH as a condition "that poses a threat of exposure to airborne contaminants when that exposure is likely to cause death or immediate or delayed permanent adverse health effects or prevent escape from such an environment."

The practical application of the IDLH designation is to provide a basis for the selection of an appropriate respirator.

Odor Threshold

Expressed in parts per million by volume in air, the odor threshold is the smallest concentration of a gas that can be detected by most individuals through the sense of smell.

It is not an absolute value as it can vary considerably among individuals. Some odors are also capable of deadening the sense of smell. It is therefore not advisable to rely on the sense of smell as an indicator of the presence of a dangerous vapor.

Knowing the odor threshold of a toxic substance is important. If, for example, a liquid has a TLV-C of 20 ppm and an odor threshold of 50 ppm, by the time an individual detects the presence of this substance by the sense of smell, harmful exposure has already occurred.

Given the number of different liquid cargoes transported on tank vessels, it is not possible for one person to know all the details concerning a particular product. It is a daunting task to become familiar with all the products a PIC might be expected to handle and transport; therefore, it is vital that the PIC know where to turn for accurate, reliable information.

Toxicity—Effect on Personnel

Poisoning by toxic liquids can occur through one or more of the following three methods: (1) ingestion, (2) skin contact, and (3) inhalation.

INGESTION

The risk of swallowing petroleum or chemical liquids in normal day-to-day operations should be minimal provided individuals always exercise good hygiene. To minimize exposure through ingestion, personnel should be reminded to wash hands thoroughly before meals and never to eat or drink on deck. If accidental ingestion does occur, guidance can be found in the MSDS or cargo information cards. Medical assistance should be sought immediately.

SKIN CONTACT

With most petroleum products, skin contact can cause irritation and lead to dermatitis. Contact with the eyes and skin can be particularly dangerous when handling corrosive cargoes such as caustics or acids. Personnel should always wear protective clothing and eye protection when there is a risk of exposure through physical contact (splash hazard). The MSDS gives recommended precautions for minimizing exposures. If a toxic or corrosive liquid comes into contact with any part of the body, guidance can be found in the cargo information sheets.

INHALATION

The inhalation of cargo vapors has long been recognized as one of the leading hazards of exposure for workers on deck. Cargo vapors are pervasive and therefore difficult to control. The effect of inhalation of petroleum vapors on an individual can vary from imperceptible to obvious signs of impairment. The acute, short-term effects of exposure to petroleum products can include headache, euphoria, eye irritation, nose and throat irritation, loss of orientation, dizziness, and a drunken appearance. Continued exposure to high concentrations may lead to paralysis and possibly death.

The toxicity of petroleum and chemical cargoes varies widely depending on the makeup of the substance. The presence of some constituents in the cargo such as benzene, lead, and hydrogen sulfide can pose a significant threat to individuals. As already discussed, the TLV provides an indication of the level of exposure to a toxic substance that is acceptable during a typical workday over an indefinite period of time. The STEL is an indication that the human body can tolerate concentrations greater than the TLV for short periods, typically no more than fifteen minutes.

The odor of cargoes varies greatly and, in some cases, can fool an individual's sense of smell. Also, with some products, the odor threshold may be much higher than the TLV. In this case, harmful exposure may occur before the individual starts to smell the cargo vapor. The impairment of one's sense of smell is especially serious if the mixture contains hydrogen sulfide. Because of these inherent dangers, the PIC should never take the absence of smell as an indication of the absence of gas, but should always test for toxicity, be aware of the TLVs, and follow the proper entry procedures for enclosed spaces such as cargo tanks.

SOURCES OF CARGO INFORMATION

There are various sources of information regarding the physical properties and hazards of cargoes. The need for current, accurate cargo information is essential for the safety of those involved in the transport of bulk liquid cargoes. Some of the more common sources of cargo specific information available to the person-in-charge include the following:

- Material Safety Data Sheets (MSDS) produced by the manufacturer of the substance (see Appendix A on enclosed disc).
- *Chemical Data Guide for Bulk Shipment by Water* (former *CG-388*), Figure 2-2, and *Chemical Hazards Response Information System* (CHRIS), Figure 2-3, from the United States Coast Guard. As of this writing, the United States Coast Guard has revised the CHRIS database and made it available to the public in a number of ways: CD-ROM, the Internet (www.chrismanual. com), and in hard copy. The CD-ROM contains physical, chemical, toxicological, and combustion properties for over 1,300 chemicals and mixtures in addition to pollution response and regulatory information. In the event that the listed sources of information do not address the substance being handled or transported, in an emergency, the PIC should contact CHEMTREC or the National Response Center.
- Tanker Safety Guide Data Sheets from the International Chamber of Shipping (ICS)

BENZENE

Synonyms— Benzol; Benzole; Coal naphtha; Coal tar naphtha; Cyclohexatriene; Phene; Phenyl hydride

United Nations Number............................ 1114

CHRIS Code .... BNZ

Formula—C₆H₆

Appearance-Odor—Clear colorless liquid with a typical, pleasant aromatic odor
Specific Gravity—0.88

Chemical Family—Aromatic hydrocarbon

Pollution Category—USEPA ___A___ IMO ___C___
Applicable Bulk Reg. 46 CFR Subchapter .... ___O___

Boiling Point .... ___80°C___ ___176°F___
                                              ___°C___ ___°F___
Freezing Point........................ ___6°C___ ___42°F___
                                              ___°C___ ___°F___
Vapor Pressure 20°C (68°F) (mmHg) .... ___75___
Reid Vapor Pressure (psia).................... ___3.22___
Vapor Pressure 46°C (115°F) (psia)........... ___4.5___
Vapor Density (Air = 1.0)........................ ___2.8___
Solubility in Water .... ___Negligible___

FIRE & EXPLOSION HAZARD DATA

Grade—C: Flammable liquid.
Electrical Group—D

General—Extremely flammable. Ignited by heat, sparks, open flame. Flashback along vapor trail may occur. Vapor may explode if ignited in an enclosed area. Precautions must be taken to prevent static electricity buildup.
Flash Point (°F) .... 12 (Benzene is solid at 12°F)
Flammable Limits........................ 1.4 to 8.0%
Autoignition Temp. (°F) 1076
Extinguishing Agents................... CO₂, dry chemical, foam, water fog
Special Fire Procedures .... Water may be ineffective on a fire. Fire parties must wear respiratory protection and rubber boots. In other respects, fight like a gasoline fire. Explosion hazard is great if ignition has not already occurred and hence civil defense authorities should also be alerted. Cool exposed tanks with water.

HEALTH HAZARD DATA

Health Hazard Ratings	Odor Threshold (ppm)	PEL/TWA (ppm)	TLV/TWA (ppm)
1, 1, 3	4.68	29 CFR 1910.1028	10

General—Benzene is a known carcinogen. Benzene vapors are severely toxic by inhalation. Benzene has a pleasant odor and narcotic effect and thus has poor warning properties.

Symptoms—Dizziness, headache, and drowsiness.

Short Exposure Tolerance—Vapor concentrations: 3000 ppm is endurable for 30–60 minutes (single exposure); 7500 ppm is dangerous in 30–60 minutes (single exposure); 20,000 ppm has been fatal in 5–10 minutes.

Exposure Procedures—Vapor—remove victim to fresh air; if breathing is difficult, administer oxygen. If breathing stops, apply artificial respiration. Skin or eye contact—remove contaminated clothing and gently flush affected areas with water for 15 minutes. Get medical help.

REACTIVITY DATA

Stability—Stable under normal conditions.

Compatibility—Material: Rubber on prolonged exposure to benzene first swells, then softens.

Cargo: Group 32 of compatibility chart.

SPILL OR LEAK PROCEDURE

Wear rubber gloves, face shield, plastic coated clothing. Wear self-contained breathing apparatus. Approach from upwind side. Avoid contact with liquid. Secure ignition sources. Small spills may be flushed away with water.

If a spill occurs, call the National Response Center, 800-424-8802.

Remarks:

Figure 2-2. Excerpt from *Chemical Data Guide for Bulk Shipment by Water* (former CG-388). Courtesy United States Coast Guard.

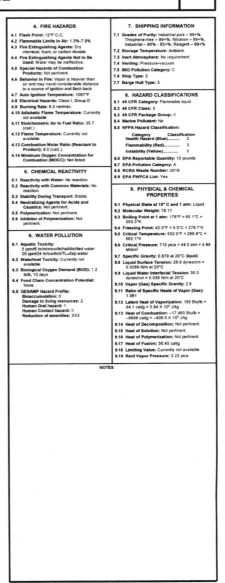

BENZENE | BNZ

CAUTIONARY RESPONSE INFORMATION

Common Synonyms	Watery liquid	Colorless	Gasoline-like odor
Benzol Benzole			

Floats on water. Flammable, irritating vapor is produced. Freezing point is 42°F.

Restrict access.
Avoid contact with liquid and vapor.
Wear goggles and self-contained breathing apparatus.
Shut off ignition sources and call fire department.
Stay upwind and use water spray to "knock down" vapor.
Notify local health and pollution control agencies.
Protect water intakes.

Fire
FLAMMABLE.
Flashback along vapor trail may occur.
Vapor may explode if ignited in an enclosed area.
Wear goggles and self-contained breathing apparatus.
Extinguish with dry chemical, foam, or carbon dioxide.
Water may be ineffective on fire.
Cool exposed containers with water.

Exposure
CALL FOR MEDICAL AID.
VAPOR
Irritating to eyes, nose and throat.
If inhaled, will cause headache, difficult breathing, or loss of consciousness.
Move to fresh air.
If breathing has stopped, give artificial respiration.
If breathing is difficult, give oxygen.
LIQUID
Irritating to skin and eyes.
Harmful if swallowed.
Remove contaminated clothing and shoes.
Flush affected areas with plenty of water.
IF IN EYES, hold eyelids open and flush with plenty of water.
IF SWALLOWED and victim is CONSCIOUS, have victim drink water or milk.

Water Pollution
HARMFUL TO AQUATIC LIFE IN VERY LOW CONCENTRATIONS.
May be dangerous if it enters water intakes.
Notify local health and wildlife officials.
Notify operators of nearby water intakes.

1. CORRECTIVE RESPONSE ACTIONS
Stop discharge
Contain
Collection Systems: Skim
Chemical and Physical Treatment: Burn
Salvage waterfowl

2. CHEMICAL DESIGNATIONS
2.1 CG Compatibility Group: 32; Aromatic Hydrocarbon
2.2 Formula: C_6H_6
2.3 IMO/UN Designation: 3.2/1114
2.4 DOT ID No.: 1114
2.5 CAS Registry No.: 71-43-2
2.6 NAERG Guide No.: 130
2.7 Standard Industrial Trade Classification: 51122

3. HEALTH HAZARDS
3.1 **Personal Protective Equipment:** Self contained positive pressure breathing apparatus; protective gloves and clothing.
3.2 **Symptoms Following Exposure:** Dizziness, excitation, pallor, followed by flushing, weakness, headache, breathlessness, chest constriction, nausea, and vomiting. Coma and possible death.
3.3 **Treatment of Exposure:** SKIN: flush with water followed by soap and water; remove contaminated clothing and wash skin. EYES: flush with plenty of water until irritation subsides. INHALATION: remove from exposure immediately. Call a physician. IF breathing is irregular or stopped, start resuscitation, administer oxygen.
3.4 **TLV-TWA:** 0.5 ppm
3.5 **TLV-STEL:** 2.5 ppm
3.6 **TLV-Ceiling:** Not listed
3.7 **Toxicity by Ingestion:** Grade 3; LD_{50} = 50 to 500 mg/kg
3.8 **Toxicity by Inhalation:** Currently not available.
3.9 **Chronic Toxicity:** Leukemia.
3.10 **Vapor (Gas) Irritant Characteristics:** If present in high concentrations, vapors may cause irritation of eyes or respiratory system. The effect is temporary.
3.11 **Liquid or Solid Characteristics:** Minimum hazard. If spilled on clothing and allowed to remain, may cause smarting and reddening of the skin.
3.12 **Odor Threshold:** 4.68 ppm
3.13 **IDLH Value:** 500 ppm
3.14 **OSHA PEL-TWA:** 1 ppm.
3.15 **OSHA PEL-STEL:** 5 ppm
3.16 **OSHA PEL-Ceiling:** Not listed
3.17 **EPA AEGL:** Not listed

4. FIRE HAZARDS
4.1 **Flash Point:** 12°F C.C.
4.2 **Flammable Limits in Air:** 1.3%-7.9%
4.3 **Fire Extinguishing Agents:** Dry chemical, foam, or carbon dioxide.
4.4 **Fire Extinguishing Agents Not to Be Used:** Water may be ineffective.
4.5 **Special Hazards of Combustion Products:** Not pertinent.
4.6 **Behavior in Fire:** Vapor is heavier than air and may travel considerable distance to a source of ignition and flash back.
4.7 **Auto Ignition Temperature:** 1097°F
4.8 **Electrical Hazards:** Class I, Group D
4.9 **Burning Rate:** 6.0 mm/min.
4.10 **Adiabatic Flame Temperature:** Currently not available
4.11 **Stoichometric Air to Fuel Ratio:** 35.7 (calc.)
4.12 **Flame Temperature:** Currently not available
4.13 **Combustion Molar Ratio (Reactant to Product):** 9.0 (calc.)
4.14 **Minimum Oxygen Concentration for Combustion (MOCC):** Not listed

5. CHEMICAL REACTIVITY
5.1 **Reactivity with Water:** No reaction.
5.2 **Reactivity with Common Materials:** No reaction.
5.3 **Stability During Transport:** Stable.
5.4 **Neutralizing Agents for Acids and Caustics:** Not pertinent.
5.5 **Polymerization:** Not pertinent.
5.6 **Inhibitor of Polymerization:** Not pertinent.

6. WATER POLLUTION
6.1 **Aquatic Toxicity:**
5 ppm/6 hr/minnow/lethal/distilled water
20 ppm/24 hr/sunfish/TL$_m$/tap water
6.2 **Waterfowl Toxicity:** Currently not available
6.3 **Biological Oxygen Demand (BOD):** 1.2 lb/lb, 10 days
6.4 **Food Chain Concentration Potential:** None.
6.5 **GESAMP Hazard Profile:**
Bioaccumulation: 0
Damage to living resources: 2
Human Oral hazard: 1
Human Contact hazard: II
Reduction of amenities: XXX

7. SHIPPING INFORMATION
7.1 **Grades of Purity:** Industrial pure – 99+%; Thiophene-free – 99+%; Nitration – 99+%; Industrial – 90% - 85+%; Reagent – 99+%
7.2 **Storage Temperature:** Ambient.
7.3 **Inert Atmosphere:** No requirement.
7.4 **Venting:** Pressure-vacuum.
7.5 **IMO Pollution Category:** C
7.6 **Ship Type:** 3
7.7 **Barge Hull Type:** 3

8. HAZARD CLASSIFICATIONS
8.1 **49 CFR Category:** Flammable liquid
8.2 **49 CFR Class:** 3
8.3 **49 CFR Package Group:** II
8.4 **Marine Pollutant:** No
8.5 **NFPA Hazard Classification:**

Category	Classification
Health Hazard (Blue)	2
Flammability (Red)	3
Instability (Yellow)	0

8.6 **EPA Reportable Quantity:** 10 pounds
8.7 **EPA Pollution Category:** A
8.8 **RCRA Waste Number:** U019
8.9 **EPA FWPCA List:** Yes

9. PHYSICAL & CHEMICAL PROPERTIES
9.1 **Physical State at 15° C and 1 atm:** Liquid
9.2 **Molecular Weight:** 78.11
9.3 **Boiling Point at 1 atm:** 176°F = 80.1°C = 353.3°K
9.4 **Freezing Point:** 42.0°F = 5.5°C = 278.7°K
9.5 **Critical Temperature:** 552.0°F = 288.9°C = 562.1°K
9.6 **Critical Pressure:** 710 psia = 48.3 atm = 4.89 MN/m²
9.7 **Specific Gravity:** 0.879 at 20°C (liquid)
9.8 **Liquid Surface Tension:** 28.9 dynes/cm = 0.0289 N/m at 20°C
9.9 **Liquid Water Interfacial Tension:** 35.0 dynes/cm = 0.035 N/m at 20°C
9.10 **Vapor (Gas) Specific Gravity:** 2.8
9.11 **Ratio of Specific Heats of Vapor (Gas):** 1.061
9.12 **Latent Heat of Vaporization:** 169 Btu/lb = 94.1 cal/g = 3.94 X 10⁵ J/kg
9.13 **Heat of Combustion:** –17,460 Btu/lb = –9698 cal/g = –406.0 X 10⁵ J/kg
9.14 **Heat of Decomposition:** Not pertinent.
9.15 **Heat of Solution:** Not pertinent.
9.16 **Heat of Polymerization:** Not pertinent.
9.17 **Heat of Fusion:** 30.45 cal/g
9.18 **Limiting Value:** Currently not available
9.19 **Reid Vapor Pressure:** 3.22 psia

NOTES

JUNE 1999

Figure 2-3. Excerpt from the *Chemical Hazards Response Information System* (CHRIS). Courtesy United States Coast Guard.

- Chemical Codes (Summary of Minimum Requirements) from the International Maritime Organization (IMO)
- *Code of Federal Regulations* (CFR) from the United States government
- *Pocket Guide to Chemical Hazards* from the National Institute for Occupational Safety and Health (NIOSH)

The information provided is of particular relevance to individuals responsible for the safe transfer and transport of bulk liquid cargoes. The person-in-charge should not only have ready access to this information but should also possess a thorough understanding of the characteristics of the substances being handled. Toward that end, the following list summarizes the various categories addressed in a typical material safety data sheet:

Cargo identification and emergency information
Components and hazard information
Primary routes of entry and emergency and first aid procedures
Fire and explosion hazard information
Health and hazard information
Physical data
Reactivity
Environmental information
Protection and precautions
Transportation and OSHA label information

A sample MSDS for commercial gasoline from ConocoPhillips can be found in Appendix A.

HAZARDS

The following list shows the main hazards associated with the handling and carriage of bulk liquid cargoes:
Fire and explosion
Static electricity
Toxicity
Oxygen deficiency
Reactivity
Corrosivity

Fire and Explosion
Tremendous strides have been made in the design of tankers to improve the safety of personnel and of the vessel. One of the primary areas of attention in the design of a modern tanker is fire safety.

HAZARDOUS AREAS OF THE VESSEL
Historically, efforts to reduce the risk of handling and transporting flammable cargoes have focused on the elimination of ignition sources from those areas of the vessel where the existence of a flammable atmosphere was likely. This led to the identification and designation of certain areas of the vessel as potentially hazardous zones.

As an additional safeguard, many tank vessels are now equipped with inert gas systems. The use of this safety system has greatly improved the fire-prevention measures in the tanker industry. As previously discussed, operators must have a thorough understanding of inerting procedures to realize the full benefit of this system. The hazards and precautions summarized in this section address the safe carriage and handling of petroleum and chemicals. For the purposes of fire prevention, a tank vessel can be divided into the following areas:

Location of flammable atmospheres: These are areas such as non-inerted cargo tanks, pump rooms, deckhouses, vent stacks, cargo and vapor manifolds, and others where the existence of a flammable atmosphere is possible.

Two general precautions cover the main methods of fire prevention in these areas:

1. Eliminate all sources of ignition:
   Smoking and open flames
   Portable electrical equipment and nonapproved flashlights
   Non-intrinsically safe electronic equipment (radios, cameras, etc.)
   Hot work
   Use of power and hand tools
   Electrostatic discharges
2. Maintain the atmospheres outside the flammable range, either by removing the hydrocarbon content (gas-freeing) or by reducing the oxygen content to 8 percent or less by volume (inerting).

Areas containing heat and ignition sources: These are working spaces such as machinery and boiler spaces, galleys, and so forth, that contain electrical equipment and other sources of heat and ignition. The main method of fire prevention in these areas is to keep them free of flammable vapors.

Living areas: Crew accommodation areas are normally free of flammable vapors but contain combustible material such as furniture, linen, paper, and more. The main method of fire prevention in these areas is to keep ignition sources to a minimum.

Pumprooms: The cargo pump room is a small, complex space that contains a large concentration of piping and equipment. Any leakage of volatile liquids has the potential to generate flammable and/or toxic atmospheres. A pump room may also contain a number of potential ignition sources such as lighting, tools, electrical equipment, and mechanical equipment. Personnel should also include the following checks as precautions against fire and explosion:

Ensure that forced draft ventilation is operating and entry procedures are followed whenever the pump room is entered. (see Chapter 15)
Make frequent rounds to find potential sources of cargo leaks, flammable vapors, and ignition.
Ensure that pumproom bilges are clean and dry.

Cargo Tanks: The prevention of fire and explosion in the cargo area of the vessel is accomplished by the following:

Maintaining an inert atmosphere in the cargo and slop tanks at all times unless they are gas free
Maintaining a positive deck pressure in the inerted tanks to prevent the ingress of air
Eliminating all possible sources of ignition

Electrostatic Hazards

Static electricity is a potential fire and explosion risk when handling certain types of petroleum and petrochemical cargoes. In some operations, the electrical charge generated is capable of igniting flammable vapors such as those found in the atmosphere of a non-inerted cargo tank.

CAUSES

The sequence necessary for the development of a static electricity hazard involves (1) charge generation, (2) charge accumulation, and (3) electrostatic discharge.

The generation of static electricity occurs at the interface of dissimilar materials. These interfaces may be between two solids, between solids and liquids, or between liquids and liquids. If the two materials are separated by some mechanical action, one will carry an excess of positive charge and the other an excess of negative charge. The separated charges typically recombine to neutralize each other.

If one of the materials is a poor conductor of electricity, recombination will be limited and a difference in charge (electrical potential) will exist between the two bodies. These electric charges can accumulate and may equalize in the form of an electrical discharge. If these electrical discharges generate a sufficient amount of heat, they can ignite flammable vapors. In cargo operations, charge separation occurs in many ways:

Friction caused by the flow of petroleum liquid through extensive piping systems, strainers, and filters
Petroleum and water mixtures in the cargo tanks
Splashing or agitation of petroleum liquids such as flow through a nozzle

When handling a static accumulator (poor conductor of electricity), the material must be given ample time for the separated charges to recombine. In practice, this is known as the relaxation time during which an accumulated charge will have an opportunity to dissipate. If the material has a high conductivity (good conductor of electricity), the recombination occurs quickly and offsets the accumulation of separated charges. Consequently, there is little generation of static electricity by materials that are good conductors. Examples are metals and water solutions, including seawater, that are incapable of holding a charge unless insulated. Oils in this category include crude oils, residual fuel oils, and asphalt.

Static accumulators are characterized as having low conductivity; they require a longer relaxation time before the charge ultimately dissipates to earth. In this case, more static electricity accumulates, increasing the possibility of an electrostatic discharge that could ignite a flammable atmosphere. Examples of known static accumulators include gasoline, naphtha, kerosene, heating oil, jet fuel, and lubricating oil.

The charge that accumulates in a liquid, solid, or mist establishes an electrical field between it and nearby earthed bodies. The strength of the electrical field is the *voltage gradient,* which is determined by the difference in voltage between the two points and by their distance apart. To prevent the possibility of an electrostatic discharge that could occur if portable equipment was introduced into a charged atmosphere, it is sound practice to bond all metal objects together. In other words, the PIC should not introduce any equipment that might be electrically insulated into a potentially hazardous atmosphere. Examples of portable equipment that usually requires bonding include portable tank cleaning machines, manual gauging equipment, temperature probes, and sampling equipment.

LOADING OF STATIC ACCUMULATOR CARGOES

To reduce the risks associated with the handling of a known static accumulating cargo, the PIC must adhere to several precautions. (It should be noted that static precautions are not necessary when the cargo tank in question is maintained in the inert condition).

Prior to the commencement of a cargo transfer, the PIC should confer with a shore representative to identify any cargoes classified as static accumulators. If in doubt, the PIC should consult the appropriate MSDS and follow the guidance pertaining to the handling of the particular product.

The PIC should take steps to minimize the presence of water in the cargo system by properly draining all cargo tanks and pipelines prior to the commencement of loading. The mixing of dissimilar liquids (oil and water) can contribute to the creation of an appreciable electrostatic charge in a space and should therefore be avoided.

It is advisable to treat all distillates as static accumulators unless they contain an antistatic additive. Distillates may carry a sufficient charge to constitute a hazard during loading and for a period of time after the completion of the loading operation.

The beginning of the loading operation is a critical point due to the risks posed by excessive initial loading rates, excessive splashing and turbulence into an empty tank, or the presence of water in the pipelines and bottom of the tank.

The PIC should minimize electrostatic generation in the early stages of the loading operation by restricting the initial flow rate to the cargo tank(s). This reduced flow rate should be maintained until the bottom framing in the tank is covered and all splashing and turbulence has ceased. The term commonly used to describe this process is known as *cushioning* a tank. According to the *International Safety Guide for Oil Tankers and Terminals (ISGOTT)*, the initial flow rate should be restricted to a linear velocity that does not exceed 1 meter per second. Table 2-7 indicates typical loading rates that correspond to a linear velocity of 1 meter per second.

Table 2-7
Flow Rates Corresponding to 1 Meter Per Second

Nominal Pipeline Diameter (mm)	Approximate Flow Rate (Cubic meters/hour)
80	17
100	29
150	67
200	116
250	183
305	262
360	320
410	424
460	542
510	676
610	987
710	1,354
810	1,782

Reprinted with permission from the *International Safety Guide for Oil Tankers and Terminals (ISGOTT)*, 5th edition. Courtesy OCIMF, ICS, and IAPH.

Throughout the loading of a cargo tank and for a period of at least thirty minutes after the completion of loading, metallic gauging and sampling equipment must not be introduced or allowed to remain in the tank. Nonconducting (nonmetallic) equipment may be used at any time; however, ropes or tapes employed with this equipment should not be made from synthetic materials.

After the relaxation period of thirty minutes has elapsed, metallic equipment may be used; however, it must be bonded and properly earthed to the vessel's structure before use.

DISCHARGE OF STATIC ACCUMULATOR CARGOES

At the discharge terminal, the PIC should consult with the shore representative concerning the proper procedure to be followed when commencing the discharge of a static accumulating cargo. In general, the initial pumping rate ashore should be limited until a sufficient cushion is developed in the bottom of the shore tank. This precaution is followed to minimize the splashing and agitation of the cargo at the initial stage of filling the shore tank.

Transport of Residual Fuel Oils

The carriage of high flash point residual fuel oil has historically been viewed as a substance that did not pose a serious flammability hazard to the vessel. Despite the fact that residual fuel oils have a measured flashpoint well above the temperature typically maintained during transport it has been found that the vapor concentrations in the ullage space may be near to or within the flammable range. The reason for this can vary from either the refining source or possible blending of various components while in storage or during loading. Compounding the problem is the fact that under present regulation the carriage of residual fuel oils is for the most part exempt from the requirement to inert the cargo tanks provided the vessel is not heating the fuel oil to a temperature within 5° C of its flash point. Therefore when transporting residual fuel oils on non inerted vessels, it is recommended that personnel monitor the vapor concentration of the atmosphere above the cargo in the tanks. Should the growth of vapors be detected in the atmosphere of the cargo tank(s), action should be taken before a flammable atmosphere is created preferably by purging the tank with low pressure air to maintain the vapor concentration of the tank at a safe level. On vessels equipped with an inert gas system it is recommended that the IG system be operated and the cargo tanks maintained in an inert condition when transporting residual fuel oil cargoes.

Toxicity

Aromatic hydrocarbons: Aromatic hydrocarbons—including benzene, toluene, and xylene—are found in varying proportions in a wide array of petroleum cargoes such as gasoline, naphtha, and even some crude oils.

The TLVs of aromatic hydrocarbons are lower than most nonaromatic hydrocarbons. For example, the TLV of benzene, a recognized carcinogen, is as low as 1 ppm. The latent effect of exposure to benzene vapors can result in potentially fatal disorders of the blood. In the United States, any cargo containing 0.5 percent or more benzene by volume is classified as a "regulated cargo," and specific rules must be followed that address handling and occupational exposure in the workplace. The detailed requirements can be found in the U.S. *Code of Federal Regulations* (Title 46 CFR Part 197 Subpart C). The person-in-charge should have a thorough understanding of the content of this regulation, as he or she is responsible for compliance with the rules.

Hydrogen sulfide: Some crude oils, described as sour, contain a high level of hydrogen sulfide. The effects of exposure to hydrogen sulfide gas can be both quick and deadly. For a more comprehensive discussion of the effects of hydrogen sulfide to individuals when exposed to concentrations in excess of its published TLV of 5 ppm, consult Chapter 15.

Precautions against toxicity by inhalation: Individuals involved in handling potentially toxic substances should avoid exposure to concentrations above the published TLV. If exposure through inhalation is possible, suitable respiratory protection should be worn to minimize the inhalation of harmful vapors. Certain operations such as the venting of cargo tanks during loading, purging, and gas-freeing may result in elevated exposure to personnel on deck. During these operations, the atmosphere exits the cargo tanks via the vent system and dilutes with the surrounding air, increasing the risk of fire and exposure to personnel. Individuals involved in such operations should wear proper respiratory protection.

Personnel are advised never to enter a compartment that contained cargo—or one that has been sealed for a period of time—without first testing the atmosphere. It should be assumed that the atmosphere of an enclosed space is incapable of supporting life until proven otherwise. All company and industry guidelines should be followed with respect to testing and entry into an enclosed space. Entry should only be permitted after a permit-to-enter or a marine chemist certificate has been issued.

Oxygen Deficiency

Air normally contains approximately 21 percent oxygen by volume. Individuals exposed to concentrations below that level are at risk of suffering from oxygen deficiency.

As the oxygen level decreases below 21 percent by volume, an individual will experience a changing breathing pattern. Unfortunately, many individuals fail to recognize the danger signs associated with an oxygen-deficient atmosphere until it is too late. This can be particularly problematic when escape involves climbing from the bottom of a space such as a cargo tank or pumproom. The degree of impairment will differ among individuals based on such variables as age, physical condition, and so forth; however, all begin to experience the adverse effects of oxygen deficiency below 16 percent by volume. The oxygen level typically maintained in the atmosphere of an inerted cargo tank—4 percent to 8 percent oxygen by volume—will result in immediate unconsciousness of an individual and irreversible brain damage within a short period of time.

The oxygen content of any enclosed space may be deficient for a number of reasons. Compartments in which even seemingly harmless liquids were carried, such as freshwater and seawater ballast, can pose a significant hazard due to a lack of oxygen. On a modern tank vessel, the most obvious cause of oxygen deficiency is an inerted cargo tank in which the oxygen level is intentionally maintained at or below 8 percent by volume.

HAZARDS ASSOCIATED WITH INERT GAS

Inert gas is used on modern tank vessels to control the oxygen content within the atmosphere of the cargo tank(s). The use of this system ensures that a nonflammable condition is maintained via oxygen deficiency within the cargo tanks unless they are gas free. On board most petroleum tankers, inert gas is derived either by using the flue gas from the boilers or an oil-fired generator. Before the flue gas is piped to the tanks it is processed (cooled and cleaned) in a scrubber. Table 2-8 shows the composition of the flue gas before and after the scrubbing process.

Table 2-8
Content of Flue Gas Before and After Scrubbing

Content	Before	After
Nitrogen	80% by volume	Same
Carbon dioxide	Approx. 14%	Same
Oxygen	2–5%	Same
Sulfur dioxide	Approx. 0.3%	Approx. 0.005% by volume
Carbon monoxide	Approx. 0.01 % by volume	Same
Nitrogen oxides	Approx. 0.02% by volume	Same
Water vapor	Approx. 5% by volume	Approx. 0.01% by volume
Soot and particulate	300 mg/m³ by volume	30 mg/m³
Heat	200°–300°C	Near ambient sea temp.

As indicated in table 2-8, the primary hazard associated with inert gas is its exceptionally low oxygen content. In addition to oxygen deficiency, exposure to an inert gas derived from a combustion process (exhaust gas) should be avoided as it contains a number of toxic constituents including carbon monoxide, sulfur dioxide, and nitrogen oxides.

The carbon monoxide content depends on the combustion conditions. The TLV of carbon monoxide is 25 ppm. At an elevated level of exposure, the blood loses its ability to carry oxygen from the lungs to the rest of the body, resulting in carbon monoxide poisoning. The symptoms are headache, drowsiness, unconsciousness, and vomiting. In extreme cases, internal suffocation may occur, followed by death. The treatment is to remove the victim to fresh air or supply oxygen and, if necessary, apply artificial respiration.

The sulfur dioxide content of the exhaust gas usually depends on the sulfur content of the fuel oil consumed in the combustion process as well as the efficiency of the scrubber.

CARGO REACTIVITY

Reactivity hazards are associated with certain liquids that tend to react to extremes of temperature, violent movement, and so forth, as well as to mixing with incompatible liquids and materials. The type of reaction will depend on the stability of the liquid and its compatibility with other liquids and materials.

STABILITY

Stability refers to the ability of a liquid product to return to a normal condition when affected by external forces. A stable liquid will not react dangerously when exposed to extreme conditions of temperature and movement. On the other hand, an unstable liquid may produce dangerous reactions.

When transporting a chemical cargo in which a vigorous self-reaction (usually resulting in polymerization) is possible, a specific chemical additive known as an inhibitor is required. Some of the more common examples of cargoes requiring the use of an inhibitor include acrylonitrile, vinyl chloride, and styrene monomer. In addition to the use of inhibitors, if a chemical cargo is potentially reactive with air, then it may be necessary to maintain an inert atmosphere (typically nitrogen) in the tank. Cargoes such as carbon disulfide, ethyl ether, and propylene oxide are examples of substances that must be maintained under a pad or blanket of inert gas.

The MSDS sheet should be consulted for details concerning the stability of a liquid and its reactivity, including the factors capable of causing violent reactions.

COMPATIBILITY

The compatibility of chemical cargoes is associated with the type of reaction that might occur when a substance comes into contact with other liquids, structural materials, and so on. Some liquids and materials are incompatible and will react violently or become contaminated when they come into contact with each other. The compatibility of liquid cargoes can be found in the MSDS and in the USCG Compatibility Chart contained in Title 46 CFR Part 150 Table 1 and Figure 1. This chart is also reproduced in the *Chemical Data Guide for Bulk Shipment by Water* (former *CG-388*). The PIC should be familiar with any unusual properties and review emergency procedures involving a casualty with each chemical cargo, also in Appendix H on enclosed CD.

U.S. Coast Guard Incompatibility Chart

X = Incompatible Groups — **Reactive Groups**

Cargo Groups	1	2	3	4	5	6	7	8	9	10	11	12	13	14	15	16	17	18	19	20	21	22
Non-oxidizing mineral acids 1		X			X	X	X	X	X	X	X	X				X	X					
Sulfuric acid 2	X		X	X	X	X	X	X	X	X	X	X	X	X	X	X	X	X	X	X	X	X
Nitric acid 3		X	X		X	X	X	X	X	X	X	X	X	X	X	X	X	X	X	X	X	
Organic acids 4		X			X	X	X	X								X	X					
Caustics 5	X	X	X	X							X	X				X	X		X	X	X	X
Ammonia 6	X	X	X	X						X	X	X	X			X	X		X			
Aliphatic amines 7	X	X	X	X							X	X	X	X	X	X	X		X	X	X	X
Alkanolamines 8	X	X	X	X							X	X	X	X	X	X	X		X			
Aromatic amines 9	X	X	X								X	X							X			
Amides 10	X	X	X			X						X									X	
Organic anhydrides 11	X	X	X		X	X	X	X	X			X										
Isocyanates 12	X	X	X	X	X	X	X	X	X	X										X		X
Vinyl acetate 13	X	X	X		X	X	X															
Acrylates 14		X	X			X	X															
Substituted allyls 15		X	X			X	X															
Alklene oxides 16	X	X	X	X	X	X	X	X														
Epichlorohydrin 17	X	X	X	X	X	X	X	X														
Ketones 18		X	X				X															
Aldehydes 19		X	X		X	X	X	X	X													
Alcohols, Glycols 20		X	X		X		X					X										
Phenols, Cresols 21		X	X		X		X			X												
Caprolactam solution 22		X			X		X					X										
Olefins 30		X	X																			
Paraffins 31																						
Aromatic hydrocarbons 32			X																			
Misc. Hydrocarbon mixtures 33			X																			
Esters 34		X	X																			
Vinyl halides 35			X																			X
Halogenated hydrocarbons 36																						
Nitriles 37		X																				
Carbon disulfide 38							X	X														
Sulfonlane 39																						
Glycol ethers 40		X										X										
Ethers 41		X	X																			
Nitro compounds 42					X	X	X	X	X													
Misc. water solutions 43		X										X										
	1	2	3	4	5	6	7	8	9	10	11	12	13	14	15	16	17	18	19	20	21	22

Figure 2-4. U.S. Coast Guard Incompatibility Chart, also in Appendix H on enclosed disc.

Corrosives

When transporting cargoes that are capable of deteriorating the structure of a cargo tank and/or the vessel, the use of suitable coatings or special construction materials is required. By their very nature, corrosive cargoes present a significant risk, not only to the vessel but also to personnel. For this reason, individuals working around or handling these cargoes must be aware of the physical dangers, emergency response, and use of proper protective gear to minimize the risks.

COATINGS

To protect the steel shell of a cargo tank, hard coatings are commonly employed to provide a protective barrier. Consult chapter 13 for a discussion of the various types of coatings used in the cargo system of a tank vessel. In certain situations, special materials such as stainless steel may be used in the construction of the tank, piping, valves, and pumps. Another alternative is the use of stainless cladding.

REVIEW

1. Define the following terms associated with the classification of liquid cargoes:
    Volatility
    Flash point
    Reid vapor pressure
2. Sketch a typical flammable range diagram and identify the following:
    Upper flammable limit
    Lower flammable limit
    Rich mixture
    Lean mixture
    Inert mixture
3. Define the following terms:
    Primary inerting
    Purging
    Gas-freeing
    Topping-up
    On inert gas equipped vessels, how is a nonflammable condition maintained during the operating life of the vessel (between shipyard periods)? Use the flammable range diagram to illustrate how this is possible.
4. Define API gravity.
5. What is the viscosity and pour point of a liquid cargo? Explain the importance of each of these terms with respect to the cargo operation on a tank vessel.
6. List and define the various terms used to express the toxicity of a substance. Why is an individual's sense of smell an unreliable indicator of the presence of a toxic vapor?
7. What is the difference between a TLV and PEL?
8. List the various sources of cargo information available to the PIC.
9. List the three primary ways a toxic substance can enter the body. Which of these exposures is the most difficult to control?
10. Any cargo classified as a static accumulator is a _____ conductor.

11. List the precautions that should be followed when loading a known static accumulating cargo into a noninerted vessel.
12. What is meant by a reactivity hazard as applied to the transport of chemical cargoes?
13. What precautions may be necessary when transporting a chemical that is self-reactive?
14. What publication would you consult to determine the compatibility of different chemical cargoes?
15. When handling corrosive cargoes, describe the precautions that should be taken to protect the vessel and personnel.

CHALLENGE QUESTIONS

16. During the discharge of a crude oil cargo you discover the main tank valve at the bellmouth in cargo tank #4S is inoperative. Describe the step-by-step process of preparing this tank for entry during the upcoming ballast leg of the voyage.
17. Prior to the widespread use of inert gas systems on tankers, what method of atmosphere control was employed to keep the cargo tank atmospheres in a non-flammable condition?
18. A properly inerted cargo tank is one with a measured oxygen content of _____ % or less and maintained at a positive pressure above _____ mmwg.
19. What concerns does the transport of high flashpoint residual fuel oils pose to operators today?
20. Why is maintaining the level of exposure to a substance below the published TLV-TWA guideline not an absolute guarantee of the safety of an individual in the workplace?
    Give two instances when it is considered necessary and acceptable to enter a compartment that is knowingly classified as IDLH?

CHAPTER 3

Chemical Tankers

By Margaret Kaigh Doyle

This chapter is devoted to the three basic elements that comprise chemical tanker operations: the cargoes themselves, the basic design and procedures, and the regulations that apply to the chemical tanker. It is important to understand what it takes to carry chemical cargoes safely and how the safety and pollution hazards are assessed. A basic grasp of the design features as well as ship operations and the difficulties encountered with everyday operations such as voyage planning, vapor control and tank washing is integral to understanding chemical tanker operations. The last part of this chapter discusses the regulations that apply to chemical tankers and the cargoes that they carry.

CHEMICAL CARGOES AND CLASSIFICATION

Chemicals

Bulk chemical shipments generally belong to one of the following family groups of chemicals: heavy chemicals, petrochemicals, coal tar products. Chemical tankers also carry a variety of products that are unrelated to chemicals including animal and vegetable oils, lube oils, molasses, juices and even wine. The carriage of these products requires as much experience, care and protection as the pure chemicals. In some trades, all these non-chemical products move in the same direction as the chemicals, or they may move in opposite directions, in either case complementing the chemical movements and making the voyages economically viable.

Heavy chemicals are "big movers" in the chemical tanker trade. Inorganic acids are the most common heavy chemicals moved. Sulphuric acid is produced from sulphur, air and water is a very popular inorganic acid because it is relatively inexpensive to produce and serves many purposes. It can be used in the production of fertilizer, explosives, in plastics such as rayon, and in the removal of oxides from metals. Phosphoric acid is used for the production of superphosphates and other products, including detergents, paints, and foodstuffs. Nitric acid is a basic ingredient of explosives, nitrate fertilizers and many dyes, and plastics. Caustic soda is also shipped in liquid form. Others include hydrochloric acid used in the steel reduction process and ore reduction, and ammonia.

Molasses is another product carried in chemical tankers. It is made from either sugar beet or sugar cane. It can then be fermented into alcohols such as rum. Alcohols are also produced by the petrochemical industry, but some can also come from the fermentation of starch, such as ethanol. Alcohols of this type include ethyl, methyl and propyl alcohols.

Edible vegetable oils are derived from soya beans, groundnuts, cottonseed, sunflowers, olives, rape and other seeds. Coconut and palm oil can be used for cooking and also in the production of soap. Industrial oils come from linseed and castor seed. Some fats are extracted from animals including lard and fish oils. Fats and oils are tri-esters of glycerol. These are sometimes called fatty acids owing to their presence in fats. The difference between oils and fats lies in their melting temperatures.

Petrochemical products form the most complex and probably the most versatile group of chemicals carried in bulk. They are all carbon compounds derived from oil or gas extensively used in the production of fibre, artificial rubber and plastics. Substances carried in chemical tankers include aromatics such as benzene that can be derived from oil or coal. Other big mover petrochemicals include xylenes (used in the production of polyester fibres); phenol (previously known as carbolic acid) and styrenes.

Basic petrochemicals, such as ethylene, propylene, and butadiene, are obtained by the cracking of gas oil from crude oil. Another source of these products is natural gas. All three of these basic petrochemicals are transported in liquid form by gas carriers. According to the Society of International Gas Tanker and Terminal Operators Ltd (SIGTTO), the world fleet of LPG carriers is engaged in the carriage of a wide range of liquefied gases, including liquefied petroleum gas (primarily propane, butane and mixtures of the two), ammonia and petrochemical gases (primarily ethylene, propylene, butadiene and vinyl chloride monomer).

The aromatic products benzene, toluene and xylenes (also known as BTX) are created using a process called catalytic reforming. These cargoes are then used as transitional agents for the production of solvents and for gasoline blending.

Coal tar is derived from the carbonization of coal. Many of the products derived from coal can be derived from oil as well because they are both fossil fuels. The derivatives include benzene, phenol, naphthalene and many more. Common products which are derived from coal include nylon, aspirin, antiseptics and herbicides.

Chemical Hazards and Problems

The hazard evaluation of chemicals is in itself a complex problem stemming from the combination of the flammability and toxicity characteristics of the chemicals themselves as well as from design and operation hazards they pose.

The International Maritime Organization (IMO) Working Group on the Evaluation of Safety and Pollution Hazards (ESPH) is a working group of the IMO Sub-Committee on Bulk Liquids and Gases (BLG) and reports to the IMO Marine Environment Committee (MEPC) and the Maritime Safety Committee (MSC). The majority of ESPH's work deals with the assignment of pollution categories and carriage requirements for products. This ensures their safe carriage and protection of the marine environment.

The overall hazard to the environment and the intrinsic hazards of the chemicals are evaluated using the hazard rating profile developed by the Joint Group of Experts on the Scientific Aspects of Marine Pollution (GESAMP). GESAMP was established in 1967 by a number of United Nations Agencies. The purpose of GESAMP is to provide advice to the agencies and, through them their Member Governments on the newly discovered safety and pollution hazards of the products being developed and subsequently traded.

The results of the environmental hazard evaluation of a given substance are laid out in the products "hazard profile". This profile acts as a fingerprint of the detailed scientific data on the intrinsic properties of a chemical substance to be widely published, without breaching the manufacturer's requirement for confidentiality. These are published bi-annually by the IMO in the form of a composite list. The GESAMP hazard profiles are based on the release into the sea of noxious substances and they fall into four categories:

- Damage to living resources
- Hazards to human health
- Reduction of amenities
- Interference with other users of the sea.

The main hazards and problems are listed below:

- *Cargo density* - the specific gravity of chemicals carried at sea varies greatly. Some are lighter than water. Others are twice as dense. Those substances which have especially high density include inorganic acids, caustic soda and some halogenated hydrocarbons.

- *High viscosity* - some lubricating oil additives, molasses and other products are very viscous, especially at low temperatures. As a result they are sticky and move very slowly, causing problems in cargo-handling and cleaning.

- *Low boiling point* - some chemicals vaporise at a relatively low temperature. This can cause containment problems, since when a liquid turns into a gas it expands, creating growing pressure. It is necessary, therefore, to provide either a cooling system or to carry the chemical in a specially-designed pressurised vessel.

- *Reaction to other substances* - some chemicals react to water, to air or to other products. Measures therefore have to be taken to protect them. Apart from the fact that an accident can lead to a dangerous reaction (such as the emission of a poisonous gas) many chemicals can be ruined if they are contaminated by other substances. Methanol, lubricating oil additives and alcohols can be spoiled by even a slight amount of water contamination. Too much oxygen can lead to a rapid deterioration in the quality of some vegetable oils. Other products can change into a different product completely.

- *Polymerization* - some substances, such as petrochemicals, do not need to come into contact with another chemical before undergoing a chemical change - they are self-reactive and liable to polymerization unless protected by an inhibitor. Polymerization is a process whereby the molecules of a substance combine to produce a new compound. The process can be accelerated by catalytic factors such as heat, light and the presence of rust, acids or other compounds. Styrene, methyl methylacrylate and vinyl acetate monomer are examples. Propylene oxide and butylene oxide are also liable to polymerization.

- *Toxicity* - many chemicals are highly poisonous, either in the form of liquid or vapor or both. The problem is sometimes made worse by the fact that toxicity can be increased when vapors from one substance come into contact with those from another.

- *Solidification* - some substances have to be kept at a high temperature, otherwise they solidify or become so viscous that they cannot easily be moved. Examples are some petrochemicals, molasses, waxes and vegetable oils and animal fats.

• *Pollution* - while many of the factors listed above present problems for the ship and crew, a considerable number of chemicals are extremely dangerous to marine and other forms of life. Although crude oil is probably the best-known pollutant of the sea, many chemicals are in fact far more poisonous and present a much greater threat - a threat which can be much more long lasting, since some of the chemicals concerned can enter the food chain and ultimately threaten humans as well as marine life.

GESAMP hazard profiles are used by the IMO to assign chemicals to pollution categories. The hazard profiles do not take into account exposure conditions and probabilities of release into the environment, i.e., considerations essential for a full environmental risk assessment of accidental spillages or intentional discharges.

The Tripartite Process

All cargoes carried in bulk are classified by the IMO and/or the vessel's flag state. The carriage requirements for a product are then determined by these entities using the guidelines set forth by GESAMP and IMO. If regulated by the IBC Code those cargoes must be authorized for carriage on that particular ship by, and listed on, the ship's Certificate of Fitness. The bulk carriage of any liquid product other than those defined as oil (subject to MARPOL Annex I) is prohibited unless the product has been evaluated and categorized for inclusion in Chapter 17 or 18 of the IBC Code (The International Code for the Construction and Equipment of Ships Carrying Dangerous Chemicals in Bulk).

Tripartite Agreement for the Provisional Assessment of Liquid Substances

When there is a need to transport a bulk liquid cargo that has not been classified, the shippers have to go to their administration and request that a Tripartite Agreement be established between the (1) shipping country, (2) the receiving country, and (3) the ship's flag state.

The Tripartite Agreement will be good for three years, before which time all outstanding or otherwise necessary data will have to have been forwarded to IMO (GESAMP) for the formal classification. Otherwise the Tripartite Agreement will expire. If the cargo is assessed as being regulated by the IBC Code or MARPOL Annex II, the ship's Certificate of Fitness and Procedures & Arrangements Manual will have to be amended (usually by the classification society, arranged by the owner or manager).

The IBC Code

The IBC Code and its predecessor, the BCH Code, were originally developed as guidance on the construction and equipment of ships carrying bulk cargoes of dangerous chemical substances. The code recommended suitable design criteria, construction standards, and other safety measures to minimise the risk involved in loading, carrying, and discharging such cargoes.

When the Code was originally developed, the main concern was not pollution It wasn't until 1973, when the IMO developed a new international convention on marine pollution known as the International Convention for the Prevention of Pollution from Ships (MARPOL), Annex II, that parts of the Code focused on the prevention of chemical pollution. Since then the IBC Code and MARPOL have been integrally linked to one another.

The international regulatory mechanism that exists for the evaluation and classification of new products is known as "the tripartite system". The tripartite system is an agreement between the cargo country of origin, destination, and ship flag state, after which the ship's certification still has to be endorsed to verify the ship's suitability for the cargo if the cargo is determined to be so regulated.

MEPC.2/Circulars

The products evaluated in accordance with these systems are then listed in MEPC 2 Circular, which is issued annually, by the IMO.

CHEMICAL TANKER DESIGN AND OPERATIONS

Chemical tankers are expensive to build, own and operate. The innate complexity of these tankers requires a major commitment from the owner in terms of capital cost, vessel management, and maintenance. Unlike the oil tanker operator the chemical tanker owner has to design a fleet which can meet the demanding requirements of a wide range of charterers and products. The chemical carrier fleet is small in terms of tonnage and numbers but is for the most part the newest and best-maintained sector of the total tanker fleet. The typical ship is a very high-value asset costing significantly more than a product or crude carrier of the same size.

Chemical tankers normally have many more tanks than oil or product carriers - thirty or more is common. This gives greater flexibility and since the amount carried of the individual cargoes is usually small (some less than 500 tons) the small size of the tank is not a disadvantage. The pipe work itself is simple not complicated, comprising one tank, one pump, and one line. What can be complicated is the flexible pipe arrangement at the manifold, especially if it links many tanks together for a common product. All procedures involving the cargo have to be carried out with great care and precision, both to avoid cargo contamination and also to ensure that cargoes owned by different shippers are kept separate. Piping, monitoring and control equipment is all highly complex.

Chemical tanker construction has to be of the highest possible standard. Because tank cleaning is crucially important to cargo quantity and purity, the stiffening is placed outside of the cargo tanks. This keeps the inside of the tank smooth allowing it to be cleaned more easily. Chemical cargo tanks are designed and constructed to avoid fatigue cracks or damage to the tank coating. The design itself has to take into consideration the properties of cargoes of varied density, some twice as dense as sea water. Other have to be carried at high temperatures to prevent solidification Both of these factors can affect the structure. Welding and other constructional features must be of the highest possible quality.

Chemical tankers make far greater use of cofferdams, double bottoms and similar devices than conventional oil tankers. To ensure that incompatible cargoes do not come into contact with each other, tanks are usually separated by a cofferdam. Most (but not all) chemical tankers have their tanks separated from the outer frame of the ship by a double bottom or double hull. If the ship is damaged in a collision or grounding this space should protect the cargo tanks from damage.

There are three basic types of chemical tankers:
- Product /chemical carriers
- General chemical carriers ("parcel tankers")
- Gas/chemical carriers

In addition to the above are specialised tankers, specifically designed for the carriage of a specific commodity (including methanol, phosphoric acid, molten sulphur, palm oil, and naphtha).

Parcel Tankers

The chemical tanker known as the parcel tanker is constructed to carry upwards of 40 different "parcels", or cargoes, simultaneously. These cargoes range from chemicals, biofuels, vegetable oils and lube oil additives to acids and other specialty chemicals that often pose a range of hazards to personal health and the environment. Such hazards include toxicity, corrosivity, flammability and reactivity, while many cargoes are also classified as marine pollutants.

In order to minimise port turnaround times and to streamline cargo-handling operations, modern parcel tankers are equipped with highly automated cargo control and monitoring systems. These are fully integrated with navigational systems and linked to a shipowner's management and communication networks. All modern parcel tankers are built with double hulls to provide the flexibility to carry oil products and to maximise the potential to carry hazardous chemicals.

Chemical/Product Tankers

In addition to the sophisticated parcel tankers, another type of chemical tanker has evolved to serve the global chemical industry. The chemical/product tanker is built to carry "simple" chemicals such as benzene, xylene, toluene, methanol and caustic soda solution, as well as a full range of petroleum products. Such chemicals are not as demanding in terms of carriage requirements as specialty chemicals and are transported in comparatively large-sized parcels. As a result, simple chemical tankers may have coated, rather than stainless steel, cargo tanks and there are fewer cargo tanks than on a parcel tanker. Simple chemical tankers can move with relative ease between oil products and commodity chemicals, and do so as market conditions fluctuate.

Gas/Chemical Carriers

Gas/chemical carriers are most likely Liquid Petroleum Gas (LPG) carriers that have been specifically designed to accommodate other liquid chemicals. Gas/chemical carriers have the following unique characteristics:
- A number of cylindrical horizontal or bi-lobe tanks constructed from mild steel or stainless steel located in a conventional hull;
- Deepwell cargo pumps in each tank;
- An insulation barrier between tanks and surrounding structures;
- Pressurization and refrigeration systems.

Design Characteristics for Chemical Tankers

The International Maritime Organization (IMO) regulations define a chemical tanker as a tanker constructed, or adopted and used, for the carriage of any liquid product listed in the IBC Code. In its simplest form, the term chemical tanker still refers to "purpose built" or converted ships that carry chemical products, either regulated or unregulated, that have special containment and handling requirements for reasons of safety, cargo properties, and cargo quality control. These ships can range from relatively simple to very complex ships, built to carry specialised single commodities, or an extensive range of cargoes of widely varying characteristics.

They are defined not only by vessel design and specific cargoes carried, but also by the simultaneous carriage of relatively small multiple cargo quantities, and by complex, labour intensive (and often simultaneous) cargo loading, discharging, and tank cleaning operations; indicative of the parcel trade.

Chemical tankers that operate in the parcel trade have numerous cargo tanks of various sizes and arrangements, have a complex cargo handling system, and are equipped for the highest degree of operational flexibility. This operational flexibility consists of two necessarily related aspects unique to these ships. The first is their extraordinary ability to interchange the use of cargo tanks between wide ranges of cargoes (of which there are over 2,000 in the parcel trade). This allows these ships to sail consistently with minimal ballast passages, utilizing the counter movements of many widely differing cargoes. It also means continuous cargo discharging, tank cleaning and loading operations throughout any one voyage.

Another aspect of operational flexibility of these purpose-built ships involves complex segregated cargo handling systems. These provide a potential for multiple simultaneous cargo operations such that sometimes simultaneous product handling can reach double figures, with products being discharged and loaded, whilst at the same time empty cargo tanks are washed and pre-washed for the next cargo, within the safe limits of the available human supervision and the capability of the shore installation to accommodate this potential. The chemical tanker today is therefore distinct, both in design and operation, and should not be confused with the less developed petroleum product tanker or, indeed, the simpler and usually smaller pure chemical carrier.

Cargo Tank Layout

All modern bulk liquid cargoes are loaded via an array of piping systems into or from tanks that are isolated with one or a series of valves. The type of system for each tanker design depends upon the tanker's intended purpose.

While chemical tankers are often required to carry full ship loads of a single commodity, small parcel sizes are often a characteristic of the trade - with parcels of less than 500 cubic metres being common. Cargo tanks are usually arranged to provide a variety of capacities to accommodate the size of the cargo shipments in the trade. Tank locations also have to reflect the IMO tank type location requirements for those cargoes.

Cargo segregation, due to cargo reactivity, toxicity, or for reasons of cargo quality, is another important design characteristic. This pertains to cargo transfer and venting systems as well as to the cargo tanks themselves - allowing cargoes to be completely segregated due to heating requirements, adjacent heat restrictions, or to preclude the possibility of cross-contamination. As cargoes have the potential for reactivity problems, and many cargoes require high carriage or discharge temperatures whilst others are heat sensitive, ship designs have been developed which utilise cofferdams to break up the cargo tank arrangements, with independent cargo transfer and venting piping arrangements for each cargo tank.

All of the valves are controlled automatically from a control room or by radio instruction to the deck crew. The cargo handling systems are augmented by systems needed for safety, environmental or operational reasons. Other systems include high/low level alarms, vapor recovery and trim indicators. These will most likely be monitored from the control room as well.

Tank Gauging, Sampling and Alarms

Remote tank level gauging is employed onboard all modern chemical tankers for all cargo tanks, along with dual independent high level alarms. In addition to tank ullage levels, the remote monitoring of all cargo parameters, such cargo temperatures and tank pressures, are featured on all chemical tankers. Closed, local tank level gauging and sampling provisions are also now universally provided for all cargo tanks.

Tank Vent Systems

Cargo tank vents on most chemical tankers are independent and fitted with pressure vacuum valves. High velocity vents are now common. Due to cargo reactivity, toxicity and cargo quality concerns, cargo tank vents are now often totally independent. Vapor control connections are fitted, and provisions are often made for tying all these vents into a common transfer line for general vapor control for multi-tank cargoes.

Cargo Handling Systems

The cargo handling systems need to reduce the possibility of cargo contamination to an absolute minimum, as well as meet the requirements for pumping systems for a wide range of cargoes. This has led to the development and introduction of submerged, usually hydraulically driven, cargo pumps as an industry standard. The submerged hydraulic pump is practically universal to the chemical tanker cargo pumping requirements.

The typical modern cargo pumping system is designed using one separate submerged cargo pump for each cargo tank - with each pump being serviced by a separate cargo line to ensure total segregation. Usually no part of the system for one tank is common with another. These designs also eliminate the need for submerged valve segregation and the consequences of malfunction, and are extremely flexible, versatile, and easy to operate. Cargo pump rooms have been largely eliminated by the wide use of individual deepwell pumps, and the absence of cargo pump rooms is a significant characteristic of the modern parcel/ chemical tanker.

Cargo loading and discharge lines are usually of six to eight inch pipe, fitted as necessary with full flow ball valves. Separate cargo rail or centre transfer lines, where fitted, are usually eight to ten inches in diameter. Loading rates are usually in the range of 150m³/hr. - 350m³/hr. per cargo line employed. When large quantities of cargo are to be loaded, the deck cargo piping can be connected via portable hoses and "jump" connections such that a good number of tanks can be loaded simultaneously. Loading rates for large volume cargoes are often thus in the 1,200m3/hr. range. However, most cargoes are loaded in relatively small parcels into one to a half dozen tanks. Numerous cargoes can be simultaneously loaded in this fashion. The restrictions on loading rates are usually due to the delivery capability of the loading pumps ashore or onboard barges or coaster vessels.

The overall advantages of individual electric or hydraulic deepwell pumping systems include:
- Infinite speed control
- No overloading possible
- Versatility of location
- Torque characteristics
- Elimination of cargo suction lines

- No separate stripping systems required
- Safe stripping
- Easy cleaning
- Elimination of pump rooms.

Cargo Tank Construction

Once cargo tank size, materials and tank distribution are determined, the structural design of the cargo tanks is considered. For high specific gravity cargoes, such as caustic soda or organic acids, the tanks must be strengthened. A basic design is for the carriage of a full deadweight of 1.5 s.g. (specific gravity) cargo (caustic soda) to s.g. 1.8 cargo (phosphoric acid). Cargo tanks are generally designed for ease of cleaning and coating. Both requirements suggest plain or corrugated bulkheads with minimum surface area. Having double bottom/double skin designs with numerous cofferdams helps place internal structures outside the cargo tanks. Deck structures are also usually inverted for this purpose.

Special Cargo Requirements

Certain special requirements for specific cargoes also have a significant impact on parcel/chemical tanker design.

Sulphuric and phosphoric acids are common cargoes and at first appear to meet the Type 3 tank carriage requirements but the Code's requirement that "ship's shell plating shall not form any boundaries of tanks containing mineral acid" necessitates double bottom / double skin or double bottom / centre tank stowage. This effectively limits these cargoes to carriage in Type 1 and 2 tanks.

For toxic cargoes, cargo transfer piping and tank vents must be fully segregated from those of non-toxic cargoes. Heating coil lines must also be separate and isolated for testing before being returned to the engine room. Products such as alkaline oxides have high vapor pressure, and are extremely flammable. Carriage of self-reactive cargoes such as propylene oxide involves the following requirements:

- all valves, flanges, fittings and accessory equipment must be alkaline oxide compatible
- PTFE ('Teflon') or materials of similar inert qualities used for all gaskets
- cargo discharge by inert gas displacement or deepwell pumps only;
- the cargo tank adjacent spaces are inerted to below 2% oxygen levels;
- cargo tanks with a specific design pressure or an installed cooling system to maintain the vapor pressure below the tank relief valve setting.

Materials and Coatings

The tanks of a chemical tanker are constructed of special materials, all designed to carry certain products. The early chemical tankers generally had tanks made of stainless steel, which resists corrosion from many products and could be cleaned relatively easily. Stainless steel is compatible with the vast majority of cargoes carried aboard such ships and it permits speedy tank cleaning operations to a very high standard. But it is expensive and may not suitable for all chemicals and so different coatings have been designed. The three typical coatings in use nowadays are phenolic epoxy (some resins), zinc silicate, and Marine Line®. Each one has advantages and disadvantages and so far no coating has been developed which is suitable for all chemicals.

Type	Suitable for
Phenolic epoxy	Alkalis, glycols, animal and veg oils, some solvents, polyurethenes
Zinc silicate	Aromatic hydrocarbons
Stainless Steel	Sulfuric acid, nitric acid, phosphoric acid, caustic soda, wine, food products
Marine Line ®	Many clean petroleum products (CPP) chemicals including benzene, MTBE, Vegetable oils, phenol, acrylonitrile, methanol, ethanol, dichloride, acetic acid and many others.

Most modern chemical tankers use stainless steel in the construction and the design of a cargo tank to maximize the benefits of cleaning, carriage and ease of tank coating. Again, stainless steel is very expensive. Different tanks may be coated with different products permitting the carriage of a full range of products. Pumps, lines, valves, fittings and associated control and monitoring equipment are usually stainless steel with chemical resistant seals.

The three most popular types of tank linings or coatings used are listed above: epoxies, phenolic epoxy and zinc silicate. The coating of cargo tanks serves three main purposes: to protect steel work from the products carried, or ballast water, to protect the product from the contamination by corrosion products present with uncoated steel, and to facilitate tank cleaning and protect the cargo from contamination by previous cargoes.

Broad chemical resistance, ease of application and long performance lifetime are three major considerations to be taken into account when choosing a coating to be applied for a tank. For chemical resistance, the choice has predominantly been phenolic epoxy coatings. This type of coating provides a cost effective option allowing the vessel to carry a broad range of chemicals together with fatty acids, latex and ammonia and other aggressive products.

Marine Line® is the best example of the advancements made in recent years as far as coating technology. It is made up of a dense, highly cross-linked molecular structure, which makes it resistant to all solvents and 98% of the chemicals carried by sea-going tankers. These include exceptionally aggressive products, such as sulphuric acid and chlorinated hydrocarbons that are normally carried in stainless steel tanks or tanks lined with expensive fluoropolymer linings. Unlike many conventional epoxy coatings, there is no tendency to absorb small quantities of cargo into the lining that can later leach out into a new cargo. For high-purity cargoes, such as technical-grade methanol, this contamination from previous cargoes can put an entire consignment off-specification.

SHIP OPERATIONS

Voyage Planning

Planning the cargo stowage for a voyage with multiple cargoes has to take into consideration exactly how much of each cargo is to be loaded and discharged in each port. Tanks of suitable sizes and coatings have to be allocated for each cargo. For cargo quality, the best tanks may be selected on the basis of last or future cargoes. As the ship goes through her discharging and loading programme the list and trim of the ship has to be taken into account because ships need to be kept on an even keel, trimmed slightly by the stern for good tank stripping and sea-keeping characteristics. If tanks are to be cleaned after a discharge to make room for cargoes to be loaded, then adequate cargo tanks for slop capacity for the wash water must be ensured. Any stowage plan must also adhere to certain cargo compatibility restrictions.

Multiple Port/Berth Operations

Chemical tanker operations usually involve multiple ports or berths. This adds a level of complexity to the stowage planning. Other ships such as oil tankers have the luxury of open ocean passages to provide for the disposal of allowable tank wash water or adequate time for tank cleaning and preparation, which a chemical tanker may not enjoy. The intensity of these operations also takes its toll on the crew. Complicating this process are instances when the cargoes are cancelled, nominated quantities are altered, or port or berth rotations are changed because of a delay beyond the control of the vessel.

Cargo Information Requirements

All cargoes for carriage in bulk have to be classified by the IMO and/or the flag state and their carriage requirements determined. If regulated by the IBC Code, cargoes must be authorised for carriage on that particular ship by, and listed on, the Certificate of Fitness. There is a mechanism in place for a relatively quick, interim classification of a new cargo and determination of its carriage requirements, referred to as a tripartite agreement. This is an agreement between the cargo country of origin, destination, and ship flag state - after which the ship's certification still has to be endorsed to verify the ship's suitability for the cargo if the cargo is determined to be so regulated.

Other information important for operational planning includes tank coating suitability, compatibility grouping, specific gravity, correction factors for temperature, heating instructions, adjacent temperature limitations, inerting or dry atmosphere requirements (purging, padding), prior cargo restrictions, prior loading cleanliness and tank testing requirements, health and safety and emergency response information. Stability and trim factors, including loadline zones, and draft restrictions, are also taken in to consideration.

Special Cargoes

There are a number of products that necessitate special handling or emergency procedures, carriage or pumping arrangements, and even crew training. These special requirements may pertain to cargo quality and transfer efficiency, as well as safety and pollution prevention. Examples include acrylic acid, cashew nut shell oil, toluene diisocyanate (TDI), latex, phenol, phosphoric acid, organic acids in general, and propylene oxide. That said, the carriage of many (what may be considered routine) cargoes involves particular tank preparation, heating, and tank cleaning procedures.

Barging & Lightering

One unique aspect of this parcel/chemical tanker trade is that many cargoes are brought to and from the larger ships by coaster vessels (in Europe and Japan) and barges. There has also been a great effort in recent years to consolidate cargoes in particular tank terminals designated as owners' berths to minimize berth calls and provide for more efficient cargo transfer operations.

Tank Cleaning

Cargo tank preparation is one of the most important parts of chemical tanker operations. All tank cleaning procedures are cargo dependent. Cargo tank cleaning usually includes the following stages: pre-cleaning, cleaning, spot cleaning, rinsing, water flushing, gas freeing, steaming, draining, and drying. Tanks are pre-cleaned by using washing machines with salt, fresh or ionized water in order to remove basic cargo residues from the tank structure. Usually, the sooner this is carried out after discharge of the cargo the easier it is to remove these residues.

Cargo tank cleaning is normally done with sea water usually heated to the temperature most compatible with the cargo residue and the tank coating. Most recent new buildings have fixed tank cleaning machines. Cleaning chemical solutions and solvents are often applied, and various combinations of steaming, fresh water rinses, etc. The types of cleaning solutions to be used are determined by the instructions of the supplier, the experience of the user and the product previously carried. All tank cleaning operations require strict adherence to safety procedures regarding crew cargo exposure, toxic vapors, static electricity, and of course, tank entry.

MARPOL Annex II requires ships to conduct mandatory in-port pre-washes after unloading certain toxic products and solidifying or high viscosity products. Pre-washes are also required in other special circumstances where the ship may not be able to comply with the regulation. After unloading a cargo tank containing a Category "X" Noxious Liquid Substance (NLS), (a cargo requiring the most care and spill prevention) the tank must be pre-washed in accordance with the procedures specified in the ship's Procedures and Arrangements (P&A) Manual. The pre-wash residue is then discharged to a reception facility prior to the ship leaving the unloading port. If facilities are not available at that berth the residue may remain onboard for discharge to a shore reception facility. Ships will normally be ready to begin pre-wash operations when the cargo tanks are empty to the limits allowed by MARPOL Annex II.

Vapor Control

Several port states require that vapor control systems be used to recover and process cargo vapors displaced during loading. In the United States these vapor control systems must be approved by the U.S. Coast Guard for the specific cargo vapors they will handle. Vapor control systems are usually approved for one or more of the three base cargoes: gasoline, benzene and crude oil. In most cases, chemical cargoes can be handled in vapor control systems approved for these base cargoes (except crude oil) at the same transfer rate, though the vessel's Certificate of Fitness must be endorsed specifically to allow the use of vapor control with that type of cargo.

Tank Inerting and Padding

Fire and explosion are among the greatest threats to any vessel but for a tanker carrying crude oil, refined petroleum, or chemicals it is an even greater threat. For this reason, an inert gas system (IGS) is designed, installed, operated, and maintained to prevent fire and explosion in an intact ship tank. Typically, this is done by adding to the tank atmosphere a gas that has less oxygen (5 percent or less) than air. Combustion is impossible without oxygen. A tank with less than 8 percent will be free of danger from explosions in that intact space.

Current regulations stipulate that chemical tankers under 30,000 dwt and with cargo tanks of less than 3,000m3 do not have to inert chemical (MARPOL Annex II) cargoes. Chemical tankers over 20,000 dwt have to inert the petroleum oil products (MARPOL Annex I) they carry. These ships usually employ independent, combustion-fired inert gas generators that operate on diesel fuel.

Many chemical cargoes require that the tanks be purged with nitrogen (N2) for dryness, and a nitrogen pad be maintained throughout the voyage for moisture protection. The N2 for purging is usually provided by the loading terminal (it can be supplied by truck or barge), and the N2 for pad maintenance is now mostly provided by onboard N2 generators or liquid N2 supply, otherwise by bottles. Only a few cargoes, such as propylene oxide, require an inert tank atmosphere for fire and explosion prevention.

REGULATIONS AFFECTING THE CHEMICAL TANKER

The International Maritime Organization

In 1958 the Intergovernmental Maritime Consultative Organization (IMCO), which later developed into the International Maritime Organization (IMO), was founded. By the time IMCO was founded, a number of important international conventions had already come into existence, including the 1948 International Convention for the Safety of Life at Sea (SOLAS), the 1954 International Convention for the Prevention of the Sea by Oil (OILPOL) and early versions of the International Conventions on Load Lines and Prevention Collisions at Sea (COLREG).

The IMO was given the responsibility for maintaining these conventions and for developing further instruments to achieve improvements in the safety of shipping and the prevention of pollution. At the latest count the IMO is responsible for 37 international conventions and agreements.

In 1971, the IMO adopted the Code for the Construction and Equipment of Ships Carrying Dangerous Chemicals in Bulk (BCH Code). The BCH Code and its successor, the International Code for the Construction and Equipment of Ships Carrying Dangerous Chemicals in Bulk (IBC Code, which applies to ships built after 1 July 1986), stipulate design and equipment standards for chemical tankers.

These international standards specify three degrees of physical protection, or ship types, i.e. IMO ship Types 1, 2 and 3. IMO Type 1 ships provide the highest degree of protection and are designed to survive a high level of prescribed damages. Those cargoes whose properties pose significant hazards, but which do not have such wide-reaching and deleterious effects as ship Type 1 cargoes when released, are specified for carriage in IMO Type 2 ships. Both these types require double hulls. IMO Type 3 ships, designed to carry products of sufficient hazard to require a moderate degree of containment to increase survival capability in a damaged condition, are only required to have single hulls. IMO ship Type 2 cargoes represent the predominant type of chemical products carried by chemical parcel tankers. The ship types are defined in the Code as follows:

- Type 1 ships must be able to survive assumed damage anywhere in their length. Cargo tanks for the most dangerous products should be located outside the extent of the assumed damage and at least 760mm from the ship's shell. Cargoes which present a lesser hazard may be carried in tanks next to the hull.

- Type 2 ships, if more than 150m in length, must be able to survive assumed damage anywhere in their length. If less than 150m, the ship should survive assumed damage anywhere except when it involves either of the bulkheads bounding machinery spaces located aft. Tanks for Type 2 cargoes should be located at least 760mm from the ship's shell and outside the extent of assumed grounding damage.
- Type 3 ships, if more than 125m in length, should be capable of surviving assumed damage anywhere in their length except when it involves either of the bulkheads bounding the machinery space. If less than 125m in length, they should be capable of surviving damage anywhere unless it involves machinery spaces. There is no special requirement for cargo tank location. (See Chapter 1, Figure 1-10)

The IBC Code also include standards for cargo transfer, materials of construction, cargo temperature control, cargo tank vent systems, environmental control, electrical installations, fire protection and extinction, ventilation in the cargo area, instrumentation and personnel protection. In recognition of the hazards associated with handling particular cargoes, the Code specifies additional special requirements.

The exemplary safety record built up by the world chemical tanker fleet over the past several decades has demonstrated the effectiveness of the BCH and IBC Codes in addressing the hazards facing such ships. The Codes are constantly revised and updated by the IMO both to reflect the introduction of new technology and to add new chemical products to the list of liquid hazardous cargoes transported in bulk by sea.

MARPOL

The major convention dealing with marine pollution is the International Convention for the Prevention of Pollution from Ships, 1973, as modified in the 1978 Protocol (MARPOL 73/78). MARPOL covers all the technical aspects of pollution from ships except the disposal of waste. The Convention has two protocols dealing with the reporting of incidents and arbitration. There are currently six Annexes that deal with specific types of pollution.

MARPOL Annex	Deals with	Entered into force
Annex I	pollution by oil	2 October 1983
Annex II	pollution by noxious liquid substances	2 October 1983
Annex III	pollution by harmful substances (those in containers or tanks).	1 July 1992
Annex IV	pollution by sewage	Not yet in force
Annex V	pollution by garbage	31 Dec 1988
Annex VI	air pollution	Not yet in force

MARPOL Annex II

The BCH and IBC Codes incorporated design and equipment requirements to implement Annex II of MARPOL 73/78. Annex II entered into force in 1987. With the inclusion of Annex II marine pollution considerations, the Codes were extended to provide a comprehensive set of standards for the safe shipment of bulk hazardous cargoes by sea. The revisions to MARPOL Annex II took effect 1 January 2007.

IMO Vessel Certifications

There are several certificates and manuals which a ship must have and maintain to carry Noxious Liquid Substances (NLSs) in bulk. The documents required depend upon the vessel's service, the NLS and other cargoes it carries, where it operates, whether the ship's flag state is a Party to MARPOL 73/78, and whether the ship must meet the IBC Code requirements.

The Certificate of Fitness for the Carriage of Dangerous Chemicals in Bulk (COF) is issued by Party nations under the IBC Codes. The COF is required for ocean-going chemical ships on foreign voyages and ocean-going tank barges when in the waters of a Party nation.

In the U.S., a Certificate of Compliance is issued to foreign ships and carries endorsements for the carriage of specific NLS cargoes for each cargo tank. The USCG issues this certificate when a foreign flag vessel has been examined and found to comply with the 46 CFR (Code of Federal Regulations). In the U.S., a ship must have an IMO Certificate of Fitness issued by its Administration and endorsed with the name of the hazardous material or NLS if the ship's Administration is a signatory to MARPOL 73/78.

The Letter of Compliance (LOC) is issued by the Officer in Charge of Marine Inspection (the local USCG in the port) after a satisfactory examination of the vessel. A foreign tanker entering U.S. waters should have on board an IMO Certificate of Fitness (COF). An IMO COF is issued by the flag state, usually through a classification society, and attests to compliance with the IMO Codes. The IMO COF includes a list of cargoes authorized to be carried by the flag administration. The LOC is endorsed to allow carriage of these cargoes in the U.S. by the local U.S. Coast Guard officer (PSC officer) endorsing the LOC.

In the U.S., the Tank Vessel Examination Letter (TVEL) is issued to non-U.S. tank vessels, which carry only flammable and/or combustible cargoes. Rather than issue a separate TVEL to a vessel with an LOC, the standard practice is to include an endorsement for flammable and/or combustible liquids on the face of the LOC. The LOC and TVEL continue to be issued and be found on board foreign tank vessels. Both the LOC and TVEL comprise the COC (Certificate of Compliance). If the ship is found in compliance with Annex II, the COC will be issued. The applicable documents must be available for inspection by boarding officers.

A Certificate of Inspection (COI) is issued under Subchapters D and O or I and O to U.S. ships and carries endorsements for the carriage of specific NLS cargoes for each cargo tank. A COI is issued by the USCG after the ship has had a satisfactory survey or inspection. The Coast Guard is responsible for issuing this certificate in accordance with 46 CFR 30-40 and 153. The specific endorsements and conditions for Annex II requirements found on the COI come from 33 CFR 151 & 157 and 46 CFR 98, 151, 153 & 172 and have the same force and effect as the regulations requiring them. Each COI is endorsed, according to the individual tanks, to show which NLSs can be carried and where they may be carried on the ship.

The Cargo Record Book (CRB) is a document required on board every ship carrying a NLS in bulk. The form of the CRB is specified in an Appendix of Annex II of MARPOL 73/78. The CRB is used to record internal and external transfers and discharges of NLS cargo or waste, information concerning inoperative cargo transfer, tank cleaning and pollution prevention equipment, actions by surveyors, and any other cargo or waste-related activities.

The signature of the USCG on the LOC will be the endorsement necessary under Codes of Federal Regulation. A separate endorsement document for MARPOL signatory chemical tankers is not necessary. An IMO COF includes a list of cargoes authorised by the flag state to be carried under the SOLAS Convention. For a chemical cargo to be carried in U.S. waters, cargo carriage must be permitted by U.S. regulations (or a tripartite agreement to which the U.S. is a party) and the cargo must be listed on the IMO COF. Separate documentation must be on board a vessel authorizing cargoes being carried under a tripartite agreement.

The Procedures and Arrangements Manual (P&A Manual) is approved by the Administration of the flag state and describes in detail the procedures for NLS cargo carriage, tank-to-tank transfers, cargo loading, unloading, stripping operations, and tank pre-washing and ventilation procedures. Ships shall have an approved P&A Manual available on board the ship for inspection by the boarding officer. Flag state administrations approve the P&A Manuals for their ships. The approval will be evident by a stamp and signature of the approving official of an authorised agency of the flag state's Administration on the cover of the P&A Manual.

REVIEW

1. Give three examples of heavy chemicals and petrochemicals considered "big movers" in the parcel trade.
2. The 'hazard profile' of various chemical substances is developed by what group under the United Nations?
3. List some of the main hazards or problems associated with the transport of chemicals at sea.
4. What IMO publication outlines the requirements for construction and equipment of ships carrying bulk cargoes of dangerous chemical substances?
5. How does the IMO define a 'chemical tanker'?
6. Purpose-built chemical tankers are often externally framed for what reason?
7. List several features characteristic of a cargo system on a parcel tanker.
8. List the various coatings used to protect the mild steel cargo tanks from attack by aggressive chemicals and corrosive cargoes.
9. What official document contains the list of cargoes that a vessel is authorized to transport?
10. When is in-port pre-washing mandated by Annex II of MARPOL?
11. What vessel specific manual contains the details of various cargo operations, ballast operation, tank cleaning, cargo residue disposal, etc. on a parcel tanker?
12. The Cargo Record Book is required on every tank vessel carrying what cargoes?
13. When is tank inerting and padding required in the transport of certain chemical cargoes?
14. Under the IBC code, what level of protection is afforded the marine environment by Type 1, 2 and 3 construction?

CHAPTER 4

Cargo Piping Systems

Cargo piping systems are an integral part of any tank vessel. The pipelines provide the path for the flow of liquid to and from the cargo tanks. The person-in-charge (PIC) of the cargo operation must have a thorough working knowledge of the piping system to avoid the possibility of spills or contaminations. Newly assigned personnel should familiarize themselves with the system by carefully tracing out the piping prior to assuming a cargo watch. Cargo piping systems can be classified in a number of ways. One approach is by location of the piping in the vessel:

1. Bottom piping (if so equipped)
2. Deck piping
3. Pump room piping (if so equipped)

BOTTOM PIPING

The cargo piping that networks through the bottom of the tanks is called bottom piping. The two piping configurations that predominate on tank vessels equipped with bottom piping are the direct piping system and the loop or ring main system.

Direct Piping System

The direct piping system is characterized by one or more main lines oriented in a fore-and-aft direction from the cargo tanks to the pump room. Each main is designed to service a certain number of tanks referred to as a cargo system or group. Figure 4-1 is a simplified view of a direct piping system on a two-cargo system (group) vessel.

Tank vessels can be designed to simultaneously handle a wide array of cargoes based on the number of cargo systems or groups. Parcel tankers are generally designed with the greatest number of segregations as seen on vessels such as the *Seabulk America* with twenty-one cargo systems.

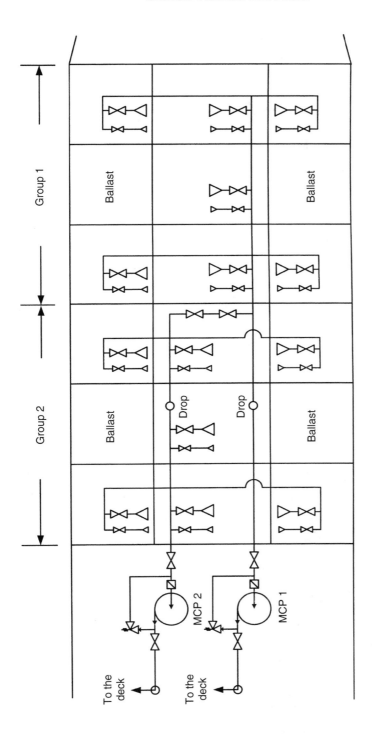

Figure 4-1. Simplified view of a direct piping system on a two-cargo system (group) vessel. Courtesy John Hanus and Mark Huber.

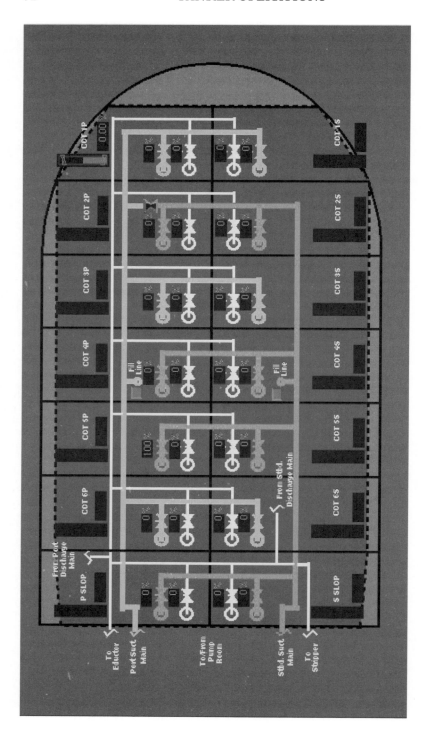

Figure 4-1a Cargo System - Bottom piping in cargo tanks.

Bellmouth

The bitter end of the piping in each cargo tank generally has a flared shape referred to as a bellmouth. Figure 4-2 illustrates a typical bellmouth found at the after end of each tank through which the cargo tanks are loaded and discharged, unless the tanks have individual drops. The cargo tanks generally have two such openings known as the main and stripping bellmouths, which differ in dimension and in location as measured in distance from the bottom of the tank. The stripping bellmouth is smaller and is placed closer to the tank bottom to facilitate draining at the end of the discharge. The bellmouth is designed to minimize vortexing, a whirlpool effect that occurs as the cargo level approaches the stripping stage in a tank.

Branch Piping

The bellmouth connects to athwartship piping known as branch piping. As shown in Figure 4-1a, the branch piping connects the port and starboard cargo tanks to the fore-and-aft main line. For example, the forwardmost cargo tanks—#1 port and #1 starboard—are connected to the port main through the branch line. Many new vessels are equipped with two bellmouths that operate through a single main, as seen in Figure 4-1.

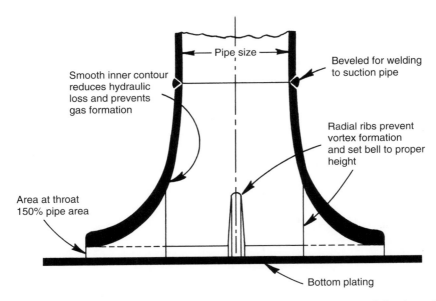

Figure 4-2. The flared bellmouth shown represents the end of the bottom piping in each cargo tank. Courtesy Hayward Manufacturing.

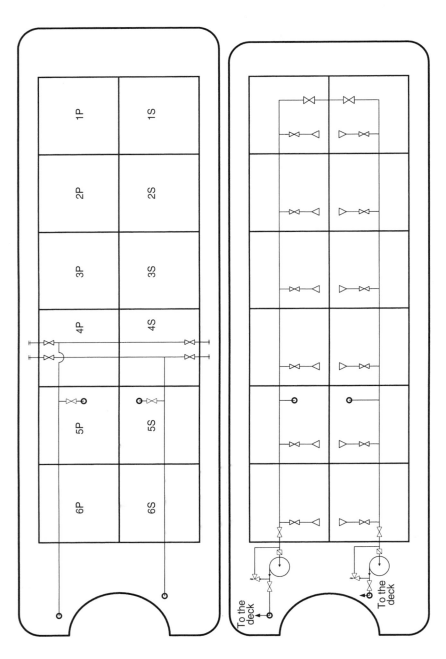

Figure 4-3. A loop or ring main piping system interconnecting the cargo tanks. Courtesy John Hanus and Mark Huber.

Loop (Ring Main)

Another type of bottom piping system commonly seen on barges and certain dedicated cargo carriers is a loop or ring main. Figure 4-3 illustrates a loop system in which the piping runs in a continuous circle throughout the bottom of the vessel, interconnecting the cargo tanks and pumps. A loop arrangement is suitable on vessels where cargo segregation is not a critical factor, such as crude-oil carriers.

Free-flow

In fact, some larger crude-oil vessels are designed with minimal bottom piping in the cargo tanks. This arrangement is commonly referred to as a free-flow vessel. Cargo flows through the bottom of the tanks via remotely controlled sluice gates in the bulkheads. As the vessel is discharged and develops a trim by the stern, cargo is directed to the bellmouths located in the aftermost tank. Only a short run of bottom piping is therefore needed to connect the bellmouths in the after cargo tank to the pumproom. This significantly reduces construction costs by eliminating a major portion of the large diameter piping that would otherwise be necessary.

Piping Details

The dimension of the piping is usually dictated by the size and capacity of the cargo pumps as well as the anticipated loading rates of the vessel. Coastal tankships and barges generally have cargo piping ranging from 10 to 14 inches in diameter, whereas the piping on large crude-oil carriers can reach 24 to 36 inches in diameter. The proper support and anchoring of the bottom cargo piping can be seen on the vessel under construction in Figure 4-4.

Figure 4-4. Large-diameter bottom cargo piping in one vessel under construction. Courtesy Jeremy Nichols.

The cargo piping is usually constructed of steel or equivalent material (carbon steel) although vessels carrying certain parcels may require the use of special material such as stainless steel. The schedule of the piping (thickness of the pipe wall) varies with the service in which it is employed—cargo, inert gas, or vapor. Federal regulation stipulates that cargo piping must be tested every year, and a record of such tests must be maintained on board. Cargo piping can fail for a variety of reasons some of which include:

- Uniform corrosion in cargo and ballast systems
- Pitting corrosion
- Metal fatigue
- Pipe alignment
- Surge pressures
- Temperature changes (thermal expansion/contraction)

Cargo piping sections are connected through the use of bolted flanges or slip-on (Dresser) couplings. Bolted flanges (Figure 4-5) are employed to connect successive lengths of cargo piping; however, they do not afford the degree of flexibility necessary to handle the bending stresses of the vessel or the thermal variances of the pipe. Slip-on couplings, on the other hand (Figure 4-6), provide a leak-free method of connecting piping without the need for flanges. These couplings accommodate changes in temperature as well as the bending of the vessel (tension and compression forces) that occurs during cargo operations and at sea. Some operators have experienced problems with slip-on couplings (Figure 4-7) due to excessive movement of the piping and deterioration of the gasket material caused by the cargo. Another approach to welded piping connections is the installation of loops or bends that allow the piping to expand and contract without causing undue strain or possible fractures.

Figure 4-5. Deck piping connected through the use of bolted flanges and slip on couplings. Courtesy Nathan Martin.

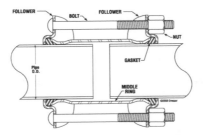

Figure 4-6. Style 38 Dresser Coupling. Copyright Dresser, Inc. All rights reserved.

Dresser Style Pipe Couplings

"Dresser" style couplings are widely used for connecting piping on board company vessels because they provide the flexibility necessary to withstand stresses caused by expansion and contraction during temperature changes, vessel motion in a seaway, and vessel vibration. These forces are absorbed without damage to the pipe or leakage at the joint so long as the Dresser couplings are properly fitted and well maintained.

Unfortunately, oil spills have occurred on board company vessels in the past due to Dresser couplings rupturing or separating.

Many of these spills happened because nearby piping was not properly supported, allowing excessive motion at the coupling.

Dresser style couplings will not maintain an oiltight seal or hold pipe sections together if pressurized without the pipe being clamped or secured on EACH SIDE of the coupling. Therefore, never pressurize the line if the pipe is not secured and firmly supported on both sides of the flexible seal. The master should be advised of any repairs requiring the removal of in-line supports so that appropriate precautions can be taken. Dresser couplings will also fail if the pipe is not straight, so always align the pipe properly when fitting new couplings.

Oil spills are extremely dangerous and costly, and every effort should be made to ensure that they do not occur on your vessel. Since Dresser style couplings are the "weak link" in the line, proper installation, support, and maintenance of these couplings is essential to safe operations.

Figure 4-7. An important message from one fleet operator concerning Dresser couplings. Courtesy Chevron Shipping Company, LLC.

DECK PIPING

The above-deck piping generally consists of one or more discharge mains connecting the cargo pumps to the athwartships manifold piping.

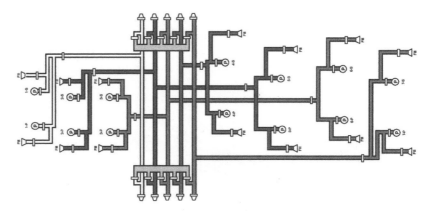

Figure 4-7a. Example of the deck piping arrangement on vessels with in-tank pumps.

The cargo manifold is the bitter end of the on-deck piping that forms the interface between the vessel and the shore facility. The manifold is generally located amidships and equipped with valves and blanks. It is considered sound practice to always close the manifold valves when there is no active transfer of cargo. Securing the vessel piping in this manner is a precautionary measure against the possibility of cargo movement (gravitation) when the transfer has ceased. The end of the manifold piping is flanged to permit connection to the shore facility via flexible cargo hoses or mechanical loading arms.

Flanges

Manifold flanges are usually constructed in accordance with a national standard to ensure a smooth operation when making the connection between the vessel and the facility. In the United States, the American National Standards Institute (ANSI) publishes the criteria for standardized flanges including such items as the following:

1. Inside diameter (ID)
2. Outside diameter (OD)
3. Bolt circle diameter (BCD)
4. Number of bolt holes
5. Thickness of the flange face
6. Raised or flush-face flange
7. Material of construction

Another recognized source dealing with standardization of flanges and manifolds on vessels is *Recommendations for Oil Tanker Manifolds and Associated Equipment* from the Oil Companies International Marine Forum (OCIMF).

When connections are made using bolted flanges, the following requirements must be met:

1. Suitable gasket material must be used in the joints and couplings to make a leak-free seal. It is not advisable to double up on gaskets or to reuse them. Gaskets are generally constructed of a fiber or neoprene material; however, Teflon may be used with certain cargoes
2. When ANSI flanges are employed, a bolt must be placed in every other hole at a minimum, and in no case should less than four bolts be used in the connection. Be aware that company policy usually specifies that a bolt installed in every hole.
3. When using non-ANSI flanges, a bolt must be placed in every hole.
4. For permanently connected flanges, a bolt must be placed in every hole.
5. Each nut and bolt should be uniformly tightened to distribute the load and ensure a leak-free seal. Any bolt exhibiting signs of strain, elongation, or deterioration should be removed from service.

The proper sequence of tightening a typical bolt connection is shown in Figure 4-8. When it is necessary to install a blank (blind flange) on the manifold, instructions can be found in Figure 4-9.

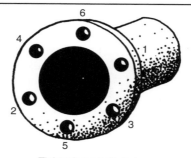

Tightening a bolt circle

The proper way to tighten up a ring is to take the slack up evenly, then tighten gradually, alternating back and forth as shown in the drawing. Don't overdo it. Remember that a riser flange can be easily cracked by tightening the bolt circle unevenly; and that a little excess pressure on the wrench can strip the bolt threads or pull out the stud.

Figure 4-8. The correct procedure for tightening a bolt circle. Courtesy Chevron Shipping Company, LLC.

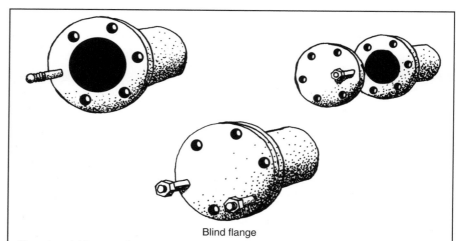

Blind flange

There is a right way and a wrong way to put a blind flange on a cargo riser. The wrong way produces damaged fingers and dented toes. This is unnecessary. Do it the right way:

1. Stick a bolt through in a bottom off-center hole.
2. Pick up the flange with both hands and hang it on the bolt using any hole in the flange. Start a nut on the bolt. The flange is now completely under control and cannot be dropped.
3. Rotate the flange and stick a bolt through in the other bottom off-center hole. These two bolts will catch the gasket.
4. Put gasket in place.
5. Insert remainder of bolts and tighten.

Figure 4-9. The correct method of installing a blind flange (blank) on a cargo manifold. Courtesy Chevron Shipping Company, LLC.

Figure 4-9a. Installing a gasket. Courtesy Stacy DeLoach.

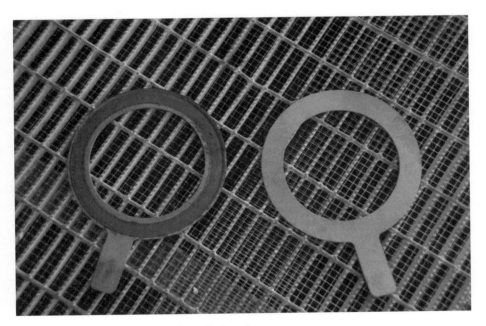

Figure 4-9b. Gaskets. Courtesy Stacy DeLoach.

Couplings

Another method of securing the hose or mechanical loading arm to the manifold is through the use of quick-connect couplings. The typical quick-connect couplings in use today are hydraulic clamps and rotating locking cams (Cam-Locks). Regardless of the type of quick-connect coupling employed, it must be acceptable to the commandant of the USCG. Figure 4-10 illustrates a bolted connection and a quick-connect coupling.

The manufacturer's step-by-step sequence of connection using Cam-Locks is illustrated in Figure 4-11.

Figure 4-10. Manifold connections are typically bolted or they employ quick-connect couplings.

1. Here a ratchet C-L Coupling in studded configuration is bolted to an existing ASA hose flange for a one-time permanent installation. The coupling is then ready to be joined at any time to a matching flanged hose, manifold, or pipe of comparable diameter.

2. The actual connection is achieved by facing the C-L Coupling to a matching ASA flange. No critical alignment of bolt holes or gasket is necessary.

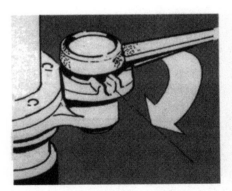

3. A simple movement with the special C-L wrench secures each ratchet cam with a wedging action, while compressing the built-in O-ring.

4. The resultant connection is positive and leakproof. It is impossible to loosen the cams, which are automatically ratchet locked, except by use of the C-L wrench.

Figure 4-11. One style of quick-connect coupling typically employed at the cargo manifold (Cam-Lock). Courtesy MMC International Corp.

Figure 4-12. Various size reducers. Courtesy George Edenfield.

Figure 4-13. Reducer. Courtesy Stacy DeLoach.

Reducers

For situations where the vessel's manifold flange differs in dimension from the flange on the cargo hose or steel arm, a reducer must be installed. Tank vessels carry many different size reducers for this purpose (Figure 4-12), to enable connection with the shore facility. Figure 4-13 illustrates a reducer that is connected to the vessel's cargo manifold.

If the flanges of the vessel and hose do not align properly, then a spool piece is frequently used to make the connection possible. A spool piece is a short section of flanged pipe that can be utilized as an adapter or an extension to the manifold. Removable spool pieces and blanks are also useful when it is necessary to provide a positive means of segregation between sensitive cargoes.

Crossovers

During a cargo transfer it is frequently necessary to cross over systems. There are several ways to interconnect cargo systems on a tank vessel. One approach commonly seen at the manifold is the use of a temporary pipe called a *runaround,* which is prefabricated pipe (Figure 4-14) designed to interconnect two or more cargo systems at the manifold.

It is usually attached (bolted) to the offshore manifold when it is necessary to interconnect the cargo systems. A number of factors may necessitate the use of a crossover, including the following:

1. Limited number of loading/discharge hoses or arms to conduct the transfer
2. Convenience when loading a single grade of cargo
3. Expediting a cargo transfer
4. Rerouting the cargo operation when a problem exists

Figure 4-14. Interconnection of cargo systems at the manifold through the use of a runaround (temporary piping). Courtesy International Marine Consultants (IMC).

Figure 4-15. Mixmaster. Courtesy Kaitlin O'Brien.

Figure 4-15a. Jumper Hose. Courtesy Christopher Adams.

A second method of interconnecting cargo systems at the manifold is the use of permanently installed piping referred to as a "*mixmaster.*" Figure 4-15 illustrates the fixed piping of the mixmaster running across the top of the manifold piping.

A third crossover uses a flexible hose, usually called a *jumper*, to interconnect cargo systems at the manifold on the vessel (Figure 4-15a). The PIC must exercise extreme caution when using a crossover between cargo systems, given the increased risk of contamination. To protect against the possibility of leakage, crossovers are typically equipped with blanks, removable spool pieces (pipe sections), or double valves for segregation.

CARGO HOSES

Flexible cargo hoses are frequently employed to make the connection between the fixed piping on the vessel and the shore facility. Barges generally carry a number of cargo hoses on board whereas tankships utilize hoses provided by the shore facility. The most common hose types used for transferring cargo are rubber, composite, and stainless steel.

Rubber Hose

Rubber hose consists of three basic components: tube, reinforcement, and cover (Figure 4-16). The PIC should check the suitability guides from the manufacturers in order to match the correct hose type to the cargoes transported. Some cargoes require specially constructed hoses such as those used in corrosive, high-temperature service and with certain types of chemicals. A number of materials are used for the inner tube of a rubber hose, the most common being Buna-N (nitrile) in oil service and Viton in aromatic and higher temperature service. For strength, several layers of polyester reinforcement and steel mesh (helix) typically surround the tube of the hose. The cover serves as the outer protective layer for the tube and as reinforcement for the hose. Covers are generally made of neoprene due to its resistance to weather, physical abuse (abrasion), and petroleum. Rubber hose has both advantages and disadvantages: it is rugged, but heavy in weight; a smooth tube provides a high flow rate, but it is inflexible; and it has good temperature resistance, but limited chemical resistance.

The end fittings of the hose are either built-in or swaged (crimped collar) steel nipples that are equipped with welded carbon steel flanges.

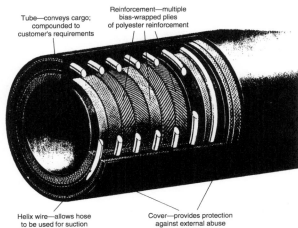

Tube—conveys cargo; compounded to customer's requirements

Reinforcement—multiple bias-wrapped plies of polyester reinforcement

Figure 4-16. Construction of a typical rubber cargo hose. Courtesy Apollo International Corporation.

Helix wire—allows hose to be used for suction

Cover—provides protection against external abuse

Figure 4-16a. Barge hose rack. Courtesy Stacy DeLoach.

Composite Hose

Composite hose is light, flexible, and resistant to most chemical cargoes. It is constructed of numerous layers of polypropylene films and fabrics with inner and outer (spiral) wire helixes. Composite hose is suitable for chemical and oil service. (Figure 4-16b)

Figure 4-16b. Composite hose cutaway view. Courtesy US Hose Corporation.

Stainless Steel Hose

Stainless-steel hose is composed of a stainless-steel corrugated tube covered by a single or double stainless-steel braid. Stainless-steel hose is suitable in situations where chemical resistance and higher temperatures may be encountered.

Markings

Cargo hoses should have the following markings:

1. Name of the product for which the hose may be used
2. For oil products, the words "OIL SERVICE"
3. For hazardous materials, the words "HAZMAT SERVICE—SEE LIST"
4. Maximum allowable working pressure (MAWP)

Vessels usually maintain a written record of the date of manufacture, latest test date, test pressure, and bursting pressure for the hoses carried on board. Cargo hoses are required to be tested annually to one and one half times the maximum allowable working pressure.

CARGO TRANSFER

During cargo transfer, the importance of visually checking the hose cannot be overstated as operators have experienced problems ranging from deterioration

 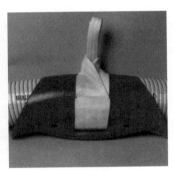

Figure 4-17. Proper support of a cargo hose requires the use of slings or saddles. Courtesy Apollo International Corp.

Figure 4-18. Flexible hoses should be supported by belt slings, saddles, or bridles. Courtesy Chevron Shipping.

of the hose lining (tube) to complete failure. Although cargo hoses are ruggedly constructed, they still represent the weak link in a cargo transfer. The person in charge of the transfer must keep a watchful eye on the hose, the connections, and in particular the vessel moorings. The moorings must be properly tended during the cargo transfer to keep the vessel in position alongside the facility. Failure to monitor the condition of the lines could result in unacceptable surging of the vessel or movement off the dock, which could place undue strain on the hoses and connections. According to federal regulation, the cargo hoses and steel arms should be long enough to allow the vessel to move to the limits of its moorings without placing a strain on the hose, arm, or transfer piping system. Several factors should be considered when determining the number of lengths of cargo hose to be connected. The PIC should account for the range of the tide in the locality, the anticipated change in freeboard of the vessel, wind and current conditions at the facility, and vessel traffic. The use of excessive lengths of cargo hose is discouraged, as there is an increased risk of a bight of hose getting pinched between the vessel and dock; this could have serious consequences. Cargo hoses are typically stopped off to bitts on deck by the manifold to facilitate securing the hose flange to the cargo manifold. Once the cargo hoses are secured to the vessel manifold piping they must be properly supported with saddles or straps to prevent kinking or damage to the hose and its coupling (Figures 4-17 and 4-18). The use of a single rope sling is not recommended when lifting or supporting the hose assembly. Figure 4-19 is a guide from one hose manufacturer illustrating some of the dos and don'ts when working with cargo hoses.

The vessel is usually equipped with a chafing rail in the vicinity of the manifold to prevent kinking or damage to hoses. The hose assembly should be drained at the end of a cargo transfer prior to its removal from the manifold. A number of methods are used to drain the hoses, including gravity, vacuum pump, or blowing the line clear with a suitable gas under pressure. Depending on the method employed it may be necessary for the valve(s) to the cargo tanks to remain open for the final clearing of the hoses. To avoid an unnecessary mess, vessel personnel should verify that draining is complete prior to breaking the connections at the manifold.

The person-in-charge is required to inspect the hose prior to use for any unrepaired loose covers, kinks, bulges, soft spots, or other defects which would permit the discharge of oil or hazardous cargo through the hose material. Additionally, there should be no gouges, cuts, or slashes that penetrate the first layer of hose reinforcement.

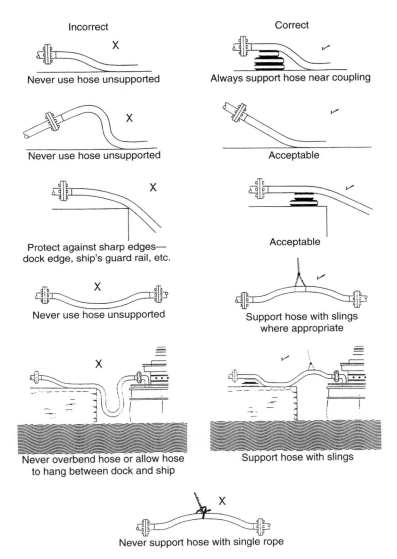

Figure 4-19. Hose handling guide from one manufacturer. Courtesy Uni-chem Hose Corp.

Mechanical Loading Arms

Modern shore facilities that accommodate large vessels and have high cargo transfer rates frequently use mechanical loading arms, which are steel pipes that telescope to make the connection with the vessel manifold (Figures 4-20 and 4-21). These are often referred to as "chicksans."

The arms are controlled hydraulically and employ swivel joints which enable them to follow the movement of the vessel at the berth. Although steel arms are capable of handling greater pressures and flow rates than hoses, the person-in-charge should be aware of several concerns with their use.

Figure 4-20. Steel Arm connected to the manifold. Courtesy Kevin Duschenchuk.

Figure 4-21. Mechanical Loading Arms. Courtesy Mark Huber.

Mechanical loading arms have a limited operating envelope, which means they are much less forgiving than cargo hoses when the vessel begins to surge or drift at the dock. The PIC should take this into account when spotting the vessel and properly tend the moorings to ensure the vessel stays in position at the berth. Most mechanical loading arms employ some form of quick-connect coupling such as hydraulic clamps or Cam-Locks when connecting to the manifold. As in the case of cargo hoses, it is imperative that suitable gasket material or O-rings be installed in the connection to maintain a leak-free seal. In the United States, mechanical loading arms must meet the design, fabrication, material, and inspection and testing requirements contained in ANSI B 31.3. Each mechanical loading arm that is constructed in accordance with this standard must have a manufacturer's certification permanently marked on the arm or recorded elsewhere in the facility. At the end of a cargo transfer there must be a means to drain or close off the arm prior to breaking the connection.

Loading Drops

The cargo piping on deck is connected to the bottom piping through vertical lines called loading drops. These drop lines are used when loading the vessel from shore. On vessels equipped with a pumproom, the drops route the cargo directly to the bottom piping and the tanks. Loading the vessel through these lines permits the operator to bypass the pumproom entirely. It is considered safe practice to avoid loading through the pumproom when a choice exists, thereby isolating the vessel cargo pumps from the loading pressure. The person-in-charge should verify the status of the pumproom valves prior to commencing the loading operation. Each cargo system or group is generally serviced by its own drop line; however, some highly segregated vessels such as those with in-tank pumps may have a drop for each tank. In the construction of new double hull product carriers, the shift toward in tank pumps has resulted in a greater number of cargo segregations (see Chapter 8, Figure 8-6).

PUMPROOM PIPING

The pumproom is a complex compartment that is the heart of the discharge operation on a tank vessel. It contains the necessary piping and pumping equipment to deliver the cargo to the shore facility. The pumproom is a comparatively small compartment usually located at the after end of the cargo section of the vessel. The aft location of the pumproom takes advantage of the tendency of the vessel to trim by the stern during a discharge. It is also close to the engine room. Locating the cargo pumps at the low point in the system enables the pumps to operate in a head condition, which results in a more efficient discharge operation. The proximity of this space to the engine room is also convenient when connecting the pump to a drive unit. Due to the complexity of the piping in this space, the person-in-charge must carefully trace the lines to become well-versed on the proper lineup of the system. Figure 4-22 is a simple pumproom sketch showing the bottom piping from the tanks directing cargo to the suction side of the pumps.

From the cargo pumps, the vertical discharge lines deliver the cargo to the manifold and onward to the shore facility. Most pumprooms are also equipped with crossovers that permit the cargo systems to be interconnected on the suction and discharge side of the pumps (see Chapter 8, Figure 8-11).

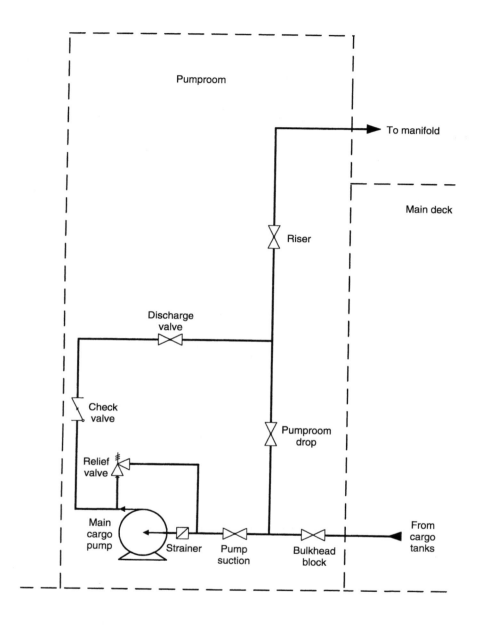

Figure 4-22. Simpified drawing of the cargo lines in a pumproom.

To prevent the possibility of contamination through the crossovers, it is advisable to maintain double-valve segregation or install blanks between the systems. Another function built into the design of the pumproom is the ability to take on ballast. The cargo pumps usually have a connection to the sea via a penetration in the hull called a sea chest. The piping that connects the sea chest to the suction side of the cargo pumps is commonly called the sea lane. As in the case of the crossovers mentioned earlier, one can expect to find double sea-suction valves separating the cargo system from the sea. To avoid a serious pollution incident, care should be exercised when taking on seawater ballast using the cargo pumps and lines. Consult chapter 12 for a discussion of proper ballasting procedures.

BASIC VALVE TYPES

Valves are the devices installed at various points in the cargo system that enable the operator to control the flow of liquid through the piping. The person-in-charge should have an understanding of the working principles and limitations of the valves in the cargo system. Valves can be categorized in the following ways:

1. *Rising or nonrising stem.* When the stem moves up and down as a function of operating the valve, it is classified as rising stem. In a nonrising stem valve, the gate or disk rides up and down on the spindle through the use of a reverse thread.
2. *Throttling or nonthrottling.* A valve is considered throttling when it is suitable for controlling liquid flow, meaning it can be operated in a partially open position. Nonthrottling valves are not designed to control flow and should only be used in a fully open or fully closed position. Operators should bear in mind that there is a risk of jamming when using a nonthrottling valve in a partially opened position.
3. *Manual or motor-operated.* If the valve wheel is physically controlled by the operator, it is a manual valve. In automated cargo systems, the valves are frequently operated through the use of hydraulic or pneumatic motors connected to the stem.

Figure 4-22a. Butterfly valve and actuator.

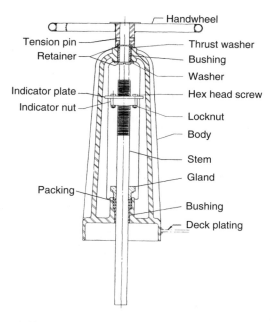

Figure 4-23. A valve operating stand on deck permits remote actuation (via a reach rod) of the valve located in the bottom of the vessel. Courtesy Hayward Manufacturing.

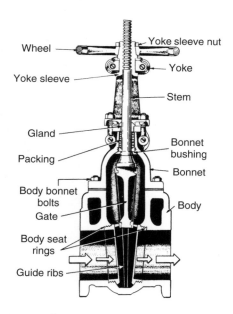

Figure 4-24. The key parts of a rising stem gate valve. Courtesy U.S. Coast Guard.

4. Local or remote actuation. Local control means the operator is at the valve and controls it directly. When the valve is located some distance from the operator such as at the bottom of a cargo tank, it is then necessary to install a reach rod for remote actuation. A reach-rod assembly is a series of steel rods coupled to connect the stem of the valve in the tank to a valve operating stand on deck (Figure 4-23).

Personnel must exercise care when operating a valve remotely with a reach-rod assembly. If it becomes necessary to use wheel wrenches or other "persuaders," the operator must guard against the application of excessive torque to the valve. This could shear the pins in the reach-rod couplings, resulting in an inoperative valve.

Gate Valve

The key parts of a typical rising stem gate valve are shown in Figure 4-24. Starting at the top of the valve is the operating mechanism called the handwheel, which is operated either manually or by a motor. The stem or spindle is the shaft that is connected to a circular gate (or disk) as seen in Figure 4-24. The packing and gland assembly makes a liquid-tight seal where the stem exits the top of the bonnet. The bonnet is the upper housing that is typically bolted to the body of the valve. When the valve is fully opened, the gate retracts into the bonnet and out of the path of the liquid. The body of the valve is the flanged portion installed in the pipeline through which the liquid flows. Within the body of the valve is a set of grooves referred to as guide ribs which keep the gate aligned when moving up and down. When the valve is in the closed position, the operating gate or disk is wedged firmly into the seat.

Gate valves are a popular choice in cargo systems given their durability and the fact that they offer the least resistance to flow in the open position. Other types of valves commonly found in cargo service include butterfly, globe, plug, and ball valves. Following are various types of valves installed on tank vessels.

Comparison of Valve Types

The following listings show some of the advantages and disadvantages associated with each type of valve shown in Figures 4-25 through 4-29.

GATE VALVE

Advantages
Simple design
Durable
Less prone to obstructions
Unrestricted flow across open valve
Suitable for local and remote actuation
Suitable for bidirectional flow

Disadvantages
Not easily automated
Prone to jamming
High cost
Numerous turns
Nonthrottling

Figure 4-25. Gate Valve. Courtesy Design Assistance Corporation (DAC).

BUTTERFLY VALVE

Advantages
Quick acting valve
Simple and compact design
Less expensive
Easy to automate

Disadvantages
Less durable
Prone to obstructions
Prone to leaking
Prone to improper seating
Nonthrottling

Figure 4-26. Butterfly Valve.
Courtesy DAC.

GLOBE VALVE

Advantages
Precise throttling
Flow control
Durable
Directional flow through valve

Disadvantages
Prone to obstructions
Prone to improper seating
Numerous turns
Not easily automated
High cost
Pressure drop across open valve
Greater resistance to flow

Figure 4-27. Globe Valve.
Courtesy DAC.

BALL VALVE

Advantages
Quick-acting
Simple and compact design
Easy to automate
Suitable for bidirectional flow

Disadvantages
Nonthrottling

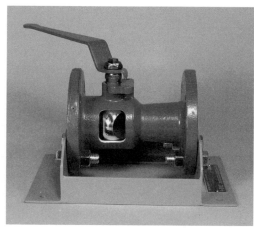

Figure 4-28. Ball Valve.
Courtesy DAC.

PLUG VALVE
Advantages
Simple and compact design
Quick acting
Less prone to obstructions
Easy to automate
Durable
Suitable for bidirectional flow

Disadvantages
Nonthrottling

Figure 4-29. Plug Valve.
Courtesy DAC.

SPECIAL VALVES

This section describes valves that serve a unique function in the cargo system of a tank vessel.

Check Valve

The check valve is designed to permit liquid flow in only one direction. It is typically found on the discharge side of the cargo pump and in the inert gas delivery line on deck to prevent return flow. The check valve opens when the discharge pressure from the cargo pump or inert gas fan lifts the operating disk in the valve, thereby allowing flow (Figure 4-30).

The check valve is designed to operate automatically either under a weight (lift check) or spring load (swing check) when the cargo pump or inert gas fan stops. The weight or tension of a spring on the disk causes the valve to seat, thereby preventing return flow. Experience has shown that check valves are prone to leaking, particularly when installed in a hostile environment. For example, the corrosive nature of inert gas can result in a check valve deteriorating to the point that it remains stuck in the open position. A similar situation can occur when handling viscous or high-temperature pour-point cargoes which are capable of gumming up a check valve to the point that it sticks open.

Figure 4-30. Check valve.
Courtesy DAC.

Relief Valve

The relief valve is also found on the discharge side of a cargo pump; it protects the piping system from the effects of over-pressurization. Relief valves are spring loaded and operate automatically when a preset pressure is reached in the discharge line of the pump (Figure 4-31).

When the relief valve opens, the cargo is returned to the suction side of the pump through a short recirculation line, preventing any further buildup of pressure. The operation and setting of the relief valves should be checked to ensure the cargo system is properly protected.

Pressure-Vacuum (PV) Relief Valve

The pressure-vacuum relief valve is specially designed to provide structural protection of the cargo tanks from the effects of over- or under-pressurization of the tank atmosphere. The valve contains two operating disks (pressure and vacuum) that are held in the normally closed position by a

Figure 4-31. Relief Valve.

Figure 4-31a. Cargo pump relief valve and check valve. Courtesy Stacy DeLoach.

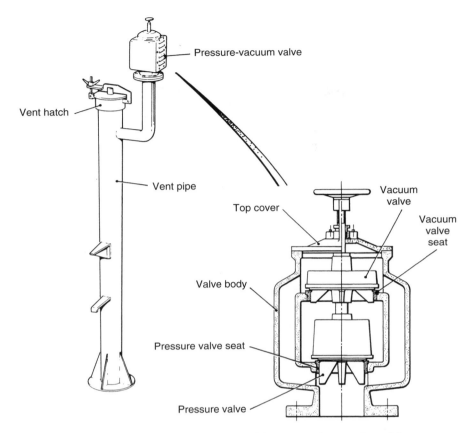

Figure 4-32. Pressure-vacuum relief valves provide structural protection of the cargo tanks. Courtesy Permea Maritime Protection.

weight or the tension of a spring. The valves are designed to open (lift) at a preset pressure or vacuum in the tank. When the pressure disk opens, the atmosphere in the cargo tank escapes, relieving the excess pressure via a flame screen and louvered vent on the body of the valve. Conversely, when the vacuum disk lifts, air rushes into the tank through the same opening to break the vacuum. Figure 4-32 illustrates a cross-sectional view of a typical PV relief valve.

Pressure-vacuum relief valves are usually installed on standpipes connected to each tank hatch or on a central vent main when designed to protect a group of cargo tanks. Figure 4-33 illustrates the typical locations of the PV valves in the cargo system. In the closed position, these valves enable the cargo tank(s) to remain sealed, thereby minimizing loss of cargo through vaporization as well as loss of inert gas pressure.

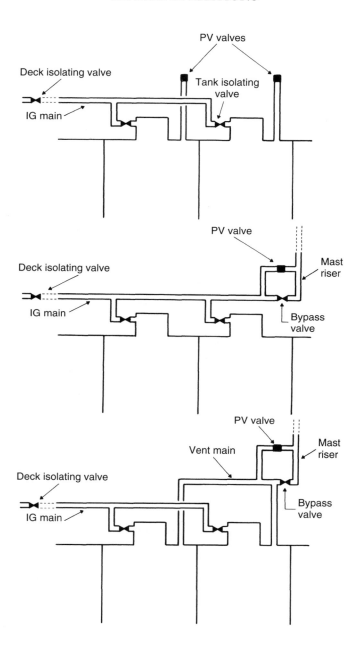

Figure 4-33. Three applications of PV valves.

As with any mechanical device, routine inspection and maintenance of PV valves is imperative to ensure proper protection of the vessel. Operators should check PV valves for the following:

1. Valves operate freely.
2. Valve seats and sealing surfaces are clean and tight.
3. Flame screens are not fouled or holed.
4. Springs or weights are in good repair.

The valves must be set to lift (operate) before the cargo tank reaches its maximum design pressure or vacuum. The pressure and vacuum relief settings are normally indicated on the body of the valve and recorded in the inert gas manual for the vessel. Relief settings vary considerably from one vessel to the next, however some typical values are as follows:

	Pressure Relief	*Vacuum Relief*
Cargo tank PV valves	1.5 psi	–0.5 psi
Mast vent PV valves	2.0 psi	–1.0 psi

BLANKS

A blank is a device that is generally inserted at one of various locations in the cargo piping system to provide a positive means of segregation between cargoes. Several styles of blanks are in use, the most common being a spectacle blank (Figure 4-34). In most cases, a blank represents a physical break in the piping; therefore, any leakage is unable to continue past this device and simply leaks out of the line. A spectacle blank consists of two disks—an open (flow) and a solid (no-flow) disk—that generally swing on a pivot point. The status of the blank is clearly indicated by the visible disk (Figure 4-34).

Figure 4-34. Spectacle blanks in mixmaster. Courtesy Stacy DeLoach.

Caution should be exercised when swinging a blank; these simple guidelines should be followed:

1. Verify that the piping is not under pressure (no active cargo transfer).
2. Prior to swinging the blank, drain the piping on both sides of the blank.
3. Ensure the pipeline in question is isolated by closing the appropriate valves.
4. Do not stand in a position that could result in contact with the cargo.
5. Carefully swing the disk out of the line and inspect all flange or mating surfaces.
6. Inspect the blank for corrosion and physical damage.
7. Inspect the O-ring or gasket material prior to installing the disk in the line.
8. Check the disk for proper alignment in the pipeline and secure the blank.
9. Inspect the blank for any leakage after it is installed.

Another style of blank is shown in Figure 4-36. In this type the blank is swung (changed) by turning the handwheel which pulls the installed disc out of the pipeline until it is clear and the opposing disc can be rotated into the downward position. The handwheel is then closed and the new disc slides into place within the pipeline.

One drawback to the use of blanks in the piping system is access for operation. Blanks require local operation, therefore they are not user-friendly when installed in such locations as the bottom piping of the vessel. Blanking devices are often found at the manifold on deck, with a mixmaster, and at crossovers in the pump room where access is not a problem.

An alternative to inserting a blank in the line is the use of a removable spool piece, a section of pipe with steel blanks on the ends of the piping. Removing the spool piece provides the ultimate assurance against any risk of contamination through the piping. Simple blanking devices, however, offer reliable protection without the labor involved with a spool piece.

Figure 4-35. Spectacle blank in pumproom.

Figure 4-36. Hamer blind. Courtesy Christopher Adams.

REVIEW

1. What are two common designs for bottom cargo piping found on modern tank vessels?
2. What is the name given to the bitter end of piping located in the bottom of each cargo tank?
3. What is meant by a "free-flow" design, as applied to a large crude carrier?
4. Nonflanged cargo piping is connected through the use of what device? Why is it used?
5. If the cargo manifold flange on the vessel differs in dimension from the cargo hose flange, what device is employed?
6. If it is necessary to cross over cargo systems at the manifold on deck, what methods are employed?
7. Prior to a cargo transfer, the hose must be visually inspected. What are the typical causes for rejection of a cargo hose? What are the required markings on a hose?
8. What is the purpose of a P/V valve on a cargo tank? List the items that an operator should check when conducting an inspection of the P/V valves.
9. Why are spectacle blanks utilized in the cargo piping of a tank vessel? List the precautions that should be followed whenever one swings a blank in the pipeline.
10. Prior to breaking the cargo hose connection at the end of a transfer, what are the typical methods of draining the line?

CHALLENGE QUESTIONS

11. On tank vessels equipped with numerous crossovers in the cargo system describe the potential concerns for vessel personnel from a quality assurance point of view.
12. During a cargo transfer with a facility a pressure surge occurs due to a faulty valve. What are the likely locations for a problem to develop in the cargo system of the vessel?
13. Describe the process of renewing a slip on (dresser) type coupling in a cargo pipeline of the vessel
14. Describe the process of hydrostatically testing the cargo piping on a tank vessel.
15. Why is it considered sound tanker practice to never load through the cargo pumproom?

CHAPTER 5

Venting Systems and
Vapor Control Operations

The method employed to vent the atmosphere of a cargo tank on a vessel is an issue that has plagued the tanker industry for many years. The earliest approaches, although simple and foolproof, presented serious concerns regarding fire safety, occupational safety and air quality.

The initial methods of cargo tank venting involved open venting through a tank top or hatch at the deck line. During loading and ballasting operations, tank atmospheres typically exited through an ullage opening in the hatch near deck level. This hatch served a dual purpose: first, as the primary venting port and second, as the gauging point for the cargo tank. Industry studies revealed that exiting cargo vapors tended to settle and accumulate around the vessel in sufficient concentrations to pose significant fire and health risks. Figure 5-1 shows the traditional approach of open-venting through the tank hatch, and Figure 5-2 illustrates the resultant vapor plume developed during a typical loading operation.

Figure 5-1. Traditional open venting of a cargo tank through the ullage opening. Shown here, the vent is equipped with a flane screen.

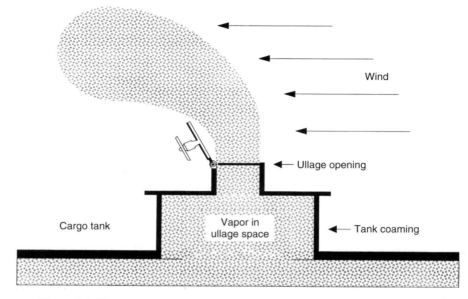

Figure 5-2. Vapor plume developed above the deck during a loading operation.
Courtesy Richard Beadon and Eric Ma.

CONTROLLED VENTING SYSTEMS

Over the past twenty-five years, the widespread installation of inert gas (IG)
systems on tankers has led to dramatic improvements in cargo tank venting, now
referred to as controlled venting systems. The deck distribution piping (the IG
main and branch lines) used to supply inert gas to the cargo tanks also serves as
the vent system during cargo loading. Through the use of this vent piping, cargo
tank atmospheres are directed aloft where the prevailing winds at the berth can
dilute and disperse the exiting vapors, thereby reducing the risks to the vessel
and personnel. Controlled venting systems are categorized in two ways: mast riser
venting and high-velocity venting. Figure 5-3 illustrates three controlled venting
arrangements in use on modern tankers.

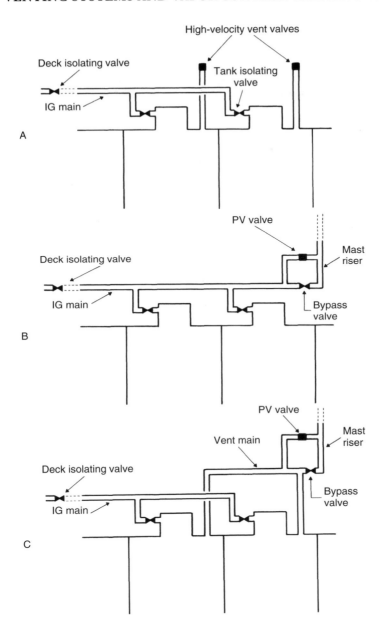

Figure 5-3. Controlled venting arrangements. A. Individual cargo tank vents employing high-velocity vents (HVVs) and standpipes. B. Common venting using a single main (IG piping) and venting through one or more mast risers or sphere through one or more mast risers or HVVs.

Flame Cell Channel

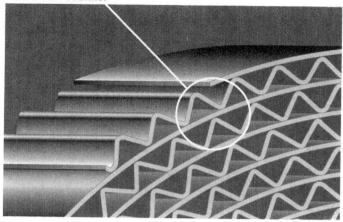

Figure 5-3a. Flame arrestor construction. Courtesy Enardo.

Mast Risers

Mast or king-post vents physically direct the exiting cargo vapors well above the deck via hard piping before releasing it to the atmosphere. The use of a tall vent stack can be seen in Figure 5-4. Controlled venting of the cargo tanks in this manner is very effective. However, when certain meteorological conditions exist (low wind and high humidity), vapor accumulation on deck and around the superstructure is still possible. These vents are equipped with flame arresters designed to prevent the passage of flame into the vent piping and cargo tanks of the vessel. A flame arrester is typically constructed of a bundle of thin metal ribbons or corrugated metal elements that act as a heat sink which has a cooling effect thereby preventing flame propagation through this device. Additionally, flame arresters aid in breaking up the pressure wave associated with the movement of a flame through a piping system (Figure 5-3a).

Figure 5-4. Vent stack method of controlled venting. Courtesy International Marine Consultants (IMC).

High-Velocity Vents (HVV)

Another approach in controlled venting systems is the use of high-velocity vent valves. Two types of high-velocity vent valves are in common use today: bullet valves (Figure 5-5) and high jets (Figure 5-6).

Figure 5-5. Cutaway view of a high-velocity pressure-vaccuum relief valve, also referred to as a bullet valve. Courtesy Ian-Conrad Bergan, Inc.

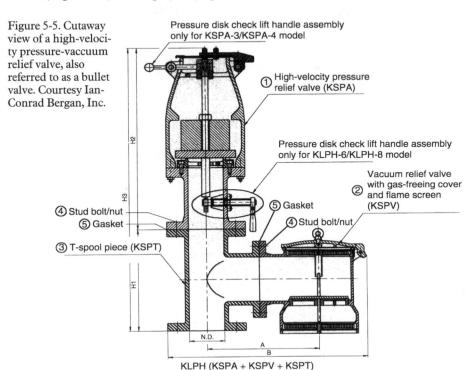

Pressure disk check lift handle assembly only for KSPA-3/KSPA-4 model

① High-velocity pressure relief valve (KSPA)

Pressure disk check lift handle assembly only for KLPH-6/KLPH-8 model

② Vacuum relief valve with gas-freeing cover and flame screen (KSPV)

④ Stud bolt/nut
⑤ Gasket
③ T-spool piece (KSPT)
⑤ Gasket
④ Stud bolt/nut

N.D.
A
B

KLPH (KSPA + KSPV + KSPT)

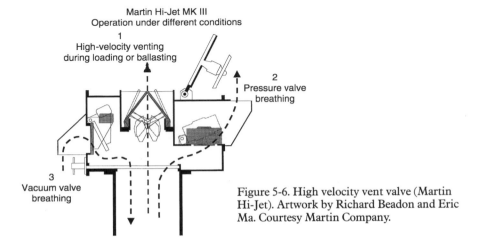

Martin Hi-Jet MK III
Operation under different conditions

1
High-velocity venting during loading or ballasting

2
Pressure valve breathing

3
Vacuum valve breathing

Figure 5-6. High velocity vent valve (Martin Hi-Jet). Artwork by Richard Beadon and Eric Ma. Courtesy Martin Company.

High-velocity vent valves operate off shorter standpipes (Figures 5-7 and 5-7a) and project the exiting vapors aloft. These devices work on a deadweight principle, preventing the release of the cargo tank atmosphere until a predetermined tank pressure is reached. According to the manufacturers of these devices, a minimum exit velocity of 30 meters per second is achieved when the valves open, ensuring that the vapors reach a considerable height above the vessel.

Figure 5-7. Bullet Valve.
Courtesy Christopher Adams.

The type of venting system an owner elects to install on a vessel depends on numerous factors, including the following:

1. Regulatory requirements
2. Vessel trade
3. Types of cargo transported
4. Health concerns
5. Venting capacity
6. Experience factor
7. Convenience/ease of operation
8. Ease of maintenance

Venting systems vary considerably from one vessel to the next; therefore, the PIC should carefully trace out the piping arrangement and be familiar with the proper lineup and use of the system.

In addition to providing improved venting of cargo tanks during loading, this piping also plays an important role in various operations associated with the proper use of the vessel's inert gas system. For example, suitable supply and venting arrangements are critical when it is necessary to replace cargo tank atmospheres.

Operations involving complete replacement of a cargo tank atmosphere include primary inerting, purging, gas-freeing, and reinerting. A detailed discussion of gas replacement methods can be found in chapter 16, "Inert Gas Systems."

Figure 5-7a. Bullet valves on standpipes. Courtesy Christopher Adams.

DEVELOPMENT OF VAPOR CONTROL SYSTEMS

In recent years, various states and localities became concerned about the affect that cargo vapors have on air quality, and this led to the development of vapor control systems. The uncontrolled release of hydrocarbon vapors from tank vessels during cargo loading contributes to the overall quantity of volatile organic compounds (VOCs) in the atmosphere, a precursor to the formation of ozone pollution. In fact, the problem became serious enough to prompt the implementation of rules limiting emissions during such operations as loading, ballasting, purging, and gas-freeing of cargo tanks.

At present, these rules apply to vessels handling crude oil, gasoline, gasoline blends, and benzene cargoes. The list of "regulated" cargoes, as they are known, can vary from state to state and in some cases even differ within a local air-quality district. Vessel operators should inquire as to the local requirements prior to arrival at the facility. The number of cargoes requiring the use of a vapor control system is expected to expand as more detailed information becomes available about the effects of these substances on the environment.

The quantity of VOCs emitted during loading and ballasting operations was documented by the U.S. Environmental Protection Agency in 1985, as shown in table 5-1.

Vapor control systems represent the next stage in the evolution of cargo tank venting designs. The basis of the system is a closed loading operation. All vents to atmosphere and deck openings to the cargo tanks are secured and remain so for the entire transfer. By means of a network of vapor collection piping connected to the manifold on deck, the vapors from each tank on the vessel are directed ashore for processing. The facility has the option of destroying or recovering the vapors. The vapors involved are those displaced by the incoming cargo during loading operations as well as those released due to cargo vaporization.

Table 5-1
EPA Emission Factors in Pounds Per 1,000 Gallons of Liquid

Emission Source	Loading Operations		Tanker Ballasting
	Ships	Barges	
Gasoline	1.8	3.4	0.8
Crude oil	0.61	1.0	1.2
JP-4	0.5	1.2	unknown
Kerosene	0.005	0.013	unknown
Distillate oil no. 2	0.005	0.012	unknown
Residual oil no. 6	0.00004	0.00009	unknown

Courtesy U.S. Environmental Protection Agency

VAPOR CONTROL SYSTEM COMPONENTS

The U.S. Coast Guard developed and published regulations (Title 46 CFR Part 39) governing the design, construction, and operation of vapor control systems on tank vessels operating in U.S. waters. It should be noted that the United States Coast Guard inspects vessels for compliance with the aforementioned regulations however enforcement of actual use of the system and reduction of emissions lies with the individual states. Internationally, emission of cargo vapors has been addressed in Annex VI of the MARPOL convention.

Vapor Control Piping

The collection of vapors is accomplished through permanently installed deck piping usually consisting of a common main, branch lines, and vapor manifold. On tankers fitted with an inert gas system, modification of the IG distribution piping permits its use for vapor control while loading (Figure 5-8). When a tanker is equipped with an IG/vapor control main, there must be a means of isolating the IG supply. The deck isolating valve required under existing IG regulations satisfies this requirement.

The vapor control piping terminates in a vapor manifold located as close as practical to the cargo manifold. To clearly distinguish the vessel vapor connection, the last meter of piping must be painted with red/yellow/red bands and labeled with the word "vapor," as seen in Figure 5-9.

As an additional safeguard against possible cross-connection of a cargo hose to the vapor manifold, a special flange is employed. The vessel vapor connection flange is equipped with a 0.5-inch stud at least one inch long, projecting outward from top dead center on the flange face. The vapor manifold must also be fitted with a manually operated isolation valve that gives clear indication of the valve's status.

If a ship is carrying incompatible cargoes, it is imperative to maintain segregation of the cargo vapors for quality assurance reasons and in some instances for safety. On a tanker with a single vapor control main, this is usually accomplished by closing valves or installing blanks in the appropriate branch lines. Other tankers (such as chemical carriers and certain product carriers) are fitted with dedicated vapor control systems for individual cargo tanks or tank groups.

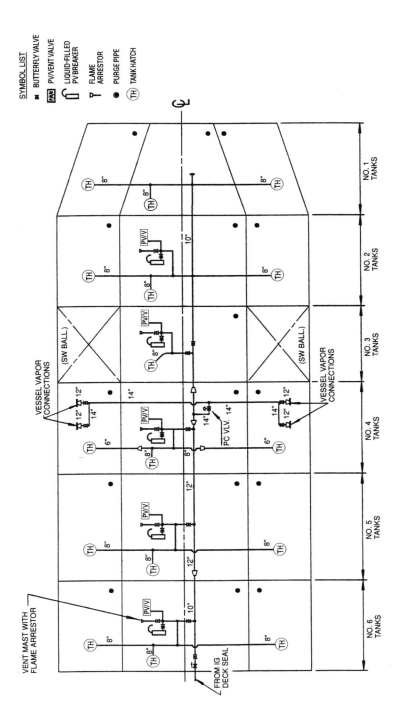

Figure 5-8. Diagram of the inert gas distribution piping on the deck of a typical tanker. Note the modifications that have been made to permit the use of IG piping for vapor control operations.

Figure 5-9. The vapor manifold is clearly marked to avoid confusion or possible cross-connection. The vapor hose is required to have the same coloring and stenciling as the vessel vapor manifold. Courtesy Mark Jones.

To further guard against contamination of dissimilar cargoes, drains must be provided for removal of liquid condensate from the vapor control piping resulting from (1) liquid carryover while loading due to mists in the vapor stream, (2) condensation in the piping due to temperature changes, (3) cargo tank overfill, or (4) cargo sloshing at sea.

Vapor Control Hose
The hose used for transferring vapors must be electrically continuous and constructed of material that is resistant to kinking and abrasion. The hose assembly should be provided with proper support to prevent excessive strain, kinking, or collapse of the hose.

The vapor hose must also meet the following *minimum* strength criteria:

Design bursting pressure	25 psi
Maximum allowable working pressure	5 psi
Vacuum (without collapsing/constricting)	–2 psi

The color and marking at each end of the hose should be similar in all respects to the vessel vapor manifold (Figure 5-10). As part of the declaration of inspection (DOI), the hose should be inspected for cuts, tears, or defects that may render it ineffective.

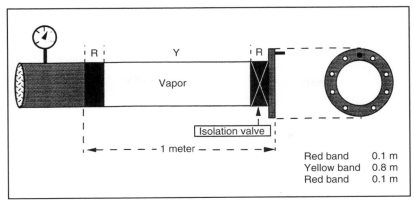

Detail at vessel manifold end of vapor control piping

Red band	0.1 m
Yellow band	0.8 m
Red band	0.1 m

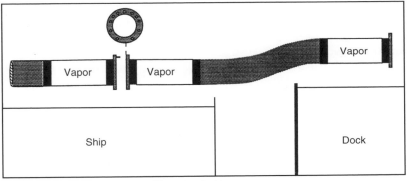

Vapor control hose

Figure 5-10a. The vapor manifold (*top*) must be fitted with a manually operated isolation valve that gives a clear indication of the valve's status. The vapor control hose (*bottom*) must be colored and stenciled as shown. Courtesy Richard Beadon and Mark Huber.

CLOSED GAUGING

In an effort to achieve a vapor-tight deck and leak-free transfer, all tankers engaged in vapor control must be equipped with a closed gauging system designed to operate over the entire tank depth. Closed gauging has been widely used for a number of years as part of the IG regulations. It originated as a result of the need to maintain atmosphere control and a positive deck pressure during cargo operations. Consequently, there are a number of closed gauging systems on the market, details of which can be found in chapter 6.

High Level/Overfill Alarms

One of many concerns associated with a closed loading operation is the risk of cargo tank overfill. Overfilling a cargo tank while topping off can result in structural damage to the vessel; also, cargo may be sent into the vapor control system. The most common cause of a spill while loading is overfilling, and the most common causes of overfilling are the following:

1. Human error—fatigue, inattentiveness, inexperience, lack of communication, etc.
2. Mechanical failure
3. Malfunctioning tank valves
4. Faulty gauging system
5. Improper lineup
6. Cargo gravitation
7. Faulty alarms

To protect against such an occurrence, alarms indicating both high level and overfill conditions must be fitted on vessels equipped for vapor control. Each alarm must be designed to provide audible and visual warning when the cargo level reaches predetermined settings. These alarms are required to be intrinsically safe and totally independent of each other. The alarm system must be equipped with a test feature to permit checking prior to the cargo transfer and to give warning in the event of a power or circuit failure (Figures 5-10b and 5-10c).

The high-level alarm is set to activate when the liquid level is between 95 percent of tank capacity and the setting of the overfill alarm. The secondary overfill alarm must be set to provide ample warning for the person-in-charge to shut down loading before the cargo tank overflows.

Figure 5-10b. High level and overfill alarm lights. Courtesy Stacy DeLoach.

Figure 5-10c. High level and overfill alarm panel.

Figure 5-10d. Overfill alarm float on the underside of the deck. Courtesy Stacy DeLoach.

Both alarms should be clearly indicated in black lettering on a white background as follows: **High Level Alarm** and **Tank Overfill Alarm**.

Additional options are available to protect the vessel structure against damage from tank overfill. One option is the use of a spill valve or rupture disk on each cargo tank. These devices generally are set to relieve at a higher pressure than the tank PV valve but certainly less than the maximum design working pressure of the tank. For vessels in ocean or coastwise service, provision must be made to prevent accidental opening due to cargo sloshing.

A second option is to install an overfill control system. In this system, generally found on tank barges, the overfill sensors on the vessel are connected to the shore facility. When an overfill occurs, the system automatically shuts down the loading operation at least sixty seconds before the tank is pressed up to capacity (Figure 5-10e).

Figure 5-10e. Overfill control system connection. Courtesy Stacy DeLoach.

Vessel Pressure-Vacuum (PV) Protection

Another consequence of closed loading operations is the possibility of over- or under-pressurization of a cargo tank. Since the advent of inert gas systems on tankers, numerous cases of cargo tank rupture and collapse have been attributed to closed operations (Figure 5-11). As a result, the need for safe operating procedures and attention to detail during cargo operations cannot be overstated.

Some of the potential causes of cargo tank over- or under-pressurization are (1) excessively high loading rate, (2) malfunctioning or undersized PV devices, (3) vapor line constriction, (4) improper vapor system lineup, (5) excessive withdrawal of vapors or cargo, (6) expansion or contraction of cargo, and (7) cargo sloshing.

To protect against the foregoing, the regulations stipulate that the system should be capable of handling vapors at 1.25 times the maximum loading rate

Figure 5-11. On the vessel shown here, a cargo tank was over pressurized during a cargo transfer resulting in significant structural damage.

of the vessel. Tank vessels are usually fitted with one or more of the following pressure-vacuum relief devices: individual tank PV valves, mast riser PV valve (if fitted), and liquid-filled PV breaker.

The set points at which these devices relieve excess pressure or vacuum should fall within the following ranges:

$$1.0 \text{ psi} < \text{pressure relief} < \begin{cases} \text{maximum design working pressure} \\ \text{or} \\ \text{setting of spill valve/rupture disk} \end{cases}$$

$$-0.5 \text{ psi} > \text{vacuum relief} > \text{maximum design vacuum}$$

The bar graph shown in Figure 5-12 further illustrates the normal operating pressures and settings for PV relief devices in a vapor control system.

It is important to realize that PV valves, like any mechanical device, can fail when needed most. Therefore, routine inspection and maintenance are essential to ensure proper structural protection of the vessel. As part of the regulations, each newly installed PV valve must have a means of checking that the valve operates freely and does not remain in the open position. For more details concerning the function of PV valves, consult chapter 4.

HIGH/LOW VAPOR PRESSURE PROTECTION

Vapor control systems must be equipped with a pressure sensing device that permits the person-in-charge to monitor the deck pressure in the vessel. High- and low-pressure alarms must give both audible and visual warning when an extreme condition exists. The alarm settings are as follows:

High pressure alarm   — not more than 90 percent of the lowest pressure relief valve setting

Low pressure alarm:

    Inerted tanker   — not less than 4 in. wg (100 mm wg) 0.144 psi

    Noninerted tanker — lowest vacuum relief valve setting

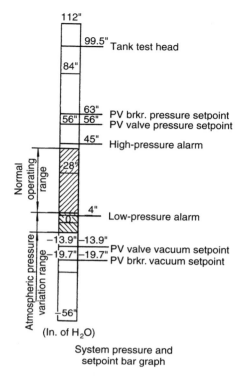

Figure 5-12. Bar graph illustrating the normal operating pressures and settings for pressure-vacuum relief devices in a vapor control system.

OPERATIONS

This section addresses a number of the operational concerns that a PIC must keep in mind during a closed loading operation.

Loading Rates

One critical element that affects the overall safety and success of vapor control operations is the determination of maximum allowable loading rates. The regulations specify that cargo loading rates must take into account the pressure drop through the vapor piping system as well as the venting capacity of the pressure relief valves on the tank. A graph reflecting the cargo loading rate versus the pressure drop for a typical installation is shown in Figure 5-13.

The maximum allowable loading rate must be determined and clearly understood by both vessel and terminal personnel prior to commencement of loading. While loading, the vessel PIC should closely monitor the loading rates and deck pressure to prevent damage to the system. When loading a "static accumulating" cargo such as jet fuel or gasoline, the initial loading rate should be limited to minimize the development of a static electrical charge. Experience has shown that agitation, splashing, and pipeline friction contribute to charge separation in certain low-conductivity cargoes. As a result, it is considered safe practice to limit the velocity to each tank to 1 meter per second until a sufficient cushion is achieved. A table of flow rates corresponding to a linear velocity of 1 meter per second through various sizes of piping and additional guidance concerning the handling of static accumulating cargoes can be found in Chapter 2.

Final Gauging

Vapor control regulations prohibit the opening of a cargo tank to atmosphere during active cargo transfer. The intent of this requirement is to keep the system vapor-tight throughout the operation. In fact, it should be totally unnecessary to open a tank to atmosphere during loading if all required equipment is functioning properly.

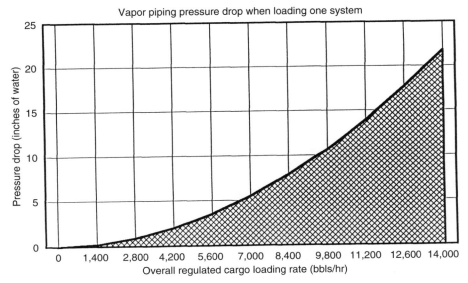

Pressure drop is from no. 1 center tank (most remote tank) to aft vapor header
1. Pressure drop is for a mixture of 45% inert gas and 55% hydrocarbon vapor.
2. Aft port vapor header is farthest from the vapor main.

Figure 5-13. This graph shows the relationship between cargo loading rates and the pressure drop through the vapor piping from the most remote cargo tank. Regulations specify that cargo loading rates take into account the pressure drop through the vapor piping as well as the venting capacity of the pressure relief valves on the tank.

Vessel personnel are permitted to open tanks for the purpose of gauging, sampling, and taking temperatures provided the following criteria are met:

1. There is no active transfer to the tank.
2. In an inerted tank, a positive pressure is maintained.
3. In a noninerted tank, the vapor pressure is reduced to atmospheric via the vapor system.
4. The cargo is not required to be "closed gauged" by regulation.
5. If a static accumulating cargo, all metallic equipment introduced into the tank is bonded to the vessel and, for noninerted vessels, a minimum thirty-minute relaxation period has elapsed since loading ceased.

Inerted Tank Vessels

The vapor control regulations are designed to complement existing requirements for inerted vessels. Despite the numerous safeguards already in effect, the need to maintain proper atmosphere control for vessel safety has never been greater. Prior to engaging in vapor control operations, vessel personnel must test each cargo tank to verify that the oxygen content does not exceed 8 percent by volume. The oxygen measurements should be taken at a point 1 meter below the tank top and at the midpoint of the ullage space in each tank. Another concern involves facilities using a blower to assist in the transport of the vapors. The use of such a blower by the shore facility could reduce the vessel deck pressure below the required minimum, which in turn could ultimately result in air being drawn into the system. The PIC of the vessel must be mindful of this possibility and carefully monitor the deck pressure during closed operations to ensure that the inert status of a tank is not compromised.

Figure 5-13a. Portable oxygen readings in the cargo tanks can be taken using this device. Courtesy MMC.

Declaration of Inspection

For vessels intending to conduct vapor control, the declaration of inspection (DOI) must include entries for critical aspects of the system. The Shell Oil addendum to the DOI in Figure 5-14 illustrates the required entries for vapor control operations. Additionally, the use of a standard checklist (Figure 5-15) similar to those used in crude-oil-washing operations should help prevent unnecessary mishaps.

S-13431 (6-91) 016241

DECLARATION OF INSPECTION ADDENDUM
VAPOR CONTROL OPERATIONS

This addendum is to be used when collecting vapors associated with the loading of oil or chemicals in bulk. The requirements are set forth in detail in 33 CFR 156.120 and 46 CFR 35.35-30 and therefore apply to both the facility and vessel unless otherwise indicated.

VESSEL NAME ▶	INITIAL	
	FACILITY	VESSEL
1 Vessel Certificate of Inspection has been endorsed as meeting the requirements of vapor control regulations. (39.30.1 (a))		
2 Initial loading rate and maximum transfer rate are determined. (156.120 (aa) 4; 35.35-30 (c) 4) _____ BPH		
3 Maximum and minimum operating pressures at the facility vapor connection are determined. (156.120 (aa) 5; 35.35-30 (c) 5)		
4 Cargo tank filling limits agreed. (39.30.1 (e))		
5 If inerted, oxygen content of the vessel cargo tanks is less than 8% oxygen by volume. (156.120 (aa) 9; 35.35-30 (c) 8)		
6 Electrical insulating flange is fitted between vessel and facility vapor connections. (156.120 (aa) 3; 35.35-30 (c) 3)		
7 All oxygen and hydrocarbon analyzers on the facility and vessel have been checked for calibration within the past 24 hours. (156.120 (aa) 7 ii)		
8 Tank barge overfill control system is connected to the facility, tested and operating properly. (156.120 (aa) 6)		
9 All cargo hatches and ullage openings are secured.		
10 All facility and vessel alarms and shutdown systems have been tested within the past 24 hours. (156.120 (aa) 7 l; 35.35-30 (c) 6)		
11 Manual valves properly aligned to collect cargo vapor. (156.120 (aa) 1; 35.35-30 (c) 1)		
12 Vapor collection hose is in good condition. (156.120 (aa) 8; 35.35-30 (c) 7)		
13 Facility vapor collection hose or articulated arm is connected to vessel vapor connection. (156.120 (aa) 2; 35.35-30 (c) 2)		

FACILITY	VESSEL
SIGNATURE	SIGNATURE
TITLE	TITLE
TIME AND DATE	TIME AND DATE

I certify that I have read the above declaration and detailed requirements and all conditions remain satisfactory

UNIT	SUBSEQUENT PERSON-IN-CHARGE	TITLE	TIME AND DATE
VESSEL			
FACILITY			
VESSEL			
FACILITY			
VESSEL			
FACILITY			

DATE AND TIME COMPLETED ▶

Figure 5-14. This addendum to the Declaration of Inspection lists the required enties that must be completed by the PICs when conducting a vapor control operation. Courtesy Shell Oil.

CHECKLIST–MARINE VAPOR CONTROL OPERATING PROCEDURES

PRE-ARRIVAL CHECKS

- High level/overfill alarms tested within 24 hrs. before operation.   ☐
- Deck pressure/tank oxygen readings checked and satisfactory.   ☐
- Closed gauging systems checked and satisfactory.   ☐
- Tanks to be loaded with regulated cargo identified.   ☐
- Tanks identified which previously contained regulated cargo.   ☐
- All cargo tank vents, P/V valves, tank hatches, tank cleaning openings tightly secured.   ☐

PRE-TRANSFER CHECKS

- Vessel properly secured at loading berth.   ☐
- Status of deck water seal satisfactory.   ☐
- Low point drains on IG/Vapor control main secured.   ☐
- Deck isolating valve secured.   ☐
- Liquid P/V breaker at proper level   ☐
- Atmospheric vent line open.   ☐
- Power supply on to cargo tank alarms.   ☐
- All alarms tested at each tank.   ☐
- Oxygen level at each tank is below 8% by volume.   ☐
- Vapor pressure alarm (audible/visual) tested.   ☐
- Overfill control connected (vessel to vessel or vessel to shore).   ☐
- All branch line valves to the vapor control main are at proper status (Consult compatability of cargoes).   ☐
- Cargo system properly lined up.   ☐
- Vapor collection system properly lined up.   ☐
- Cargo hoses/Vapor hoses connected.   ☐
- Insulating flange or non-conductive hose in use.   ☐

Figure 5-15a, b, c. The use of a standard checklist similar to those used in crude-oil-washing operations should assist in preventing unnecessary mishaps. Courtesy Richard Beadon and Mark Huber.

CHECKLIST–MARINE VAPOR CONTROL OPERATING PROCEDURES

• Pre-transfer conference required by DOI completed.

• Terminal consulted regarding dropping the deck pressure.

• If terminal operates a vapor line vacuum assist, have operating limits been determined?

• Certificate of Inspection or Certificate of Compliance endorsed.

• All oil transfer procedures complied with.

• Manifolds open - loading operations commenced.

LOADING/TRANSFER OPERATIONS

• Initial loading rates observed.

• Cargo/Vapor connections checked and found satisfactory.

• Liquid/Vapor flow checked and found satisfactory.

• ISGOTT recommendations observed regarding static accumulator oils.

• Maximum allowable loading rates not exceeded.

• Vapor pressure on deck monitored.

• Loading rates adjusted as necessary for topping-off.

• All applicable regulations pertaining to Inert Gas systems complied with.

POST TRANSFER CHECKS

• Finished loading and gauge out. (Note: IG vessel to maintain positive deck pressure.

• Cargo hoses drained, disconnected, and cargo system valves shut.

• Vapor manifold shut and vapor hose disconnected.

• Branch line valve status on IG vessel checked and satisfactory.

• IG plant run up and deck pressure topped up if necessary.

Figure 5-15b

CHECKLIST–MARINE VAPOR CONTROL OPERATING PROCEDURES

CRITICAL FAULTS INVOLVING VAPOR CONTROL SYSTEMS

- Vapor hose constricts/collapses/kinks or damaged in any way which renders it ineffective.
- Tank overfill alarm is tripped.
- High level alarm/tank overfill alarm fault.
- Tank overfill control (auto-shutdown) inoperative.
- Inoperative gauging system.
- Mechanical failure of branch line/vapor manifold valve.
- High/Low vapor pressure condition.
- Inoperative P/V relief valves.
- Inability to maintain less than 8% oxygen by volume in cargo tanks.
- Inability to maintain a positive deck pressure throughout transfer.
- Faulty line-up of the vapor collection system.

ALWAYS CONSULT COMPANY POLICIES AND OIL TRANSFER PROCEDURES FOR YOUR VESSEL.

Figure 5-15c

Lightering

When vessel-to-vessel cargo transfers occur in a locality where emission limits are in effect, the vapors are generally handled through vapor balancing. Simply stated, vapor balancing is a closed transfer between the service vessel and the vessel to be lightered (VTBL). The cargo vapors displaced from the service vessel during liquid cargo transfer are returned to the VTBL via a vapor connection. When inerted vessels engage in a closed lightering operation several additional requirements must be met including the following:

1. The service vessel must have a means to inert the vapor transfer hose prior to commencement of the lightering operation.
2. The service vessel must have an oxygen analyzer fitted within 3 meters of the vessel vapor connection. The analyzer must have a suitable connection for a calibration gas to enable the testing of the instrument.
3. The service vessel must have a visual and audible alarm that sounds when the oxygen content in the vapor system exceeds 8 percent by volume.
4. The service vessel should be equipped with an oxygen indicator located where the cargo transfer is controlled (i.e., cargo control room or on deck).
5. An electrical insulating flange or a length of nonconductive hose must be installed between the vessel vapor connection on the service vessel and the VTBL.

Vapor balancing requires careful coordination between the PICs of each vessel. To date, vessel experience with vapor balancing has revealed some interesting facts:

1. Cargo pumping rates are critical when conducting a closed lightering operation. The PICs must discuss and carefully monitor the pumping rates throughout the transfer.
2. Given the volatile nature of the cargo, vessels have experienced dramatic increases in deck pressure due to vapor growth resulting from cargo pumping (agitation) and increasing ambient temperatures (vaporization). In some instances, vessels have reached a high deck pressure condition necessitating a shutdown of the lightering operation, as venting off the excess pressure to atmosphere is no longer an option.
3. To ensure the safety of the lightering operation, it is important to verify that each vessel is properly inerted prior to commencement of the transfer. Each vessel must check the oxygen content of the cargo tanks following the guidelines mentioned earlier in this chapter. A measured value of 8 percent oxygen or less by volume is considered acceptable and it is the responsibility of both PICs to monitor the level throughout lightering.

Maintenance

The venting and vapor control system of the vessel requires periodic inspection and maintenance. Consult the manufacturer's manual for the recommended intervals and details of the service. If no guidance exists, create a preventive maintenance (PM) schedule for the vessel based on operator experience with the installed system. Critical areas requiring attention in the vent system generally include the following:

1. Low point drains (liquid condensate/cargo)
2. Flame screens/arresters (fouling/holes)
3. Cargo accumulation (polymerization/solid residue/scale)
4. PV relief devices (see chapter 4)
5. High-velocity vents (smooth operation of high jets/bullet valves)
6. Vent valve seating surfaces (gummed up with residues/damage due to chattering)
7. Stop valves (branch lines/mast riser/vapor manifold)
8. Gauges/alarms (deck pressure/high-level/overfill)

The increased complexity of cargo tank venting and vapor control systems requires ongoing training and diligence on the part of the vessel PIC. Sound tanker practice dictates that everyone involved in the operation should fully understand the proper use of these systems and carefully check the lineup before commencing a transfer. There is no room for complacency or second-guessing as serious damage can quickly result to the vessel, personnel, facility, environment, and cargo. Remember, the goal of the PIC should always be to strive for a safe, efficient, and environmentally sound cargo transfer.

REVIEW

1. List the advantages of a controlled venting system over traditional open venting of cargo tanks.
2. What are the typical methods of controlled venting employed on a tank vessel?
3. What is the principle of operation of a high-velocity venting device?
4. What are the most common types of HVV found on tank vessels today?
5. What is the manufacturer's stated exit velocity from these devices?
6. What fire protection, if any, is afforded by the use of these devices?
7. When loading a vessel using a controlled venting system, is it still possible to experience vapor accumulation around the deck and superstructure?
8. What causes physical liquid carryover through the venting system?
9. Why is it important to check low-point drains in the vent/vapor control piping?
10. What types of cargo must be loaded without venting to atmosphere today? Where?
11. Describe the typical piping configuration for a closed loading system (vapor control) on each of the following vessels: crude oil, clean oil, and parcel carrier.
12. Sketch the vessel vapor manifold showing all required details.
13. Why is it imperative to check the integrity of the P/V relief devices? Describe the items that should be checked when performing an overhaul of the P/V valves on the vessel.
14. What is a pressure drop calculation/loading rate curve?
15. What additional entries must be completed on the DOI for a vapor control operation?
16. When conducting a closed lightering operation, what is the method typically employed?
17. What is the purpose of an insulating flange? Where is it typically found?
18. When an inerted vessel conducts a closed load, oxygen readings must be taken at what locations in the cargo tanks? What oxygen reading is considered acceptable?
19. List the minimum strength criteria for a vapor hose.
20. When performing closed lightering operations, what additional equipment is required on the service vessel?
21. While conducting a closed loading operation a cargo tank is overfilled. List the problems confronting the vessel and the checks that must be performed before resuming operations.

CHALLENGE QUESTIONS

22. Describe the line-up of the vent/vapor piping when conducting the simultaneous loading of regulated and non regulated cargoes.
23. When conducting a closed lightering operation list some of the issues the PIC's of each vessel may confront.
24. During an electrical storm, a lightning strike at the top of the mast causes a vent fire despite the fact that the PIC had shutdown the loading operation and secured the deck. Discuss the possible causes and actions to be taken.
25. a. While conducting a closed loading operation on an inerted vessel you notice the deck pressure is steadily dropping and approaching the lower acceptable limit (low deck pressure alarm). List the possible causes and your actions as the PIC.
    b. While conducting a closed loading operation on an inerted vessel you notice the deck pressure is steadily increasing and approaching the upper acceptable limit (high deck pressure alarm). List the possible causes and your actions as the PIC.

CHAPTER 6

Cargo Measurement and Loss Control
John O'Connor

The various parties who have an active interest in the methods used to quantify oil have standardized cargo measurement procedures over the past several decades. Those who wish to familiarize themselves further with the theoretical and practical aspects of cargo measurement can refer to Chapter 17, "Marine Measurement," in the *Manual of Petroleum Measurement Standards* published by the American Petroleum Institute. This publication has several sections, the most pertinent being Section 1, "Guidelines for Marine Cargo Inspection, and Section 2, "Measurement of Cargoes On Board Tank Vessels." These and other related publications have been formalized during many years of discussion and technical critique in order to arrive at an industry consensus of what can be considered standard methods to gauge and sample petroleum and other liquid cargoes on board tank vessels.

THE PURPOSE OF CARGO MEASUREMENT

The purpose of a tank vessel is to transport bulk liquid cargo, which earns revenue for the vessel's owners. In order to properly care for and handle the cargo while it is being loaded, carried, and discharged, accurate measurement of the cargo is important to determine quantity and quality. This is accomplished by physical gauging, temperature measurement, sampling, and calculation of the quantities in each cargo tank, the vessel pipelines, and slop tanks. Special circumstances may require measurement of cargo in ballast tanks, void spaces, or the vessel's fuel tanks.

To understand the significance of accurate measurement, it is necessary to have a clear picture of how the cargo on board the vessel is purchased and sold. In many instances, the charterer of the vessel has arranged to purchase the cargo from a supplier at the loading port. The quantity (or volume) of cargo is supplied to the vessel from shoreside storage tanks. The volume of cargo that is delivered from one or more shoreside tanks is listed on documents such as a bill of lading or a certificate of quantity. This is the volume of cargo that the charterer has purchased. As vessel personnel are not part of the measurement process for the shoreside tanks, the crew (ultimately the master) accepts the cargo on board the vessel with no direct knowledge of whether the volume as listed is representative of the actual volume supplied. The crew can measure the volume of cargo contained in the vessel's tanks, and the two volumes can then be compared. If they are close, the master could be reasonably confident that the volume listed on the bill of lading and/or the certificate of quantity is accurate. Should there be a large discrepancy between the two volumes, the master has the obligation to have the cargo on the vessel measured a second time to make sure that all potential errors have been eliminated. If the volumes still cannot be reconciled, then the master will usually sign the bill of lading under protest, noting the discrepancy and any steps taken to account for the difference.

PARTIES INVOLVED

The following list identifies all the parties involved in cargo transfer and measurement:

The *vessel owner* is an individual or corporation that holds title to the vessel.

The *vessel operator* may be the vessel owner, although in most cases, the operator is a different party contracted or employed by the owner to conduct day-to-day operations of the vessel.

A *charterer* is an individual or corporation that employs the vessel for one or more voyages to carry cargo.

The *supplier* is the party providing the cargo, which is normally stored at a terminal or in another vessel.

The *receiver* is the party that accepts the cargo. The vehicle used to accept or receive the cargo could be the vessel, a storage tank or tanks, or the metered pipeline receipt that is provided after cargo passes through a pipeline.

A *terminal* is a shoreside facility capable of storing, receiving, and/or supplying cargo.

An *inspection company* is a third party employed equally by the supplier and the receiver to measure the cargo. Measurement will routinely consist of gauging and measurement of the cargo and sampling, followed by sample analysis at the inspection company laboratory, and finally calculations to determine the cargo quantity and quality.

TERMS OF SALE

In most circumstances, someone other than the vessel owner or operator has title to the petroleum that is transported on a tank vessel. In most cases the oil is owned by the charterer. The charterer takes possession of the oil at the point of custody transfer, which normally is the interface between the terminal's cargo hose and the vessel's manifold flange.

Following are the three most common terms used in the sale and purchase of oil cargo:

1. Free on board (F.O.B.): Under these terms, risk passes to the buyer (receiver) at the F.O.B. point, which is normally the loading port. The buyer purchases the quantity as listed on the bill of lading and accepts any risk of loss during the voyage. Quality of the cargo under F.O.B. terms is based upon the quality of the cargo in the supplier's shore tank or tanks.

2. Cost, insurance, freight (C.I.F.): Usually the C.I.F. term is followed by the name of the discharge port or ports. The C.I.F. cost to the buyer includes the price

(cost) of the quantity of cargo as listed on the bill of lading, plus insurance and the freight payment. In cases where the buyer procures his own insurance, the terms of sale would then be cost and freight (C.&F.). The quality of the cargo in both C.I.F. and C.& F. cases is based upon the quality of the cargo in the supplier's shore tank or tanks.

3. Delivered: When cargo is purchased or sold on delivered terms the cargo owner, who would be called the supplier in this arrangement, agrees to be compensated by the receiver (buyer) based upon the volume of cargo that is measured in the receiver's shoreside tanks. Quality of the cargo is determined by the vessel's composite sample, as found when the vessel arrives at the discharge port.

These different terms allow suppliers and receivers to limit their exposure or liability, as in each set of terms, the risk or loss is accepted at a different point.

MEASUREMENT EQUIPMENT

While many vessels have been fitted with a remote tank gauging system which is primarily used to monitor cargo tank levels during loading and discharge operations, the equipment used to conduct final measurements is portable and manually operated. The choice of equipment is mainly determined by the oil carried, and whether or not the tanks are under positive pressure with inert gas.

Based upon the vessel's design and regulatory requirements, gauging of the cargo tanks will be classified as open (open to the atmosphere), restricted, or closed, as seen in Figure 6-1.

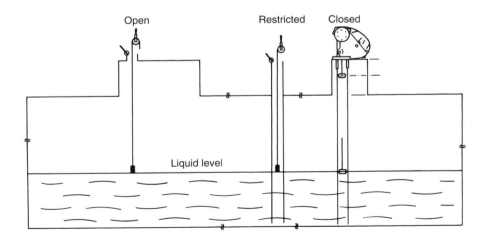

Figure 6-1. Gauging methods on a tank vessel are typically classified as *open, restricted,* or *closed,* based on the amount of atmosphere released from a tank during the process. Copyright © International Maritime Organization (IMO), London.

Cargoes that are not required to be inerted or carried under a nitrogen blanket are loaded into tanks that are open to the atmosphere and therefore at atmospheric pressure. Cargoes that require inerting are introduced into tanks that are maintained at a positive pressure slightly higher than atmospheric pressure. Under most conditions, these tanks should not be opened to the atmosphere. Should it become necessary to open an inerted cargo tank, follow industry-recommended safety precautions and maintain a positive pressure (see Chapter 16 for further information). Tank vessels that routinely carry inerted cargoes or load at terminals with restrictions on the emission of cargo vapors are fitted with closed measurement devices. These devices permit gauging, temperature reading, and sampling of the cargo to be accomplished without the release of atmosphere or loss of IG pressure.

Many tank vessels are equipped with standpipes and vapor control valves (Figure 6-2) that permit the operator to gauge each tank with a portable sonic tape (Figure 6-3). Portable tapes are typically used when topping off cargo tanks to compare readings with the permanently installed closed gauging system. Cargo surveyors (gaugers) frequently use portable sonic tapes to "gauge out" a vessel at the completion of loading or prior to the discharge of cargo.

Figure 6-2. Gauging standpipe and vapor control valve. Courtesy Stacy DeLoach

Figure 6-3. Vapor control valve and manual sonic tape. Courtesy Marine Moisture Control (MMC)

Many types of closed gauging systems are found on modern tank vessels. Figures 6-4 and 6-5 show two types—electric resistance and radar.

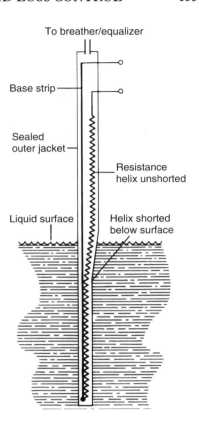

Figure 6-4. The Metritape level sensor is an "electric tape measure" that hangs from the top to bottom in the tank. Two wires out of the sensor top carry an electric resistance signal that is directly proportional to the tank ullages. When the sensor is submerged in the liquid, the weight of the liquid compresses the sensor and causes a short circuit in the submerged portion of the helix windings thus changing the electrical resistance. A change in ullage of 1 meter causes a corresponding change of 1 meter in the length of unshorted resistance helix and an electrical resistance change of 100 ohms. An ullage readout is displayed locally at the tank top and remotely in the cargo control room. Courtesy Metritape.

Detail of Metritape sensor

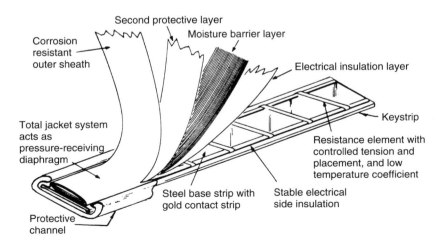

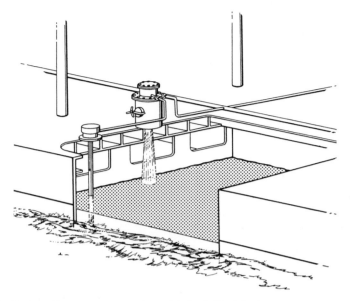

Figure 6-5a. Radar ullages systems such as this Saab TankRadar measure ullages by bouncing radar waves off the surface of the liquid in a tank. Such systems are intrinsically safe and extremely accurate (+/- 5mm). Courtesy Saab Marine Electronics.

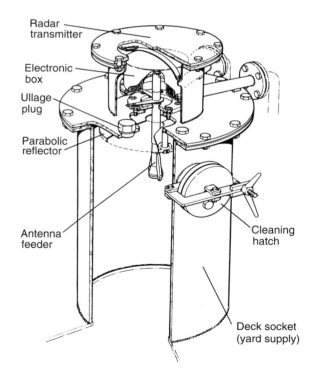

Figure 6-5b. Transmitter unit for Saab TankRadar. The radar transmitter measures the tank ullage. With specific data stored in a permanent memory, the computer can calculate the tank volume. This system also monitors cargo temperature and inert gas pressure, which, along with the tank ullages are available as a direct console or terminal. Courtesy Saab Marine Electronics.

Figure 6-5c. Saab closed gauging system radar dome. Courtesy George Edenfield

Typical cargo measurement equipment carried on board a tank vessel consists of the following:

Open tank gauging tapes
Gauging bobs
Water indicating paste
Product indicating paste
Thermometers
Closed tank gauging devices
Temperature probes
Sample bottles
Sample bottle container assembly

For tanks that do not require inerting, and where the ullage hatches can be opened the simplest equipment used to measure the liquid level in a cargo tank would be a gauging tape and plumb bob (Figures 6-6 and 6-7). The gauging tape can range in size from 15 meters to 30 meters, and is typically graduated in centimeters and millimeters. The plumb bob is also graduated in a similar fashion, and when paired with a tape, can be used to determine the liquid level in a cargo tank, either by the innage or ullage method.

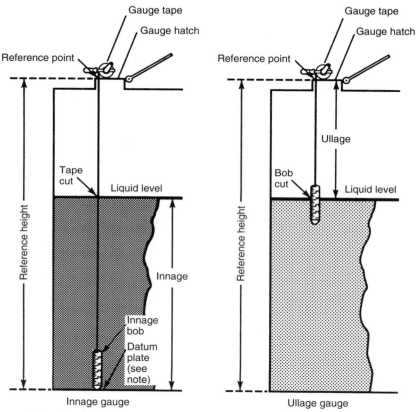

Note: The datum plate may actually be the ship's bottom, a striking plate, or another point from which the reference height is measured.

Figure 6-6. Manual tank gauging. From *Manual of Petroleum Measurement Standards*, First Edition, July 1990, "Measurement of Cargoes on Board Tank Vessels." Reprinted courtesy of the American Petroleum Institute.

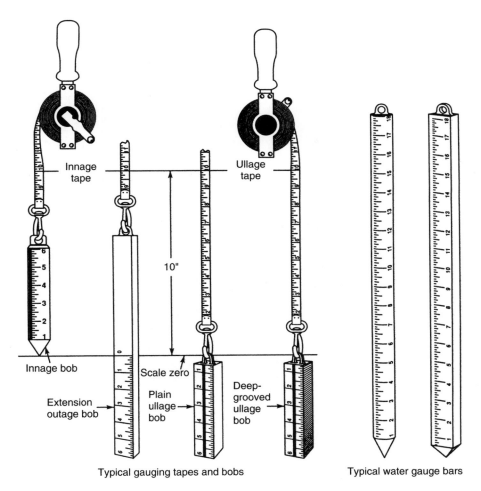

Typical gauging tapes and bobs   Typical water gauge bars

Figure 6-7. Several types of bobs connected to handheld tapes. From Manual of Petroleum Measurement Standards, First Edition, July 1990, "Measurement of Cargoes on Board Tank Vessels." Reprinted courtesy of the American Petroleum Institute.

For tanks that are under positive pressure, and where the tank or ullage hatches have been fitted with a vapor control valve (Figure 6-8), a closed system gauging device (Figure 6-9) needs to be utilized.

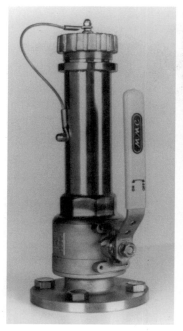 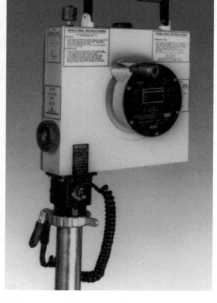

Figure 6-8. Vapor control valves permit gauging a tank without the need to open the hatch. Courtesy Marine Moisture Control (MMC).

Figure 6-9. Closed system gauging device. Courtesy Marine Moisture Control (MMC)

Closed system gauging devices can be coupled to the vapor control valve on the tank top or ullage hatch, thereby permitting physical gauging of the tank without the escape of inert gas.

Cargo tank sampling can also be accomplished via either a glass bottle sampler (Figure 6-10) or closed system sampler (Figure 6-11).

Determining the temperature of the cargo can be accomplished through the use of either a cup case thermometer (Figure 6-12) or a portable electronic thermometer (Figure 6-13).

Locating water in the cargo can be accomplished by using water finding paste on the plumb bob and gauging tape or by a portable ullage-interface device that can be used with either an open or closed gauging system. When a plumb bob coated with paste is lowered to the tank bottom and then retrieved, you will note that the paste changes color upon contact with water. Some cleaning away of the paste from the bob may be necessary to see the graduations on the bob so as to determined the water's depth in the tank. Closed system devices use a probe on the end of the tape which senses and reports on the water/cargo interface, thereby providing the location (innage or ullage) of the water in the tank.

Figure 6-10. Glass
bottle sampler

Figure 6-11 Closed System Sampler. Courtesy Kevin
Duschenchuk

Figure 6-12. Cup case
thermometer

Figure 6-13. Portable electronic thermometer

EQUIPMENT USE

There is often a debate about the equipment to be used in the measurement process, namely, whether to use the equipment that is carried on board the vessel or the equipment provided by the petroleum inspection company. The crew should keep in mind that the inspection company is paid by the cargo owner, or by both the supplier and receiver, to gauge and sample the cargo. Inspection companies are routinely audited to determine compliance with industry standards and to verify that their equipment is maintained in good condition and calibrated on a regular basis. All the equipment needed to conduct and complete the inspection must be in the possession of the inspector. Should a measurement discrepancy arise, the parties with interest in the cargo will look to the inspection company to clarify or reconcile the difference. If the inspection company uses the vessel's equipment, it has utilized devices beyond its control and cannot attest to the material condition or accuracy of those devices; its position is thereby compromised. Inspectors should use their own equipment, and it would be prudent for the vessel's personnel to compare the inspectors' equipment to their own. This allows the vessel to establish a baseline comparison, which may be used after the conclusion of the gauging process to reconcile differences. As an example, suppose that the vessel has experienced an in-transit difference in the cargo quantities between the load and the discharge port. If the vessel's loaded quantity was determined using the vessel's gauging tape and cup-case thermometers, and the predischarge quantity was determined using the inspector's gauging tape and temperature probe, the two different sets of equipment will yield differing volumes. As this situation would normally be uncovered just prior to the start of the discharge operation, it would be best to delay the start of the discharge and compare the vessel's equipment to the inspector's equipment. There may be a situation where the inspector's tape differs from the vessel's tape by one-eighth inch. Possibly the vessel's cup-case thermometer reads 3° lower in temperature than the inspector's electronic thermomometer. In any event, it is best to reconcile discrepancies in measurement while the cargo is still in the vessel's tanks. This affords all interested parties a second opportunity to regauge the vessel.

TANK STRUCTURE AND MEASUREMENT TERMS

Many things can affect the measurement of cargo—the structure of each cargo tank (including the location of the gauging point), internal framing members, deadrise, turn of the bilge, and cargo pipelines. Prior to undertaking any gauging or sampling of a cargo, the PIC should become familiar with the actual physical structure of the cargo tanks. Information may be obtained from the vessel's drawings or plans, from a tank arrangement diagram, and from crewmembers with a working knowledge of the vessel.

Common terms used in the measurement process include the following:

Calibration tables (ullage/innage) are tables developed by recognized industry methods that represent the volumes in each tank according to the liquid (innage) or empty space (ullage) measured in the tank. The tables are entered with linear measurements (for example, feet, inches, meters, or centimeters) to obtain calibrated volumes such as gallons, barrels, cubic meters, or cubic feet.

Reference height is the distance from the tank bottom and/or datum plate to the established reference point or mark (see Figure 6-14).

Observed reference height is the distance that is actually measured from the tank bottom or datum plate to the established reference point.

Reference point (gauging point) is the point from which the reference height is determined and from which the ullages/innages are taken. Historically, most tank vessels use the rim of the ullage opening in the hatch as the reference point for gauging the tank (refer again to Figure 6-14).

Ullage (also referred to as "outage") is the measured distance from the surface of the liquid to the reference point. In other words, it is the measurement of free space above the liquid in a tank (refer again to Figure 6-14).

Innage (also referred to as "dip" or "sounding") is the measured distance from the surface of the liquid to a fixed datum plate or to the tank bottom (refer again to Figure 6-14).

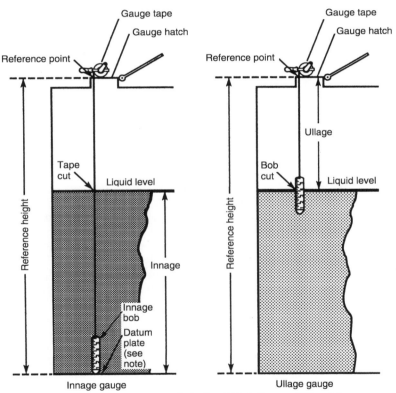

Note: The datum plate may actually be the ship's bottom, a striking plate, or another point from which the reference height is measured.

Figure 6-14. The measurement of free water using a bob coated with water sensitive paste. For accurate determination of free water, it is important not to lay the bob on its side at the bottom of the tank. From *Manual of Petroleum Measurement Standards*, First Edition, July 1990, "Measurement of Cargoes on Board Tank Vessels." Reprinted courtesy of the American Petroleum Institute.

In addition to these measurement terms, there are volumetric terms that are also important to understand. These terms include the following:

Total observed volume (TOV) is the total measured volume of all petroleum liquids, sediment, water in suspension, and free water at the observed temperature.

Gross observed volume (GOV) is the total measured volume of all petroleum liquids, sediment, and water in suspension, excluding free water at the observed temperature. Gross Observed Volume = Total Observed Volume – Free Water or, GOV = TOV – FW.

Free water (FW) is the volume of water present in a container that is not in suspension in the contained liquid.

Gross standard volume (GSV) is the total volume of all petroleum liquids, sediment, and water in suspension, excluding free water, corrected by the appropriate volume correction factor for the observed temperature and API gravity, relative density, or density to a standard temperature such as 60°F or 15°C. Gross Standard Volume = Total Observed Volume – Free Water X Volume Correction Factor or, GSV = (TOV – FW) X VCF.

Total calculated volume (TCV) is the total volume of all petroleum liquids, sediment, and water in suspension, corrected by the appropriate volume correction factor for the observed temperature and API gravity, relative density, or density to a standard temperature such as 60°F or 15°C, and all free water measured at the observed temperature. Total Calculated Volume = Gross Standard Volume + Free Water, or TCV = GSV + FW.

Net standard volume (NSV) is the total volume of all petroleum liquids, excluding sediment, water in suspension, and free water, corrected by the appropriate volume correction factor for the observed temperature and API gravity, relative density, or density to a standard temperature such as 60°F or 15°C. Net Standard Volume = Gross Standard Volume – Sediment and Water, or NSV = GSV – S&W

Sediment and water (S&W) is the nonhydrocarbon solid material and water in suspension in a petroleum liquid.

Onboard quantity (OBQ) is the material remaining in vessel tanks, void spaces, and/or pipelines prior to loading. Typically, the onboard quantity includes water, oil, slops, oil residue, oil/water emulsions, sludge, and sediment.

Remaining on board (ROB) describes the material remaining in vessel tanks, void spaces, and/or pipelines after discharge. The quantity remaining on board includes water, oil, slops, oil residue, oil/water emulsions, sludge, and sediment.

MEASUREMENT PROCEDURES

Prior to the start of any measurement procedure, it is prudent to refer to the vessel's tank calibration tables. The calibration or strapping tables (as they are sometimes called) are generated when the vessel is built. The tables will be separated into a number of sections, one section for each tank. The tables will list the location of the reference point for each tank as well as the total gauge height. The simplest form of calibration table assumes that the tank is a cube with no internal framing members or pipelines. The shipyard specialist who generates the tables does so simply by calculating the total volume of the tank, then dividing that figure by the number of feet and inches or meters and centimeters between the gauging point and the tank bottom. Measurements are normally shown to the nearest quarter-inch, although there are tables for larger vessels that may be broken down into increments of one-eighth inch.

For accurate measurement to take place, it is important to have what is termed repeatability during the measurement process. Repeatability is the ability to take multiple ullage or innage measurements over time and get the same value. It is recommended that each measurement be done twice within a short period of time. To illustrate: a tank holds 30,000 barrels when full, and the distance between the gauge point and the tank bottom is 50 feet. The total volume of the tank divided by the gauging distance would equal the number of barrels per foot, or in this case, 600. To extrapolate, each inch would equate to 50 barrels and each quarter-inch would account for 12.5 barrels. Some calibration tables take certain aspects of the tank's shape into account such as the turn of the bilge, the volume of space occupied by any cargo or ballast lines that run through the tanks, and internal framing members.

After reviewing the calibration tables to determine the reference point and height, gauging of the tanks can commence. Prior to taking each set of measurements, the gauging equipment should be checked for wear. Handheld gauging tapes (Figure 6-15) should be checked for kinks and excessive wear at the clip that holds the gauging bob to the tape.

Cup-case thermometers (Figure 6-16) should be checked to make sure that the thermometer glass has not been cracked and that the mercury in the glass has not separated. "Standard" or lab-approved thermometers can be obtained that are certified by independent laboratories to read temperatures within a small tolerance (usually 0.5°) over a defined range, for example, 50° to 110° Fahrenheit. If a standard

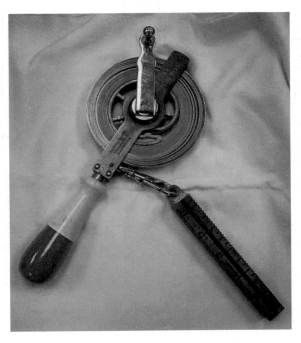

Figure 6-15. Manual gauging tape

thermometer is carried on board, it should be used only to check the readings of the working thermometers. If a standard thermometer is not available, then the vessel's temperature device should be checked against the petroleum inspector's device as these are routinely calibrated in a laboratory setting and certified to be accurate.

An electronic temperature probe, whether a single device or incorporated into a tape device, should also be checked against a standard thermometer when available.

Prior to proceeding onto the deck of the vessel to begin the measurements, the trim and list of the vessel should be checked. If the vessel has noticeable trim or list, the PIC can correct it by shifting cargo fore and aft or port and starboard. If transfer of the cargo is not possible, it may be necessary to shift some of the vessel's fuel oil or ballast. In any event, it should be noted that even a small amount of trim or list could greatly affect the cargo volumes calculated from the measurements taken.

One of the most important factors to consider is the location of the gauging point. Ideally, the gauging point should be centered directly above the center-point of the tank. Having the gauging point situated in this location negates any adverse

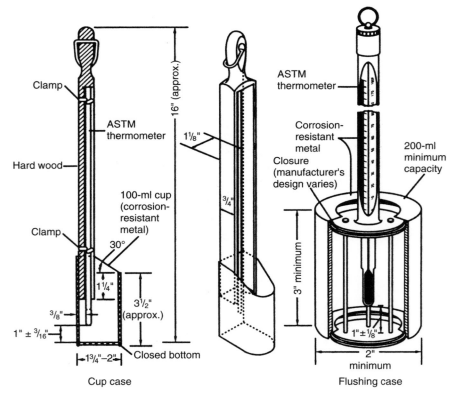

Figure 6-16. Typical thermometer employed when measuring temperature of cargo in a tank. From *Manual of Petroleum Measurement Standards*, First Edition, July 1990, "Measurement of Cargoes on Board Tank Vessels." Reprinted courtesy of the American Petroleum Institute

effects of the vessel's trim or list. The following diagrams illustrate the effect of trim on the measurement of cargo when the gauging point is located at the after end of the tank (Figure 6-17).

In Figure 6-17a, the liquid level of the cargo remains a constant distance from the tank top when the vessel is on an even keel. The ullage measurement would yield the same results regardless of the location of the gauging point on the tank.

Figure 6-17b shows the vessel trimmed by the stern. Note how the distance between the liquid level of the cargo and the tank top changes. Although the volume of cargo in the tank is the same as in Figure 6-17a, the measurement from the reference point (gauge point) to the surface of the cargo has changed.

To correct or adjust the ullage measurements, it is necessary to apply trim corrections when the vessel is not on an even keel. These corrections are normally found in the vessel's calibration or strapping tables. The corrections themselves can take various forms, with the two most common being a correction applied to the measured (observed) ullage or a volume correction applied to the volume measured with the uncorrected ullage. In the first instance, the calibration or strapping tables may provide a correction that corresponds to each foot or meter of trim, and interpolation is required. In the second instance, a volume, given in barrels or cubic meters, is provided to apply to the tabulated volume.

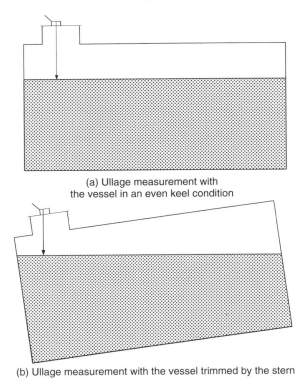

(a) Ullage measurement with
the vessel in an even keel condition

(b) Ullage measurement with the vessel trimmed by the stern

Figure 6-17. The effect of vessel trim is evident on the observed (ullage) measurement of cargo in a tank. (a) Ullage measurement with the vessel in an even keel condition. (b) Ullage measurement with the vessel trimmed by the stern. Courtesy John Hanus and John O'Connor

PRELOADING INSPECTION AND MEASUREMENT

Before cargo is loaded into a tank, it is necessary to determine the contents of the tank, if any. This can be accomplished by following certain procedures. If the cargo tank was washed after the last cargo, a determination must be made as to the presence of any water that may be on the bottom of the tank. If the tank is not inerted, it may be possible to determine this by a visual inspection through the ullage opening in the hatch, using a bright flashlight. If it is not possible to sight the bottom of the tank through this opening, then another access point may need to be opened, such as the trunk or Butterworth plate (tank cleaning opening). If liquid is observed on the tank bottom, then attempts should be made to obtain a measurement. Prior to using the gauging tape, it is best to place a coating of product or water-indicating paste on the surface of the gauging bob. The indicating paste will change color when in contact with petroleum or water (depending on the type of paste used) and will facilitate reading the measurement once the bob has been withdrawn from the tank. If the liquid is below the gauging hatch, then the gauging tape must be extended into the tank until the tip of the bob makes contact with the tank bottom. Once the bob makes contact with the tank bottom, the total height up to the reference point should be recorded and compared with the tank's total gauge height as listed in the calibration tables for the vessel. Performing this comparison will confirm the tank bottom was reached and the tape or bob was not caught on one of the tank's internal framing members or some other obstruction on the tank bottom. The bob should be left in this position for several seconds, with appropriate steps taken to ensure that the tape and bob remain still. When the bob is withdrawn from the tank, the petroleum/water measurement should be read to the nearest one-eighth inch or centimeter. In practice a minimum of two measurements should be obtained from each tank. If the two measurements coincide, then the readings should be recorded. If the two measurements do not match, then a third and possibly a fourth measurement should be taken until consecutive readings provide matching results.

Should it be determined (from the visual observations) that the liquid is in contact with all four bulkheads of the tank, trim corrections can be applied and the vessel's calibration tables can provide the tank volume. However, if the liquid surface does not make contact with all bulkheads—for example, if the liquid was contained at the after end of the tank and did not reach the forward bulkhead—and trim corrections were applied, it is possible that the resultant number would indicate that there was less than zero volume in the tank. In this situation, the volume in the tank can be calculated by using the wedge formula. The volume of the liquid wedge at the after end of the tank can be determined from certain physical dimensions of the tank combined with the observed measurements from gauging. The dimensions required are tank length, tank width, distance between the reference (gauging point) and the after bulkhead, and the measurement of the liquid. It will be necessary to ensure that all measurements are in the same system (either English or metric), so conversion of one or more measurements may be required before calculations can begin. For ease of use, a wedge formula work sheet is provided at the end of this chapter. It includes instructions on the calculation process.

Another situation may arise during the preloading (OBQ) measurement if the last cargo carried was a viscous or high pour-point cargo. Due to the peculiar properties of these cargoes, they require specialized handling such as heating in order to make them pumpable. When cargoes such as these are discharged, it is normal for the internal surfaces of the tank to be coated with a film (clingage) or thicker layer that is most often encountered along the bottom of the tank. Volume measurement in this situation requires additional steps and some common sense. In a situation where thick, viscous cargo such as vacuum gas oil (VGO) is encountered, multiple measurements should be obtained to create a profile of the layer on the tank bottom. In some cases, the layer on the bottom may be thicker at the forward end of the tank than in the after section. As a rule of thumb, the more measurements that can be obtained and averaged, the more representative the calculated volume will be of what is actually contained in the tank. While not exact, this method is currently the most practical.

POSTLOADING INSPECTION AND MEASUREMENT

After the cargo has been loaded, the vessel may be trimmed to an even keel (if possible), and it should be placed in an upright position, eliminating list. Ullage measurements should be taken from the reference point to the surface of the liquid. Manual ullaging is accomplished by lowering the gauging bob into the liquid until part of the bob is covered. The bob should not be immersed, as a reading cannot be obtained if the entire bob is covered with oil. When gauging light (volatile) cargoes such as naphtha, gasoline, and jet fuel, the use of a product-indicating paste on the bob will facilitate reading the measurement. Once again a minimum of two readings should be obtained from each tank. For two readings that do not match, subsequent measurements and readings should be taken until confidence in the readings is obtained. In certain situations, such as a lightering operation in an open seaway, the vessel may be in motion, slowing rolling from side to side, and measurements may need to be taken several times until an average reading can be determined. In situations such as this, a notation should be made on the cargo calculation work sheet that the vessel was observed to be rolling at the time of gauging, which might introduce errors into the measurement process.

One situation that is frequently encountered after loading is the presence of free water in the cargo tanks. The source of this water may be residual ballast water that could not be stripped out of the tank, or it may be ballast water that was reintroduced to the tank when loading through the same vessel or shore pipeline that was used to discharge the ballast to the terminal. It may also be that the vessel has a leak in a ballast line or a heating/steam coil passing through the cargo tank. In a worst-case situation, there may be a breach of the tank-shell plating to a ballast tank or, in the case of a single-hull vessel, the sea. Free water is measured using the same method used to determine the OBQ prior to loading (Figure 6-18).

The measurement of free water (also referred to as water cuts or thieving) is accomplished from the gauging hatch with water-indicating paste applied to the bob. The bob should be lowered into the tank, ensuring that the bob reaches the tank bottom by checking the gauge height. The tip of the bob should rest on the tank bottom for several seconds before it is withdrawn and the measurement is read. There are times when the paste does not fully turn color and appears speckled or spotty. (Speckled or spotty traces of water are the result of the water

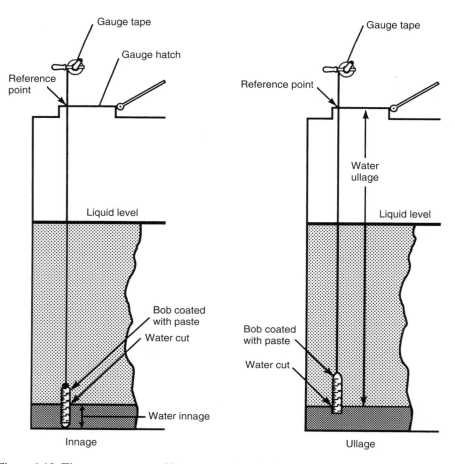

Figure 6-18. The measurement of free water using a bob coated with water sensitive paste. For accurate determination of free water, it is important not to lay the bob on its side at the bottom of the tank. From *Manual of Petroleum Measurement Standards,* First Edition, July 1990, "Measurement of Cargoes on Board Tank Vessels." Reprinted courtesy of the American Petroleum Institute.

not settling out due to insufficient time or possibly due to the density of the cargo.) A second measurement can be attempted, leaving the bob tip on the bottom for a longer period of time to see if more of the paste changes color. If not, then it should be recorded that traces of water were observed up to that point as measured on the bob.

The various cargo measurements are recorded on an ullage or cargo report that is sent to all the parties involved with the movement. Figure 6-19 is one such form that provides a detailed accounting of the cargo loaded or discharged at a particular terminal.

VESSEL ULLAGE/SOUNDING AND CAPACITY REPORT

	Before	After
Load	☐	☐
Discharge	☐	☐
Lightering	☐	☐

☐ To ☐ From Vessel ______

Sea valve seal numbers
Port ______
Starboard ______
Overboard ______

Vessel	Port	Product/cargo	Voyage no.	Date/time gaged

Tank Number	Ullage (ft/m)	Trim/List Corrected Ullage (ft/m)	Total Observed Volume (bbl/m³)	Free Water		Gross Observed Volume (bbl/m³)	Temperature (°F/°C)	°API @ 60°F or Density @ 15°C	Volume Correction Factor Table ____	Gross Standard Volume (bbl/m³) @ 60°F/15°C
				Innage/Ullage (ft/m)	Volume (bbl/m³)					
Capacity of vessel lines (bbl/m³)				When this form is used for on-board quantity/remaining on board, transfer this number to OBQ/ROB Report						

				SIGNATURES:
Gross standard volume (bbl/m³)		Draft forward (ft/m)		
Plus free water (bbl/m³)		Draft aft (ft/m)		Vessel representative
Total calculated volume (bbl/m³)		List (degrees)		
API gravity 60°F/relative density 15°C				Port representative
Sediment and water (volume percent)				
Weight correction factor (API Table 11)				Measurement representative
Long tons/metric tons (GSV)				

Figure 6-19. Sample cargo (ullages) report that is completed after each cargo transfer. This report provides an accurate accounting of the cargo on the vessel for all parties concerned with its movement. From *Manual of Petroleum Measurement Standards*, First Edition, July 1990, "Measurement of Cargoes on Board Tank Vessels." Reprinted courtesy of the American Petroleum Institute.

In order to determine the quality of the loaded cargo, samples must be obtained which are analyzed in a laboratory. The quality of the cargo is as important as the quantity, and in some cases even more so. In most cases, the vessel's crew will not be directly involved in the sampling process, but steps can be taken to prevent problems. First, unless special requirements dictate, the inspector should not be allowed to obtain a sample of the cargo until the cargo tank, or tanks, have finished loading. Second, the inspector should know the contents of each tank and where the liquid level of the cargo falls within the tank. This may be accomplished by providing the inspector with an ullage sheet of recorded measurements and an escort from the vessel's crew. Third, the inspector should properly label each sample container. Finally, a "retain" sample should be obtained from the inspector that can be sealed and retained on board in case quality differences arise during the voyage or after its completion.

Samples must be representative of the cargo loaded, and therefore special sampling equipment and techniques or procedures have been developed. It is routine for the inspector to provide his or her own equipment. Sample containers are usually one-quart glass bottles that can be sealed with a plastic cap or cork (Figure 6-20).

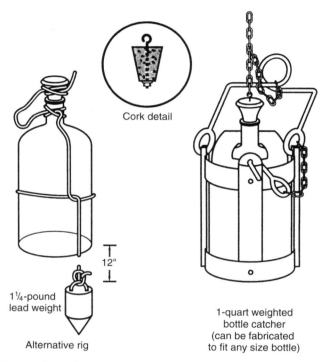

Cork detail

1¼-pound lead weight

12"

Alternative rig

1-quart weighted bottle catcher (can be fabricated to fit any size bottle)

Figure 6-20. One method of sampling the cargo in a tank. Sampling is an important function in the movement of cargo to maintain quality assurance. From *Manual of Petroleum Measurement Standards*, First Edition, July 1990, "Measurement of Cargoes on Board Tank Vessels." Reprinted courtesy of the American Petroleum Institute.

Each bottle is held in an assembly that allows it to be lowered into the tank to a predetermined level before the container is opened and the bottle allowed to fill. There are several different types of samples. A spot sample is a sample that is taken at a specific location (depth) in a tank. Upper, middle, and lower samples are spot samples taken at the midpoint of the upper third, middle third, and lower third of the cargo tank. A grab or line sample is a sample that is obtained at the header or manifold at a specific time during the loading or discharge operation. An all-levels sample is obtained by lowering a weighted, stoppered bottle or beaker to a point 1 foot (0.3 meter) above either the tank bottom or the free water level, opening the container, then raising it at a rate that allows it to emerge from the cargo with the container being about 75 percent full (maximum 85 percent). A running sample is obtained by lowering a weighted, unstoppered bottle or beaker to a point 1 foot (0.3 meter) above either the tank bottom or the free water level; the open container is then raised at a rate that allows it to emerge from the cargo with the container being about 75 percent full (maximum 85 percent). To obtain a composite sample in the laboratory, the sample is volumetrically blended from all tank samples of the same cargo, according to the volume contained in each tank. The type of sample to be taken is normally specified by the cargo supplier or receiver, and the vessel's crew should note which type of sample was pulled from each tank. Bottom or 1-foot samples are often requested when initially loading a tank with a cargo that is sensitive to contamination. The tank is loaded until a sample can be reasonably taken and the loading operation is normally stopped pending the results of the laboratory analysis.

PREDISCHARGE INSPECTION AND MEASUREMENT

The procedures used to measure the cargo prior to the discharge are the same as those used after loading. Prior to gauging, the inspector may inquire as to any rough weather that was encountered during the sea passage as this information may become useful at some later time if it is determined that a physical loss of cargo occurred during the voyage, or that an excessive and unaccountable volume of free water is found along with the cargo.

Normally the inspector will wish to gauge the cargo tanks and then sample them. By following this sequence, the inspector can determine the volume in each tank, which will make the sampling procedures easier. Since ullage and free water measurements were obtained at the loading port, it is best if these measurements are kept handy for easy reference during the gauging process. In this way it will be immediately possible to determine if there has been an error in gauging, and follow-up gauging can be accomplished on the spot. It should be noted that in some cases, it is discovered that the error in gauging occurred at the loading port. Appropriate notations should be recorded on the cargo documents, as these will assist the vessel's crew and inspector with reconciling any in-transit differences.

POSTDISCHARGE INSPECTION AND MEASUREMENT

Gauging the vessel after discharge for ROB (cargo remaining on board) is similar to the gauging for OBQ, although in some cases the petroleum inspectors will appear to be conducting a much more thorough job. This may be attributed to the fact that the cargo owner wishes to be informed if all cargo has been discharged from the vessel. Any cargo that is left on the vessel will most certainly be part of a physical loss of cargo, and someone will suffer economically. Measurements will most certainly be taken from each cargo tank's reference point and, in many cases, from a second or third location in the tank as well. These multiple measurements are an attempt to locate and quantify the remaining cargo. In situations where there is a substantial volume in one or more tanks, the inspector will attempt to retrieve a sample. The sample will be used to determine the quality of the ROB, thus placing it in one of these categories; cargo, sediment, or sludge.

Petroleum Inspectors typically seek to obtain innage gauges for ROB measurement. During the gauging procedure the inspector will try to insure that gauging is being done from the tank's reference point, and that the plumb bob has been lowered to the bottom of the tank. Tank reference heights are provided in the vessel's gauging, or strapping tables, and indicate the total height from each tank's official gauge point to that point on the tank bottom directly below the gauge point. Inspectors will look to place the tip of the plumb bob on the tank bottom, being careful to not let the bob lean over which would introduce error into the measurement. Having verified that the plumb bob has reached the tank bottom, the bob is withdrawn from the tank and the level of the ROB is read directly off the bob, or in cases of large ROB quantities where the bob is covered, then at that point where the ROB ends on the gauge tape.

Difficulties seemingly always arise during ROB measurements, and often times this leads to disagreements on how the ROB should be located, measured, and calculated. This is where it becomes important for the vessel's crew to understand the intricacies of the different measurement methods. Of the most importance is to determine whether the ROB should be trim or wedge corrected. Cargoes that are viscous or have a high pour-point and that have cooled during the discharge and tank stripping process may set up in a wedge form along the tank bottom. Determining whether the cargo lies in a wedge form can only be accomplished via multiple measurements along various points of the tank bottom. As a minimum, two (2) measurements at different points obtained from forward and aft in the tank are required to determine and prove that a wedge exists.

Many newer cargo loading programs possess the ability to calculate wedge volumes when the correct measurements and parameters are entered. Wedge volumes are also easily calculated manually, and will be discussed in greater details later in this chapter.

LOSS CONTROL

There are many difficulties that are encountered during the shipment of petroleum around the globe. One of these difficulties is the accurate quantification of the petroleum cargo that is loaded and discharged. Some of the factors that contribute to measurement error are:
- Use of various calculation methods
- Application of different volume correction tables
- Use of uncalibrated or defective equipment
- Poor sampling procedures
- Vessel construction and tank arrangement
- Weather

In addition to these factors, there are other possible causes whereby oil may be lost or gained during either the shipment or cargo operation. Some of these include:
- Use of different volume correction tables
- Under or over reporting of OBQ and ROB volumes
- Errors in quantifying free water or Sediment and Water (S&W)
- Evaporative losses
- Diverted cargo to ballast and void spaces
- Inefficient Crude Oil Washing operations
- Poor cargo discharge planning

It should be noted that a loss or gain of cargo can either be apparent, or actual. An apparent loss or gain would be when different sets of volume correction tables are used between the load and discharge ports. The quantities calculated at both ends would be different, while no physical loss or gain of cargo was realized. An actual loss or gain of cargo might be the result of cargo leaking into a ballast tank, or an evaporative loss where the light ends of the cargo are vented from the tank and leaked to the atmosphere.

All tank vessel officers, including engineering officers when it comes to fuel oil, should develop an awareness of good measurement procedures and record keeping. The purpose of the vessel is to transport cargo, and the Officers are responsible for insuring that the cargo is loaded, transported and discharged in a manner that insulates the vessel owner from cargo claims. While the Chief Officer is routinely deemed to be the cargo officer, Junior officers must begin to take partial responsibility for the cargo operation, including but not limited to; adequate segregation, cargo measurement in the form of tank gauging and temperatures, sampling, P/V valve operation and maintenance, cargo heating, and crude oil washing. By attending to these tasks and insuring they are carried out with the utmost care, all Officers can contribute to the safe and secure transport of the cargo.

Since petroleum is a relatively high value cargo, many of the interested parties employ what are referred to as loss control specialists to monitor cargo loading and discharge operations in an effort to pinpoint and mitigate losses due to the numerous factors previously named. These individuals may possess a marine background, or may have been a measurement specialist with training in all aspects of cargo measurement and troubleshooting not only on board vessels but in refineries and terminals as well.

As discussed earlier in this chapter the accuracy of cargo measurement depends upon controlling the variables that arise during the measurement process. The vessel's crew, and primarily the vessel's Chief Officer, can contribute significantly to the loss control process. They should at all times be aware of the procedures and equipment used by others who are involved in cargo measurement but are not part of the crew including independent inspectors, terminal personnel, and P&I representatives.

Simply put, loss control is a reconciliation process followed by those interested parties involved in the movement of petroleum cargoes. The importance of loss control with respect to the transportation of petroleum is that it:

- Pinpoints the sources contributing to cargo measurement error.
- Assists the charterer and vessel owner in reconciling measurement differences and discrepancies
- Assists a vessel owner in defending against erroneous cargo and ROB claims
- Reduces insurance costs for both the charterer and vessel owner

At it's best, loss control should be a preventative process, being aware and observing what is to happen and taking proactive steps to reduce or eliminate measurement errors. While loss control personnel often have to deal with large discrepancies in cargo quantities, these are typically the easiest to reconcile as they are usually the result of a large calculation error or significant measurement discrepancy. The harder ones to handle are the smaller discrepancies, usually associated with quantification of OBQ, ROB, line displacements and water in the cargo.

Attention to good cargo measurement procedures as described earlier in this chapter, as well as maintaining accurate cargo records will certainly benefit the loss control process. Recordkeeping, in the form of cargo documents, ullage reports, voyage summary reports and load / discharge data can make all the difference when attempting to successfully reconcile a cargo contamination or claim for short delivery.

VESSEL EXPERIENCE FACTOR

The Vessel Experience Factor (VEF) is a tool that can be used by vessel personnel to assess the accuracy of cargo quantities that are delivered from shore tanks or other vessels. It is a tool that should always be employed by the vessel's staff for each cargo load and discharge. A full description of the development and use of VEF's can be found in the API's *Manual of Petroleum Measurement Standards*, Chapter 17, Section 9.

Cargo quantities that are delivered from refineries or petroleum terminals are determined from measurements of shore tanks or meter systems. The vessel's crew do not partake in these measurements, and as such, they have no way of determining if the quantities are accurate. The quantity delivered by the shore facility is the quantity recorded on the Bill of Lading, and for custody transfer purposes this is the quantity that was 'shipped' and for which the Master accepts responsibility.

By collecting and maintaining accurate data the vessel's crew can generate a VEF that will assist them in assessing the accuracy of the reported quantities from the shore facility. With this tool, the Master may determine that the Bill of Lading quantity is not representative of the quantity of cargo that was actually shipped onto his vessel, and can then sign for the Bill of Lading under protest.

First, it is important to make sure that the VEF data obtained was based upon accurate measurements. As discussed earlier in this chapter, utilization of calibrated equipment which is employed in the proper manner will generate data that is the most accurate. Next, accurately recording the data for each voyage builds up a historical comparison of shore quantities vs. vessel quantities. Using this historical comparison a calculated "experience factor" can be determined. Figure 6-21 shows a recommended Vessel Experience Factor calculation worksheet with sample data. The factor is generated by dividing the total vessel quantities by the total shore quantities.

From inspection of the data it can be clearly seen that for those voyages that qualify for inclusion in the calculations the vessel routinely shows quantities in excess of the shore quantities. Therefore, application of the VEF factor shown above of 1.0011 would be applied to the vessel's receipt on the current voyage to provide an indication of what the shore quantity should be. Let's use this sample VEF of 1.0011 to a current voyage to see how it can be used to expose a shore quantity that may be accurate.

Vessel Loaded (US bbls):	864,722
Actual Shore Quantity:	865,491
Difference:	- 769
Percentage	- 0.09

APPENDIX C - VEF CALCULATION FORM EXAMPLE

Vessel Experience Factor - Calculation
Load or Discharge

Vessel: **M/T CONSENSUS**
Date: **December 1, 2004**

1	2	3	4	5	6	7	8	9	10	11	12	13	14
	List all voyages				BBLS / M³/MT (Use same units for all entries)				Step 1		Step 2	Qualifying Voyages	
Cargo	Voyage Number	Cargo Description	Terminal - Port	Date	Vessel Sailing / Arrival TCV	OBQ ROB	Load / Discharge TCV	B/L or Outturn TCV	Gross Error > 2%?	Vessel Load/Discharge Ratio	Qual. Voyage. (>0.30%) Y/N?	Vessel TCV	Shore TCV
Last	35	Arab Med	Ras Tanura	30-Oct-04	849,442	840	848,602	845,100		1.00414	N		
2nd	34	Mercy	Puerto La Cru	10-Sep-04	496,330	150	496,180	495,200		1.00198	Y	496,180	495,200
3rd	33	Cusiana	Covenas	20-Aug-04	~~325,289~~	~~196~~	~~325,093~~	~~310,494~~	Y	~~1.04702~~	N		
4th	32	Schiehallion	Sullom Voe	26-Jul-04	903,214	310	902,904	901,350		1.00172	Y	902,904	901,350
5th	31	Gullfaks	Mongstad	21-May-04	877,236	246	876,990	877,473		0.99945	Y	876,990	877,473
6th	30	El Sharara	Zawia	28-Mar-04	853,115	121	852,994	851,625		1.00161	Y	852,994	851,625
7th	29	Xikomba	Xikomba	15-Jan-04	605,052	232	604,820	606,981		0.99644	N		
8th	28	Forties	Hound Point	11-Dec-03	705,878	115	705,763	705,692		1.00010	Y	705,763	705,692
9th	27	Rabi Light	Cap Lopez	30-Sep-03	855,504	294	855,210	852,941		1.00266	Y	855,210	852,941
10th	26	Doba	Kome Kribi 1	28-Jul-03	881,892	392	881,500	880,427		1.00122	Y	881,500	880,427
11th	25	Belanak	Belanak	15-May-03	688,938	217	688,721	689,314		0.99914	Y	688,721	689,314
12th	24	Oriente	Esmeraldas	1-Apr-03	652,238	146	652,092	650,748		1.00207	Y	652,092	650,748
13th	23	Cabinda	Malongo	26-Feb-03	872,491	338	872,153	871,387		1.00088	Y	872,153	871,387
14th		Note: Vessel in dry dock Jan 1-28, 2003 for below deck piping changes											
15th													
16th													
17th													
18th													
19th													
20th						Totals:	9,237,929	9,228,238				Totals:	7,784,507	7,776,157
						Average TCV Ratio:		1.00105				TCV VESSEL	1.00107
												TCV SHORE	

Qualifying Range (excluding Gross Errors)
L: 0.99805   H: 1.00405

Vessel Experience Factor: 1.0011

Notes:
- List last voyage first
- Do not include load and discharge information on the same form
- Cross out either "load" or "discharge" and other inapplicable title information
- The average TCV ratio is the total vessel loaded TCV divided by total shore TCV

Figure 6-21. Vessel Experience Factor Worksheet

Based upon this comparison, a difference of 769 barrels is not very much, and without further review could be deemed to be acceptable. Now, let's use the VEF to conduct an additional comparison.

Vessel Loaded (US bbls):	864,722
VEF:	1.0011
Expected Shore Quantity:	863,763
Actual Shore Quantity:	865,491
Difference	- 1,728
Percentage	- 0.20

As the vessel's data indicates the vessel's quantity is routinely more than the shore quantity, the VEF is used as a divisor. By dividing the vessel's quantity by the VEF, you generate an "expected" shore quantity, which can then be compared to the actual quantity from the shore terminal. In this case, the shore quantity would appear to be overstated by an amount of 1,728 barrels. This being the case, it may be prudent for the Master to have the vessel regauged to make sure the vessel's quantity is accurate and no mistakes were made. If nothing on the vessel is uncovered, then the Master would issue a letter of protest to the shore terminal stating the quantities and highlighting his comparison which indicates the shore quantity may be in error.

Another source of disagreement, especially between vessel and inspection company personnel, is the quantification of small quantities in the cargo tanks. While the calculation of small volumes in the cargo tanks can sometimes be handled by the vessel's strapping tables, there are circumstances where the tables may not be applicable. When strapping tables are generated, it is most often assumed that the vessel will be on an even keel, and in an upright position. Some tables also come with trim factors, to be applied when the vessel is either trimmed by the head or stern. List corrections are uncommon, and the simplest solution is to remove the list prior to gauging.

Situations where the strapping tables may not be applicable are when there is very little cargo, and the cargo does not touch all four sides of the tank. This situation is usually encountered when the tanks have been emptied, and stripping fails to remove all of the cargo. In this circumstance, and when the vessel is trimmed by the stern, the cargo may lie in a wedge shape, touching the sides and back wall of the tank, but not the front. The most accurate method to determine the volume remaining in the tank in this situation is by using a wedge formula. This formula utilizes the obtained cargo gauges along with the tank dimensions, and volumes from the vessel's strapping tables to determine the cargo remaining in the tank.

It is most important to determine whether the cargo reaches (or touches) all four of the tanks' sides. A single gauge at the gauge hatch will not prove whether the cargo does in fact touch all sides of the tank. To ascertain this, there is a simple formula as shown in the worksheet that will assist you in determining whether the cargo is in contact with all four sides, or whether the cargo lies in a wedge shape. In conjunction with this formula, it is always advisable to obtain at least one or two additional gauges from other points in the tank. This may not always be possible, and in cases of high ROB quantities the removal of butterworth plates may be an option.

The wedge formula is laid out below, in Figure 6-22, and provides all of the information necessary to perform the calculations.

Figure 6-22. Wedge Calculation Worksheet

Let's take a look at an example where at the conclusion of the cargo tank stripping operation, tank No. 1 Center is found to have a measured ROB innage gauge of 9.5 centimeters at the gauge point. Refer to the example sheet below, shown in Figure 6-23.

CARGO TANK WEDGE CALCULATIONS

Tank No.	U	-(D x F)	x F	+ S	= A	A²	V¹	+(S x L)	= W¹	A² x W¹	+(F x 2)	= V
1 Center	1.95	0.28222	0.017839	0.095	0.124752	0.015563	16.4	1.634	10.03672	0.156203	0.035679	4.38
2 Center												
3 Center												
4 Center												
5 Center												
6 Center												
1 Port												
1 Starboard												
2 Port												
2 Starboard												
3 Port												
3 Starboard												
4 Port												
4 Starboard												
5 Port												
5 Starboard												
6 Port												
6 Starboard												

A = Adjusted Product Innage at the Aft Bulkhead   0.12475
D = Mean Total Gauge Height   15.82 meters
F = Trim Factor   1.13974
L = Length of Tank   17.2 meters
S = Measured Product Innage at Gauge Point   0.095 meters

U = Distance from AFT Bulkhead to Gauge Point   1.95 meters
V = Volume Calculated by Wedge Formula   16.4 cu m
V¹ = Volume extracted from the Vessel's Tables   14.0 meters
W = Width of the Tank   14.67 meters
W¹ = Theoretical Width of the Tank

To determine if the measured R.O.B. at the cargo tanks' gauge point reaches all four of the tanks' bulkheads, solve the following equation for "X":

$X = A \div F$
$6.99 = 0.12475 / 0.01784$ L = 17.2 meters
                                        6.99 < 17.2 therefore use wedge
If "X" is less than the tank length (L), use the wedge formula, or
If "X" is greater than the tank length (L), use trim corrections.

Comments :

Draft AFT   : 5.7
Daft FWD   : 2.1
Trim   = 3.6
LBP   + 201.8
"F"   = 0.017839
"D"   x 15.82
"DF"   = 0.28222

Cargo :
API :
Pour Point :
Ambient Air Temp. :
Ambient Sea Temp. :

Figure 6-23 Wedge Calculation for Cargo Tank #1

In our example if the vessel's strapping tables for tank No. 1 are entered with an innage of 9.5 centimeters, the resulting cubic meters are read as 16.4 cubic meters. As stated previously, since strapping tables are generated with an even keel vessel in mind, this quantity of 16.4 cubic meters may not be representative of what is actually in the vessel's tank. Additionally, application of a trim factor to the innage of 9.5 centimeters when the vessel has a trim is not reliable as trim factors only account for cargo that is in contact with all four sides of the tank. This is where the wedge formula comes in handy.

Using the simple formula shown in Figure 6-23, by dividing the adjusted cargo innage at the aft bulkhead by the Trim Factor, it is easy to determine whether to use the trim tables or wedge formula. In the example, the resultant answer of 6.99 is less than the overall tank length of 17.2 meters, and therefore the wedge formula should be used to quantify the cargo in tank No. 1 Center.

The wedge formula yields a quantity of 4.38 cubic meters, which is less than the quantity from the strapping tables of 16.4 cubic meters, and is a more accurate representation of what is in the cargo tank. Given that there may be measurable ROB in some or all cargo tanks, you can see how using the wedge formula will aid the vessel's crew in reducing the ROB quantity that will be reported to the charterer.

In summary, good cargo measurement procedures and loss control practices should be employed by the vessel's staff to reduce cargo discrepancies and subsequent claims from the terminal and charterer.

REVIEW

1. What role does an "inspection company" play in the shipment of bulk liquid cargoes?
2. How are tank vessel gauging methods classified?
3. What equipment is necessary when performing cargo measurement on a tank vessel?
4. What information is derived from the calibration tables of a tank vessel?
5. Explain the difference between ullage, innage and free water measurements in a cargo tank.
6. Explain the difference between ROB and OBQ.
7. What effect does vessel trim have on the accuracy of cargo measurement?
8. How are the effects of trim and list addressed during cargo volume calculations?
9. List some of the factors that contribute to cargo measurement errors.
10. What is a "vessel experience factor" and how is it determined in practice?
11. What is a "wedge formula" and when is it applied?
12. Who is ultimately responsible for accepting the measured quantity of cargo that is loaded or discharged on a tank vessel?

CHAPTER 7

Cargo Calculation
Kelly Curtin and Kevin Duschenchuk

O nce the vessel has been "gauged out" and all of the necessary information is
recorded, the calculation of the cargo volume begins. The goal of the calculation process is to determine the volume/tonnage of cargo transferred (loaded/
discharged) at the terminal. Cargo tonnage must also be determined to enable
personnel to verify that the vessel's stability, draft, trim, and stresses are within
safe limits. The cargo tonnage, often referred to as the cargo deadweight, is combined with the fuel, water, stores, and light-ship tonnages to arrive at the overall
displacement of the vessel.

It is essential that accurate readings be obtained for the following:

Volume of cargo (ullage/innage/water cuts)
Temperature of the cargo
API gravity/specific gravity of the cargo

For the convenience of the reader, the most common units of measure employed
in the industry are listed below as well as in the appendix of the text.

Barrel = 42 gallons (U.S.)
Cubic meter = 6.2898 barrels
Metric ton = 1,000 kilograms = 2,204.6 pounds
Long ton = 2,240 pounds
Observed (gross) barrel = 42 gallons (at the observed temperature)
Standard (net) barrel = 42 gallons (at the standard temperature 60°F)

The first step in the calculation is to correct the volume of cargo to a standard volume. Liquid cargoes expand and contract with changes in temperature;
therefore, the petroleum industry established a standard temperature to be used
when calculating the standard volume of cargo in a tank. In the United States,
the American Petroleum Institute (API) set the standard temperature at 60°F
(15.6°C); in countries that use the metric system, 15°C (59°F) is the standard
temperature.

The volume correction factor is found by entering the appropriate API tables with the observed temperature and API gravity of the cargo. Separate API tables are to be used depending upon the cargo itself. For example Table 6A is used for Generalized Crude Oils, Table 6B is for Generalized Petroleum Products, and Table D1555 is for cargoes such as Benzene, Styrene, and Toluene. Using the following formula, the standard (net) volume can be determined by multiplying the observed (gross) volume by the volume correction factor.

Standard (net) volume = observed (gross) volume × volume correction factor (VCF)

The tonnage in a tank is found by entering the appropriate conversion table (also found in the API tables) with the API gravity of the cargo to determine the stowage factor. There are two ways of finding the tonnage:

Standard (net) volume × long tons per barrel = long tons
Standard (net) volume / barrels per long ton = long tons

The following example illustrates a typical cargo calculation including the excerpts from the API tables:
A parcel of #6 fuel oil is loaded and the following information is determined:
Observed temperature = 135°F
API gravity = 14.0
Observed (gross) volume = 15,000 barrels
Note: To determine the actual quantity of cargo loaded at the terminal, all water and OBQ must be taken into account.

PROBLEM: What is the standard (net) volume and how many long tons are in the tank?
1. Determine the volume correction factor by entering the API tables with the observed temperature of 135°F and API gravity of 14.0. In this case the volume correction factor is 0.9707 (see table 7-1).
2. Multiply the observed (gross) volume (barrels) by the volume correction factor to find the standard (net) volume.

15,000 barrels × 0.9707 = 14,560.5 barrels

3. Enter the (volume-to-weight) conversion table with the API gravity of the cargo to find the appropriate conversion factor. In this case:

Long tons per barrel = 0.15186 or barrels per ton = 6.585 (see tables 7-2 and 7-3)

4. Either multiply the standard (net) volume by the long tons per barrel or divide the standard (net) volume by the barrels per ton.

Standard (net) volume X long tons per barrel = long tons
14,560.5 X 0.15186 = 2,211.2 long tons
Standard (net) volume / barrels per long ton = long tons
14,560.5 / 6.585 = 2,211.2 long tons

TABLE 7-1
Excerpt from API Table 6B—Generalized Products
Volume Correction Factors for Generalized Products

API Gravity at 60°F

Temp. (F)	13.0	13.5	14.0	14.5	15.0
135.0	0.9709	0.9708	0.9707	0.9705	0.9704
135.5	0.9707	0.9706	0.9705	0.9703	0.9702
136.0	0.9705	0.9704	0.9703	0.9701	0.9700

TABLE 7-2
Excerpt from API Conversion Table 11

API Gravity	Long Tons Per Barrel
13.8	0.15207
13.9	0.15196
14.0	0.15186
14.1	0.15175
14.2	0.15165
14.3	0.15154
14.4	0.15144

TABLE 7-3
Excerpt from API Conversion Table

API Gravity	Barrels Per Long Ton
13.8	6.576
13.9	6.580
14.0	6.585
14.1	6.590
14.2	6.594
14.3	6.599
14.4	6.604

Tables 7-1, 7-2, and 7-3 reprinted courtesy of the American Petroleum Institute, *Petroleum Measurement Tables—Volume Correction Factors*, volume 2, 1980.

CARGO PLANNING

Vessel personnel normally develop a detailed plan (prestow) of the upcoming load based on tentative orders and anticipated cargo values received from the owner/operator and terminal. Some of the factors considered when drawing up the cargo plan include the following:

Number of grades and quantity
Limiting draft (seasonal load line)
Vessel trim
Bending stresses and shear forces
Tank preparation (cleaning/pipeline flushing/drying)
Cargo segregation

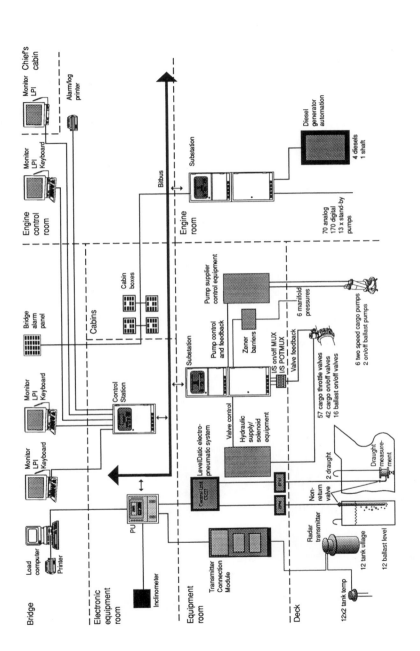

Figure 7-1. Schematic of an integrated monitoring and control system installed on a tanker (Saab MaC/501 system). Note that because of an interface to the contralized microcomputer system, the loading computer can be used not only to precalculate the load, but also to provide virtually instantaneous updates of draft, trim, and stress throughout all phases of the ship's operation. Courtesy Saab Marine Electronics.

Cargo compatibility
Regulatory requirements
Loading port and discharge port sequence
Number of cargo hoses/arms
Special cargo handling requirements

Cargo planning can be accomplished in a variety of ways. Modern tank vessels are frequently equipped with a cargo calculator or a software program that contains the hydrostatic data and characteristics of the vessel. The use of these devices has improved the cargo planning process by increasing accuracy and by saving time over the laborious hand calculations of the past. Many of the cargo loading programs in use today still require vessel personnel to input certain key information when performing a prestow or final calculation. Therefore, the PIC should have a thorough understanding of the cargo calculation process to avoid any claims against the owner as well as improper stowage of the vessel. As of this writing, owners are increasingly integrating the vessel computers with various inputs (tank ullages, cargo temperature, IG deck pressure, oxygen content, draft readout, and so forth) directly from the cargo system (Figure 7-1).

The reduction of crew size and the heightened complexity of systems on modern vessels has necessitated the increased use of computers and automated cargo systems as seen in Figures 7-2, 7-2a and 7-3.

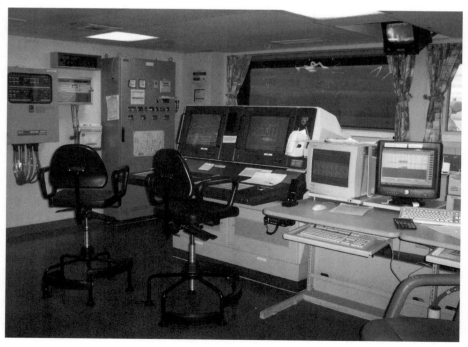

Figure 7-2. Cargo Control Room. Courtesy Mark Huber.

Figure 7-2a. CCR screen. Courtesy Mark Huber.

Figure 7-3. Depending on the vessel, the cargo operation may be controlled on deck, in the cargo control room, or, as shown here, on the bridge. The installation shown enables the PIC to monitor cargo tank ullages, temperatures, stress, draft, trim, and inert gas pressures on a single screen. Cargo system valves can be operated by touching a light pen to the screen. Courtesy Saab Marine Electronics.

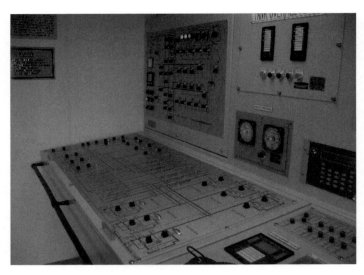

Figure 7-3a. Cargo
Control Console.
Courtesy George
Edenfield.

LOAD LINES AND ZONE LIMITATIONS

The load line of a vessel (its "Plimsoll mark") indicates the maximum permissible tonnage that can be loaded based on the zone (location in the world) the vessel will pass through and the season of the year. From the earliest seafaring times, attempts were made to ensure the safety of personnel and the vessel by imposing strict limits on the cargo carried each voyage, thereby reducing the risks from overloading (Figure 7-4). Loadlines were designed so that a vessel would have sufficient freeboard, or reserve buoyancy, to safely transit a geographic zone in weather conditions that might be expected.

Figure 7-4. American Bureau of shipping inspectors check a vessel's load line markings. Courtesy American Bureau of Shipping.

Today, international standards governing load lines have been implemented by the maritime nations of the world and incorporated into U.S. rules. When a vessel is constructed, the appropriate load line markings are calculated and permanently etched on the hull by an authorized classification society. In the United States, the load lines are usually assigned by the American Bureau of

Shipping (Figure 7-5). Each mark corresponds to a given displacement, or the total tonnage of water displaced by the vessel. This tonnage is exactly equal to the weight of the loaded vessel. The various marks are identified as tropical, summer, winter, and winter North Atlantic zones including salt water and freshwater conditions. Due to the fact that salt water is more buoyant than fresh, an allowance is made for the extra sinkage of the vessel in freshwater; this is known as the freshwater allowance (FWA). This system of marking permits vessels to load more cargo in regions of predominantly fair weather and during seasons when good weather can be expected.

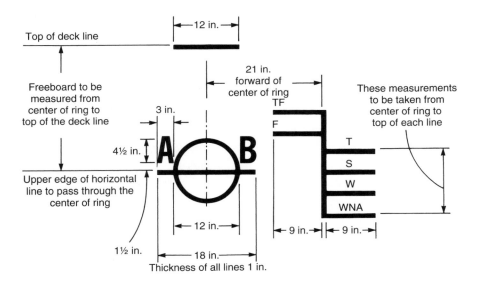

The center of the ring is to be placed on each side of the vessel at the middle of the length as defined in the Load Line Regulations. The ring and lines are to be permanently marked, as by center punch, chisel cut, or bead of weld.

A B	American Bureau of Shipping
T F	Tropical Fresh Water Allowance
F	Fresh Water Allowance
T	Load Line in Tropical Zones
S	Summer Load Line
W	Winter Load Line
W N A	Winter North Atlantic Load Line

Figure 7-5. Load line markings for oceangoing vessels are placed amidships on both sides of the hull. The American Bureau of Shipping is authorized to assign load lines to vessels registered in the United States and other countries. Courtesy American Bureau of Shipping.

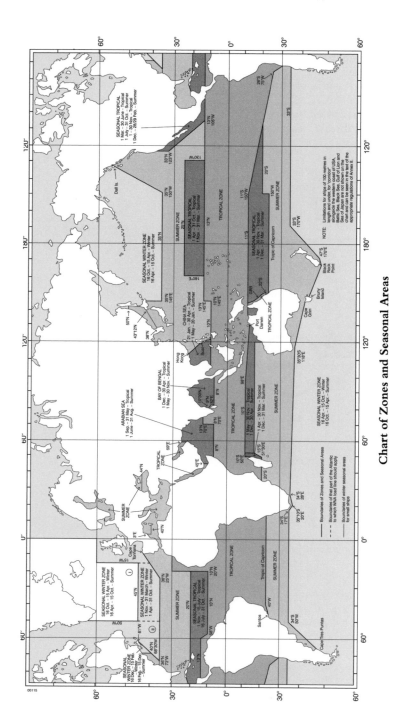

Chart of Zones and Seasonal Areas

Figure 7-5a. Loadline chart of the world. Copyright International Maritime Organization (IMO), London.

For example, a tanker loading in Valdez, Alaska, on December 1 (winter seasonal zone) would load to the winter marks. In contrast, a vessel loading on the same date in Rio de Janeiro, Brazil (tropical zone), would be allowed to load to the tropical marks.

Zone Allowance

When a vessel is loaded to the tropical marks in December, such as the example at Rio de Janeiro, it cannot cross into the summer zone without violating the law. In this instance, the vessel is not only governed by the zone in which it is loaded, but also by the zones through which it transits. If the loaded passage of the vessel proceeds into the summer zone, then its draft must not exceed the summer marks at any time while inside that zone.

The trick is to load the maximum allowable tonnage without being overloaded as the vessel proceeds from one zone to the next. Allowance can be made for the burn-off of fuel and other consumables while proceeding to the *controlling zone*, which is the zone that imposes the greatest restriction on the loading of the vessel.

The extra tonnage a vessel may load beyond that permitted by the controlling zone is known as the zone allowance. The following example shows how to determine zone allowance:

A tanker loads at a terminal within the tropical zone. It is estimated that the vessel will enter the summer zone nine days after departing the loading port. The approximate burn-off of fuel at sea speed is 50 tons per day and water consumption is 10 tons per day. How many tons can the vessel be loaded beyond the summer load line?

$$\text{Zone allowance} = 9 \text{ days} \times 60 \text{ tons/day} = 540 \text{ tons}$$

Note that a vessel sailing in the other direction (from summer to tropical zone) would not require this calculation, due to the fact that it was loaded in the controlling zone.

CARGO TERMS

The following section is a review of the key terms used in cargo calculations:

Deadweight is the total weight of cargo plus crew, stores, water, fuel, and ballast on board at a given time. In the tanker industry, deadweight is used as a rough measure of the cargo carrying capacity of a vessel.

Displacement is the weight of the water that is displaced by the hull; it is exactly equal to the weight of the vessel when floating freely. Thus the term displacement is used to denote a vessel's weight in tons at a given draft. When the displacement of the vessel is known, a number of important hydrostatic values can be determined by inspection either from the tables or curves of the vessel.

Light ship is the displacement, or weight in tons, of a vessel minus cargo, crew, stores, fuel, water, and ballast; in other words, it is the weight of the empty vessel.

Tons per inch immersion (TPI) shows the number of tons required to submerge a vessel 1 inch amidships. The change in draft on a vessel is proportional to the amount of weight loaded or discharged. The TPI varies with the draft and shape of the vessel at the waterline. The values for TPI can be found in the hydrostatic tables or on the deadweight scale of the vessel.

This information is valuable when computing changes in mean draft caused by the loading and discharging of weight. The following formula is used:

Weight loaded (or discharged) / TPI = change in mean draft

EXAMPLE: A vessel with a mean draft of 25' 00" has a TPI of 150 tons/inch. What will be the new draft after loading 900 tons?

900 tons/150 tons per inch = 6 inch increase in draft

$$
\begin{array}{r}
25'\ 00" \\
+\ \ 06" \\
\hline
25'\ 06"
\end{array} = \text{new mean draft}
$$

Trim is the difference between the forward and after drafts of the vessel. The calculation of trim is closely linked to the following terms:

Longitudinal center of buoyancy (LCB) is the center of volume of the underwater portion of a vessel's hull and is the point through which it is assumed all upward (buoyant) forces act. The LCB of the vessel is found in the hydrostatic tables or curves of the vessel.

Longitudinal center of gravity (LCG) is the counterpart of LCB, or the point in the hull through which it is assumed all the downward (gravitational) forces act.

Trimming moment. When a weight is loaded or discharged at a given distance forward or aft of the tipping center, a trimming moment is created.

Likewise, a moment is created when a weight on the vessel is shifted forward or aft. Moments are expressed in foot-tons and computed using the following formula:

Trimming moment = weight (tons) × distance (feet)

Trim arm or **lever** is the numerical (longitudinal) difference between LCB and LCG. The position of LCB and LCG relative to each other determines the amount of trim and whether it will be by the head or stern. The LCB can be found by entering the hydrostatic tables or curves supplied by the naval architect with the displacement of the vessel. LCG is determined by first multiplying the LCG for each cargo tank by its tonnage to yield the longitudinal moments. The same calculation is made for fuel, water, stores, miscellaneous tanks, and light ship. The total moments (sum of the longitudinal moments) divided by the total displacement will give the location of the LCG. Once the LCG and LCB have been determined, numerically compare these values to find the trim arm or lever. The trim of the vessel can then be found using the following formula:

$$
\text{Trim (inches)} = \frac{\text{displacement} \times \text{trim arm}}{\text{MT1"}}
$$

Tipping center or **center of flotation** is best described as the hinge about which a vessel rotates longitudinally. This hinge is not fixed in a single position but moves forward and aft with changes in draft and trim.

Moment to change trim 1 inch (MT1) is used in conjunction with the trimming moment to determine the change in trim of the vessel. MT1 varies with the draft of the vessel, and the exact values can be derived by inspection from the hydrostatic tables.

EXAMPLE: A vessel has a draft of 25' 00" forward and aft (vessel is on an even keel). The MT1 at this draft is 1,000 foot-tons. If cargo weighing 100 tons is moved aft a distance of 100 feet, what are the new drafts?

$$\text{Trimming moment} = 100 \text{ tons} \times 100 \text{ feet}$$
$$= 10,000 \text{ foot-tons}$$

$$\text{Change in trim} = 10,000 \text{ foot-tons}/1,000 \text{ foot-tons}$$
$$= 10\text{" by the stern}$$

$$\begin{array}{l} 25'\ 00" \\ -\ \ 05" \\ \hline 24'\ 07" \end{array} = \text{new draft forward}$$

$$\begin{array}{l} 25'\ 00" \\ +\ \ 05" \\ \hline 25'\ 05" \end{array} = \text{new draft aft}$$

Note that the new drafts were determined by applying half the change in trim to arrive at the forward and after drafts. Due to shifting the weight aft, half of the change in trim (5 inches) was added to arrive at the after draft and the same amount was subtracted to arrive at the forward draft.

When a weight is loaded or discharged, a slightly different problem is encountered. The first step is to determine the new mean draft produced by the changed displacement of the vessel. The change in trim can then be computed and applied to the new mean draft to find the forward and after drafts.

EXAMPLE: The initial draft of the vessel is 25' 00" forward and aft (even keel) and the MT1 is 1,000 foot-tons. TPI at this draft is 50 tons per inch.

If 100 tons are loaded 100 feet aft of the tipping center, what are the new drafts?

$$\text{Increase in mean draft} = \frac{100 \text{ tons}}{50 \text{ tons per inch}} = 2"$$

$$\begin{array}{l} 25'\ 00" \\ +\ \ 02" \\ \hline 25'\ 02" \end{array} = \text{new mean draft}$$
$$\text{Trimming moment} = 100 \text{ tons} \times 100 \text{ feet}$$
$$= 10,000 \text{ foot-tons}$$

$$\text{Change in trim} = \frac{10,000 \text{ foot-tons}}{1,000 \text{ foot-tons}}$$
$$= 10\text{" by the stern}$$
$$\begin{array}{l} 25'\ 02" \\ -\ \ 05" \\ \hline 24'\ 09" \end{array} = \text{new draft forward}$$

$$25' \ 02"$$
$$\underline{+ \quad 05"}$$
$$25' \ 07" \ = \text{new draft aft}$$

Freshwater and dock-water allowances: Freshwater is less buoyant than salt water; therefore the load line regulations allow for this by assigning the vessel a freshwater mark. The freshwater allowance is the number of inches the draft will change when moving from freshwater to salt water and vice versa. In many ports, the harbor water may be classified as brackish (a mixture of salt and freshwater) requiring the calculation of a new allowance. Here the term dock-water allowance comes into play.

Dock-water allowance is the number of inches a vessel may load below its marks in water of a known specific gravity. To find this value, a sample of the water alongside the dock is obtained and the specific gravity is measured with a simple hydrometer. The specific gravity of freshwater is 1.000 and of salt water 1.025. If the sample at the dock reveals a specific gravity of 1.010, how can the dock-water allowance be found?

First, the freshwater allowance should be found by referring to the hydrostatic tables or vessel characteristics. In this case, the vessel has a freshwater allowance of 10" and the specific gravity alongside is found to be 1.010.

Second, the dock-water allowance can be determined since it is already known that at a specific gravity of 1.000, the allowance is 10", and at a specific gravity of 1.025, it is zero. The value at the dock is 1.010 or 15/25 of the way between the zero and 10" allowance. Thus:

$$\text{Dockwater allowance} = 15/25 \times 10"$$
$$= 6.0"$$

To eliminate confusion, vessel personnel often develop a table of dock-water allowances based on the different specific gravities that may be encountered.

TABLE 7-4
Dock-water Allowances
(Freshwater Allowance = 10")

Specific Gravity	Allowance
1.000 (fresh)	10"
1.005	8"
1.010	6"
1.015	4"
1.020	2"
1.025 (salt)	0"

Bending stresses and shear forces. One of the major concerns in the safe transport of bulk liquid cargoes by tank vessel is the stress on the hull. The two primary problems are bending stresses and shear forces. First, longitudinal bending occurs when there is a concentration of weight at a particular location, such as in the midsection of the vessel or near the ends. The bending point(s) of a vessel can be calculated with a cargo-loading computer program or by using a traditional trim/stress form. Additionally, the PIC should check the bending stresses on the hull during a cargo operation by comparing the mean and mid-ship drafts.

Sagging (Figure 7-6a) occurs when there is a concentration of weight in the midsection of the vessel, causing the deck to be subjected to compression forces while at the same time the keel is under tension.

Hogging (Figure 7-6b) occurs when there is a concentration of weight at both ends of the vessel, causing the deck to experience tensile forces while the keel is under compression.

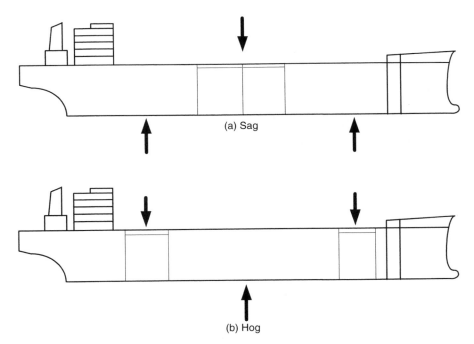

Figure 7-6. Bending stresses (a = sag, b = hog) exerted on the hull due to a concentration of weight at certain locations in the vessel.

A well-conceived cargo plan should minimize the bending stresses of the vessel through proper distribution of the cargo and ballast over the length of the vessel. It is important to remember that sea conditions (Figure 7-7) can amplify these stresses, ultimately causing permanent damage to the vessel's structure (fractures) and in the most extreme case, loss of the vessel (structural failure) (Figure 7-7a).

Shear force occurs when two forces act in opposite directions (parallel to each other) such as at a bulkhead between an empty ballast tank and a full cargo tank. The weight (gravitational) and buoyant (upward) action experienced on either side of the bulkhead causes a shear force.

Figure 7-7. Heavy weather amplifies the bending stresses on a tanker's hull. Courtesy Kevin Duschenchuk.

Figure 7-7a. Excessive bending stresses (hogging). Courtesy gCaptain.

LOADING TO FINAL DRAFT

The cargo plan (prestow) serves as a guide for the PIC during the actual cargo transfer. The cargo information used in developing the prestow does not necessarily match the values derived during the actual loading operation. For example, when the cargo temperature, API gravity, or volume differ slightly from the prestow, it is often necessary to make adjustments during the final stages of the loading operation. At the final stages of the loading operation, it is common practice to check the draft marks from the dock and compare them to the precalculated values. As discussed in chapter 9, it is recommended that the loading operation should end in slack (partially full) tank(s), often called the "trim tank(s)," to accommodate any last minute changes in cargo quantity or trim.

LOADING AND DISCHARGING RATES

Loading and discharging rates vary considerably with different vessels, terminals, and cargoes. For example, crude-oil carriers routinely load at rates in excess of 100,000 barrels per hour at the Alyeska Terminal in Valdez, Alaska, while a small product or parcel carrier might load at less than 1,000 barrels per hour at a refinery.

Some of the factors that influence cargo transfer rates include the following:

Number and capacity of cargo pumps
Pipeline dimensions (vessel and shore facility)
Temperature and viscosity of the cargo
Capacity of the venting or vapor control system
Capacity of the inert gas system
Location of (distance from) and elevation of shore tanks
Use of shore booster pumps
Experience of vessel personnel
Number of cargo tanks open

In order to determine when the cargo operation will finish, the PIC should calculate the loading/discharging rates at periodic intervals. On most vessels, it is a common practice to calculate hourly rates to confirm that the loading rate has not changed from the agreed value and to estimate the time of completion of the cargo operation (sailing time). The loading/discharging rates are typically expressed in barrels-per-hour, or cubic-meters-per-hour depending on the vessel. To calculate a rate, the PIC should perform the following steps:

Gauge the vessel's tanks at periodic intervals and refer to the calibration (ullage) tables to determine the total cargo on board.

Find the difference between this figure and the last total; divide this difference by the number of hours elapsed between the measurements. The result is the loading or discharging rate per hour.

To determine the estimated time of completion, divide the amount of cargo remaining to be loaded or discharged by the hourly rate to yield the number of hours remaining.

CALCULATING VESSEL EXPERIENCE FACTOR (VEF)

A tanker's VEF, or vessel experience factor as described in Chapter 6, is a historical compilation of ship-to-shore cargo volume variations and was designed to be used, primarily, as a loss control tool to help assess the validity of quantities derived from shore tank measurements. Vessel experience factors are also frequently used to determine custody transfer quantities when shore-based measurements are not available.

Since vessel capacity tables are typically derived from the vessel's building plans, and not from (more accurate) tank calibration measurements, there are usually significant differences between the quantity of a cargo measured in a calibrated shore tank and the same cargo measured in vessel tanks. These differences may be as high as ± 2%.

Using the following cargo figures, determine the vessel's VEF for the past 10 loads.

Date	Port	TCV	OBQ	Loaded TCV	Shore TCV	Difference	Voyage Load VEF	Total Load VEF
10/13/08	Valdez, AK	902610	1659	900951	899531	1420	1.001579	1.001579
10/20/08	Valdez, AK	900364	280	900084	896006	4078	1.004551	1.003062
11/02/08	Valdez, AK	901396	157	901239	896351	4888	1.005453	1.003858
11/26/08	Valdez, AK	905216	1587	903629	897871	5758	1.006413	1.004497
12/05/08	Valdez, AK	905768	2799	902969	897448	5521	1.006152	1.004828
12/15/08	Valdez, AK	904697	3397	901300	896894	4406	1.004913	1.004842
12/23/08	Valdez, AK	899322	483	898839	893936	4903	1.005485	1.004934
1/04/09	Valdez, AK	902376	2714	899662	897401	2261	1.002519	1.004632
1/15/09	Valdez, AK	651988	411	651577	647883	3694	1.005702	1.004720
1/28/09	Valdez, AK	901810	180	901630	899542	2088	1.002321	1.004473

- Voyage Load VEF = vessel TCV / shore TCV

- Total Load VEF = sum of all previous vessel TCV / sum of all previous shore TCV

1.004473 is the tanker's VEF after 10 loads.

Following the previous example, and formulas, figure out the total VEF for the given discharge quantities below:

Date	Port	Open TCV	ROB	Disch TCV	Shore TCV	Difference	Voyage Discharge VEF
10/16/08	Ferndale, WA	903040	542115		361484		
10/17/08	Anacortes, WA	542040	339		538829		
10/24/08	Ferndale, WA	899938	649446		249646		
10/28/08	Richmond, CA	649570	249409		400751		
10/29/08	Martinez, CA	249332	144		246111		
11/10/08	Long Beach, CA	901087	451253		447529		
11/17/08	Barbers Pt., HI	451279	801		449973		
12/01/08	Tacoma, WA	904993	0		899966		
12/09/08	Ferndale, WA	905111	453889		450168		
12/11/08	Tacoma, WA	453870	11		453610		

ANSWER:

Date	Port	Open TCV	ROB	Disch TCV	Shore TCV	Difference	Voyage Discharge VEF	Total Discharge VEF
10/16/08	Ferndale, WA	903040	542115	360925	361484	559	0.998454	0.998454
10/17/08	Anacortes, WA	542040	339	541701	538829	2872	1.005330	1.002569
10/24/08	Ferndale, WA	899938	649446	250492	249646	846	1.003389	1.002747
10/28/08	Richmond, CA	649570	249409	400161	400751	590	0.998528	1.001657
10/29/08	Martinez, CA	249332	144	249188	246111	3077	1.012502	1.003142
11/10/08	Long Beach, CA	901087	451253	449834	447529	2305	1.005151	1.003543
11/17/08	Barbers Pt., HI	451279	801	450478	449973	505	1.001122	1.003138
12/01/08	Tacoma, WA	904993	0	904993	899966	5027	1.005586	1.003751
12/09/08	Ferndale, WA	905111	453889	451222	450168	1054	1.002341	1.003594
12/11/08	Tacoma, WA	453870	11	453859	453610	249	1.000549	1.003287

The tanker's Total Discharge VEF for the 10 discharges is 1.003287

CARGO LOADING SOFTWARE

For years cargo/stability software has been available for the marine industry to assist with cargo planning and stability calculations. CargoMax©, LoadMaster©, Capstan3©, and ShipMaster© are just a few examples of this type of software. All of these programs not only assist in the planning of loading and discharging operations, but can also give real-time calculations and graphical representations of the condition of the vessel in regards to strength and stability. The prudent tanker mate or PIC does not just load their vessel and hope that everything will work out. With the aid of this type of software, an individual can plan each evolution of the cargo/ballast operation to ensure that the vessel stays within regulatory and operational requirements, and at the same time ensure the safety of the crew, vessel, and cargo. The remainder of this chapter will explain the basics of working with one such program, CargoMax from Herbert Software Solutions, Inc. (HSSI). By working with the enclosed CargoMax demo CD, the reader will become familiar with the operation and limitations of such a program. With this knowledge a tanker mate or PIC will be able to move from vessel to vessel, and understand the information provided by such software and how to apply it.

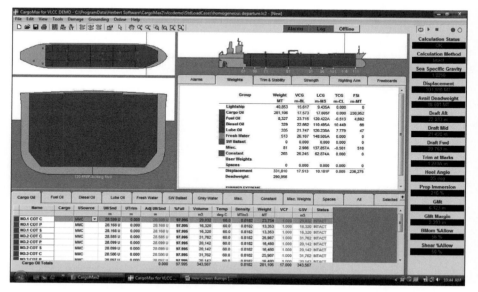

Figure 7-8. Courtesy of HSSI©.

For over 25 years CargoMax has set the standard for shipboard stability and load management software. This proven system combines rigorously tested algorithms with a user friendly interface to provide quick and consistent results; ensuring the safety of the vessel is maintained at all times.

CargoMax is used worldwide by the industry's highest quality operators for Crude Oil Tankers, Product Tankers, LPG / LNG Tankers, Chemical Tankers, RoRo, General Cargo, Container Vessels, Passenger Vessels, Barge / ATB / ITB, Drill Ships, FPSO / FSO, Research Vessels and Military Vessels.

After installation of the enclosed disc on your computer you will have full operational capability of the included CMax demo software for a generic VLCC. It is recommended that before starting the following sample exercises that you work with the software and become familiar with the capabilities of the program.

We will begin by looking at what the default screen shows. In the upper left, the plan and profile window displays the locations of all the included tanks and weights that are used for calculations. In most cases only the active deadweight entry tanks will be displayed. The upper right window shows the strength summary for the current condition. The window shows the shear force and bending moments along the length of the vessel at predetermined points, mainly tank boundaries. Finally the bottom window is the deadweight entry table. This is where you will be doing all of your data entry, to determine the stability condition of the vessel.

We will start off by going to the File drop down menu and then "Open Standard Loading Condition" and select "homogeneous departure.lc2". This is a sample load case for the VLCC with a maximum (98%) departure load of a homogenous cargo, see Figure 7-8. First off, depending upon your vessel and location, either Metric or US units can be used for results. Right click in the "Ull/Snd" column and you can "Change Ull/Snd Units" to any format the user chooses. "m" (meter) will be used for all further references. Individual units can be changed in order to accommodate the user's needs. By right clicking in the "Temp", "Ull/Snd", and "Volume" columns we can choose between metric and US units such as Celsius/Fahrenheit, meter/feet, and m3/barrels.

There are a few basic operations that need to be understood in order to fully utilize the ship stability program. Since CargoMax (CMax) is in use aboard many different types of vessels, it needs to also handle and calculate various cargoes, fuels, and liquids as well. All liquids have defined parameters; two of the most important include API gravity (density), and temperature. These parameters are usually given by either the loading terminal or the manufacturer of the product. In this practice we will say that the homogenous cargo that is loaded is Alaskan North Slope Crude Oil, ANSCO, with an API gravity of 31.5° at a temperature of 13.1° C. In order to define the cargo in the deadweight entry table as ANSCO, we need to click the "Tools" menu and then "Cargo Library". Once the "Cargo Library" window is open select "Alaskan Crude". Click on the "Method" box and choose "API Table 6A" from the drop down menu and then update the API to 31.5°. Once these are input, click "OK" and "YES" to save changes. Now that CMax has the needed info for calculations, we need to identify the cargo as ANSCO by double clicking any cell under the "Cargo" column in the deadweight window. Using the drop down menu select "Alaskan Crude", and notice how the weights change. As stated before, temperature is also crucial in calculations, not for weight but for volume. Now click on any cell under the "Temp" column and enter "13.1° C". (Enter it all the way down) Notice how the "GSV" volume change. Once these two parameters are entered we can begin manipulating the deadweight table for our needs. Ullages can now be input and calculated. The correct way to input ullage is to type "3.15U" for a 3 meter 15 centimeter ullage measurement. If a sounding is to be used instead, then an "S" is typed after the numbers instead of a "U". Now that we have learned how to define cargoes and liquids, all of the deadweight entries can be specified for such products as HFO, MDO and even ballast water. If the product you are looking for is not listed in the cargo library click on the "add new" button and name the product and define its specifications.

It is important to understand that not all cargo operations, and vessels, operate in pure salt water environments. Sometimes vessels will operate in rivers and harbors that might be fresh or brackish water. CMax needs to be told what that environment is in order to properly display the calculated results. To accomplish this, click on the "Tools" menu and then "Options". In the "Calculations" tab from the "Options" menu, the "Seawater Specific Gravity" can be entered. For this example the water at the terminal has been tested, and is found to be slightly brackish at 1.018, this is where you would input that salinity.

Now that the cargo has been defined as ANSCO and the vessel is in brackish water, we can look at what CMax can tell us. Looking at the results bar on the extreme right of the screen, we can see that the calculated drafts for this loaded condition are; Fwd: 22.318 meters, Mid: 22.550 meters, and Aft 22.761 meters. By studying these drafts the user can see that the vessel is in a sagged condition. Also of importance in the results bar is the "Deadweight" calculation. CMax has calculated that this particular load condition in brackish water is causing the vessel to be overloaded by 841 MT's. The "GMt", "Shear % Allow", and "BMom % Allow" are also very important results. The tanker mate does not want to load or discharge their vessel in any way that might endanger the safety of the ship. In this particular load case, it is apparent that both the shear and bending stresses are very low.

CMax can display these results in graphical and tabular views as well. By clicking on the "Strength" tab in the "Results Window", more detailed results can be viewed. As shown in Figure 7-9.

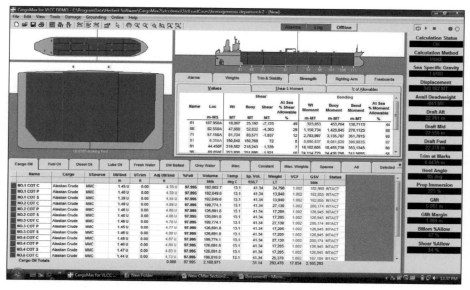

Figure 7-9. Courtesy of HSSI©.

Greater detail of the stability condition can be seen by choosing the "Righting Arm Summary" view, as seen in Figure 7-10.

The UTrim column is used to correct the volume in the tank from the observed ullage when the vessel is not on an even keel. By entering the trim of the vessel in this column the actual volume in the tank will be shown. Remember, that unless the vessel is on an even keel (has no trim) the actual volume of cargo in the tank will differ from what is stated in the ullage tables. By entering the trim in the UTrim column, the calculations in the subsequent columns are updated automatically to show corrected volumes, weights, and drafts. Some vessels may have CMax integrated with the cargo gauging system, which will automatically update the ullages and trim over a given time period. The prudent operator will always compare these values with the trim derived from the observed drafts, and it may be necessary to manually input the trim.

When a trim correction is manually inputted in the UTrim column new volumes and weights are calculated which may result in a change in draft and new trim. It may be necessary to renter the new trim for greater accuracy.

Perhaps the greatest advantage of using a ship stability program like CargoMax is the ability to manage emergencies, such as collisions, groundings and structural

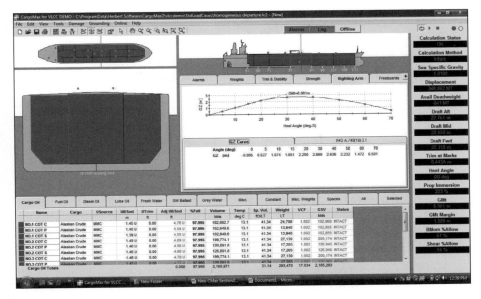

Figure 7-10. Courtesy of HSSI©.

failure. With the proper inputs of the current load condition CMax can predict what will happen if certain compartments are flooded. In this application you can chose which compartments are "damaged" or flooded by clicking the desired space in the "Deadweight Entry" window, and then choosing "Damage Selected Items" from the "Damage" drop down menu. For example, in this loaded condition, what would happen if the Fore Peak, #1 WBT Stbd, and #2 WBT Stbd were holed and flooding? After selecting these three compartments, view the results, as shown in Figure 7-11.

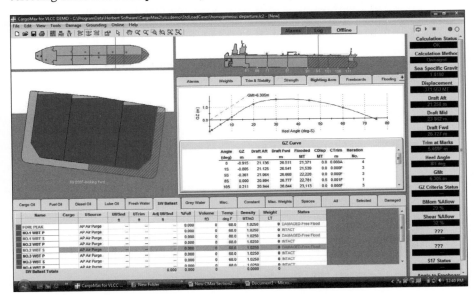

Figure 7-11. Courtesy of HSSI©.

This chapter should in no way be viewed as a comprehensive user guide. With the included "CargoMax Users Manual" file and practice, the tanker mate can become a skilled user.

In the following exercises it is important to remember that there are many different ways to load and discharge a vessel. In some of these exercises you will need to "add new" products to the cargo library. These examples are worked out using a specific approach. After completing the exercises, you should go back and try different methods using different numbers. The different aspects of tanker operations, such as stability and operational needs will become more understandable as you work with a vessel stability program.

For the new student (or people studying this subject for the first time) the answers provided in the following exercises were found without entering a trim correction in the "UTrim column". This was deliberately done to make it easier to understand how to use the software and solve the exercises. We encourage the reader who has experience with Cargo Loading software to try these calculations utilizing the UTrim column.

CargoMax Exercise #1

Your VLCC is preparing to load at an offshore terminal in Ras Tanura, Saudi Arabia. Begin by opening up a "New Loading Condition." This will give you an empty vessel to begin loading.

Before any ship sails it needs fuel, lubes, and potable water to get somewhere.

Load the following items:

All Fuel Tanks with HFO to 90% at a temperature of 62° C and an API gravity of 12.5°.
All Diesel Tanks with MDO to 80% full at 15.6° C and an API gravity of 34.0°.
All Lube Oil Tanks are all at 50% with the default characteristics.
All Fresh Water Tanks at 75% with the default characteristics.
All Misc Tanks at 30% with the default characteristics.

Finally, load the cargo tanks with SLCO (Arabian Light Crude Oil) with an API gravity of 33.5° at a temperature of 28.1° C, to the following ullages (U):

>   1X (1P, 1C, 1S): 4m 55cm
>   2X: 1m 55cm
>   3X: 6m 10cm
>   4X, 5X: 1m 55cm
>   Slops: 1m 85cm

After inputting the previous information, save the load case as "ex1.lc".

QUESTIONS:
1. What is the total observed volume of the load?
2. What are the vessel's calculated drafts in salt water at the draft marks?
3. What is the calculated deadweight of the vessel?
4. Does the vessel have sufficient transverse GM to be stable?

5. What is the bridge visibility, and does this meet the IMO requirement?
6. What would be the vessel's calculated drafts at the draft marks if in fresh water?
7. Is this load condition acceptable to sail with in both Salt and Fresh water voyages? If not, why?

ANSWERS:
1. 2,037,365 bbls
2. Fwd: 19.605m   Mid: 20.949m   Aft: 22.170m
3. 282,573 MT
4. Yes, with a GM margin of 3.596m.
5. 253.631m, yes with a margin of 246.369m.
6. Fwd: 20.172m   Mid: 21.414m   Aft: 22.543m
7. Yes

CargoMax Exercise #2

The vessel has now departed Ras Tanura in the loaded condition, saved in the previous exercise. You will eventually be headed to Long Beach, CA, to off load your cargo. However due to your loaded drafts you can not enter the harbor. The plan is to first lighter off of Mexico near the Cortes Bank. During the voyage the main engines will be initially supplied by the #2 HFO Port and Starboard tanks until they are empty and then evenly from the #1 HFO Port and Stbd tanks for the remainder.

The vessel will average 14.5 kts for the entire voyage. The average fuel burn per day for the vessel will be 1,050 bbls for the 13,815 nautical mile voyage. After figuring the fuel burn off, save the load case as "ex2.lc".

QUESTION:
What will be the vessel's arrival drafts at the lightering area? (ignore time/date changes)

ANSWER:
1. 14.5 kts × 24 hrs = 348 nm per day
   13,815 nm ÷ 348 nm/day = 39.7 days
   1,050 bbls/day × 39.7 days = 41,685 bbls used

   #2 HFO P: 18,342.8 bbls Empty
   #2 HFO S: 15,218.0 bbls Empty
   #2 HFO total fuel used = 33,560.8 bbls

   41,685 bbls − 33,560.8 bbls = 8,124.2 bbls
   8,124.2 bbls ÷ 2 = 4,062.1 bbls (in order to maintain even list, equal
                                      amounts taken from each tank)

   #1 HFO P: 7,360.2 bbls - 4,062.1 bbls = 3,298.1 bbls
   #1 HFO S: 5,209.6 bbls - 4,062.1 bbls = 1,147.5 bbls

   Arrival Drafts:
       Fwd: 20.198m   Mid: 20.577m   Aft: 20.921m

CargoMax Exercise #3

The VLCC needs to discharge enough SLCO to the lightering vessel in order to be able to dock at the Pier T terminal which has a controlling depth of 17.7m. The terminal also has a minimum under keel clearance requirement of 1m.

The cargo plan calls for discharging only from 2×, 4× and the Slop tanks to the lightering vessel. During the lightering, the vessel will burn a total of 1,000 bbls of HFO from the #1 HFO Port and Stbd evenly, and 125 bbls of MDO from the storage tank. The planned discharge rate is to be an average of 52,500 bbls/hr.

After computing the stop ullages save the load case as "ex3.lc".

QUESTIONS:
1. What will be the finish ullages (U) of the discharged tanks in order to meet the draft restriction and maintain an even keel?
2. What is the TOV to be discharged to the lightering vessel?
3. How long will the discharge take?

ANSWERS: The following answer is one of many solutions to this problem.
1. 2C: 12.95m   2P&S: 12.94m   4X: 16.16m   Slops: Empty
2. 485,206 bbls will be discharged.
3. Total discharge time will be approximately 9 hours 15 minutes.

CargoMax Exercise #4

The vessel has now docked at the Pier T terminal to discharge the remaining cargo. While discharging its cargo, the VLCC will also be loading segregated ballast and fuel, and use 200 bbls of MDO.

QUESTION:
1. After loading HFO, with an API of 11.3° at a temperature of 46.1° C, into the #2 HFO tanks, you gauge out with the following ullages: 1.90m Port, and 1.35m Stbd. How much HFO was loaded, corrected for temperature?

ANSWER:
1. 33,357.6 bbls

We then load the following segregated ballast tanks (from the saltwater harbor at a temperature of 20.0°C) to the following soundings (S):

       Fore Peak: 15.25m
       #1 WBT P&S: 30.50m
       #2 WBT P&S: 30.00m
       #3 WBT P&S: 30.00m
       #4 WBT P&S: 30.00m
       #5 WBT P&S: 21.35m

After loading the segregated ballast tanks to the soundings listed above, the Captain asks to load minimal dirty ballast into tank 1 for the return leg of the voyage. The Captain has specific sailing requirements which must be maintained: Shear Stresses no greater than 60%, Bending Moments no greater than 75%, and a maximum trim of 2.60m by the stern.

QUESTION:
2. How much dirty ballast at 15.6° C has to be loaded, and where, in order to meet his sailing requirements of Shear Stresses no greater than 60%, Bending Moments no greater than 75%, and a maximum trim of 2.60m by the stern. Calculate to the nearest centimeter.

ANSWER:
2. Load dirty ballast into 3 Center to an ullage of 10.34m, which equals
    137,704.8 bbls, or 22,441 MT of dirty ballast.

*Don't forget that dirty ballast is loaded as a new cargo, therefore the tank has to have it
selected as such in the "Cargo" column.

After computing the dirty ballast needed save the load case as "ex4.lc"

REVIEW QUESTIONS

1. What information (derived from observation) is needed by the PIC to
    perform a cargo calculation? What is the standard temperature used by
    the American Petroleum Institute in the creation of the API tables?
2. What are calibration tables? What information is derived from the calibra-
    tion table of the vessel?
    How does trim of the vessel affect the accuracy of tank gauges? How is
    trim accounted for in a cargo calculation?
3. Describe the difference between ullage, innage, and water cuts
    (thievage).
4. What is the difference between an observed (gross) volume and a standard
    (net) volume of cargo measure? How does one convert from the observed
    volume in the tank to a standard volume?
5. What information is needed when entering the API tables to determine
    the volume correction factor? How is the standard volume converted to
    a weight, namely tons?
6. List the major considerations when planning the cargo load.
7. What is the freshwater allowance of a vessel? When loading a vessel in
    brackish water, how is the allowance determined? What is meant by the
    controlling zone when discussing loadlines?
8. Define the following terms:

Deadweight	TPI	LCG
Displacement	MT1	Trim arm
Light ship	LCB	Trimming moment

9. When planning a cargo load, what stresses should be taken into account?
    Explain the difference between hogging and sagging. Where is a shear
    force experienced on a tank vessel?
10. Explain how cargo transfer (loading/discharging) rates are determined.
    Why is it necessary for the PIC to frequently check the rates?

REVIEW PROBLEMS

The following cargo problems can be answered using the MT PETROLAB
reference tables found in Appendix B in the enclosed CD only. Unless otherwise
specified the vessel is assumed to be on an even keel. Refer to the formulas in
Chapters 6 and 7 to assist you in solving these problems.

PROBLEM:
  1. Using the sample ullage tables in Appendix B, determine how much cargo was loaded using the following gauges:

     #1 Center: 9'-10"
     #2 Center: 6'-01"
     #3 Center: 6'-03"
     #4 Center: 6'-03"
     #5 Center: 8'-07"

ANSWER:
     #1C = 172,743.3 bbls
     #2C = 197,225.6 bbls
     #3C = 196,654.1 bbls
     #4C = 196,835.9 bbls
     #5C = 179,032.9 bbls
     Total = 942,491.8 bbls

PROBLEM:
  2. The vessel is currently at a mean draft of 60'-00". If the vessel can only load to a mean draft of 60'-05" due to a minimum under keel clearance of 3'-00", how many additional barrels of crude oil can be loaded? The crude oil has a temperature of 53.5° F, and an API gravity of 32.8°. Use the appropriate tables in Appendix B.

ANSWER:
  Remaining cargo which can be loaded (in tons) = TPI × 5"
  Remaining cargo which can be loaded (in tons) = 552.6 LT/in × 5in
  Remaining cargo which can be loaded (in tons) = 2,763 LT

  GSV of cargo which can be loaded (in bbls) = 2,763 LT / 0.13446 (Long Tons per bbl. from API table 11)

  GSV of cargo which can be loaded (in bbls) = 20,548.9 bbls

  GOV of cargo which can be loaded (in bbls) = GSV (in bbls) / Volume Correction Factor (VCF) for temperature 53.5°F and API gravity 32.8°.

  GOV of cargo which can be loaded (in bbls) = 20,548.9 bbls / 1.0030

  GOV of cargo which can be loaded (in bbls) = 20,487.4

PROBLEM:
  3. For the following trim conditions, and using the enclosed sample ullage tables in Appendix B, determine how much cargo was loaded using the following gauges:

     #1 Center: 10'-06" 4 foot trim by the stern
     #2 Center: 7'-06" 3 foot trim by the stern

#3 Center: 93'-02" 4 foot trim by the stern
#1 Port: 6'-03"   5 foot trim by the stern
#5 Stb: 9'-07"   1 foot trim by the stern

ANSWER: (Interpolate for accuracy)
  #1C = 10'-06"
  Trim correction +1.513"   = corrected ullage   10'-07.513"   = 171,200.8 bbls

  #2C = 7'-06"
  Trim correction +0.194"   = corrected ullage   7'-06.194"   = 194,171.7 bbls

  #3C=93'-02"
  Trim correction -1.253"   = corrected ullage   93'-00.747"   = 11,730.9 bbls

  #1 Port: 6'-03"
  Trim correction +1.650"   = corrected ullage   6'-04.650"   = 100,574.2 bbls

  #5 Stb: 9'-07"
  Trim correction -0.208"   = corrected ullage   9'-06.792"   = 77,734.1 bbls

PROBLEM:
  4. Use the tables found in Appendix B You are gauging out a load of Number
      6 fuel which you have loaded into #5 wing tanks. The API gravity of the
      product is 14.6° and the average temperature in the tank is 131°F. The
      vessel is even keel when you observe the following ullages:

      #5 Port: 7'-03"   #5 Stb: 7'- 04"
  No water or OBQ is present in the tanks. What is the GSV loaded in tons?

ANSWER:
      #5P = 80,046.9 bbls
      #5S = 79,951.7 bbls
      Total = 159,998.6 bbls

  TOV =159,998.6 bbls
  GOV=159,998.6 bbls (No water present)

  GSV aboard = (GOV) × Volume correction factor (VCF) for temperature & API
  GSV aboard = 159,998.6 bbls × 0.9721 (Table 6B)
  GSV loaded = 155,534.6 bbls
  TCV loaded = 155,534.6 bbls (No water present)

  GSV loaded in tons = GSV loaded × conversion factor of long tons per barrel
  (Table 11)
  GSV loaded in tons = 155,534.6 bbls × 0.15123
  GSV loaded = 23,521.5 LT

PROBLEM:
  5. Upon arrival at the load port you determine the vessel has 1,048.9 bbls.

OBQ. You load the vessel with a crude oil at a temperature of 68.5° F and an API gravity of 35.0°. At the completion of loading you gauge the tanks and observe the following ullages. . Referencing the tables found in Appendix B, determine in barrels the TOV aboard, TOV loaded, GSV aboard, GSV loaded, TCV aboard, TCV loaded, and the GSV loaded in tons.

#1 Center: 7'-01"   1 Port: 10'-02"   #1 Stbd: 9'-11"
#2 Center: 7'-06"
#3 Center: 6'-09"
#4 Center: 6'-11"
#5 Center: 10'-07"   #5 Port: 8'-05"   #5 Stbd: 8'-05"

While gauging you find water in the following tanks at the indicated ullages.
#1 Port: 96'-10"
#4 Center: 96'-01"
#5 Port: 96'-10"

ANSWER:

#1C = 178,094.0 bbls		
#1P = 96,207.9 bbls		126.2 bbls water
#1S = 96,447.4 bbls		
#2C = 194,206.1 bbls		
#3C = 195,589.8 bbls		
#4C = 195,417.1 bbls		5,482.6 bbls water
#5C = 174,987.7 bbls		
#5P = 78,886.9 bbls		219.1 bbls water
#5S = 78,876.9 bbls		
Total = 1,288,713.8 bbls		5,827.9 bbls water

TOV aboard =1,288,713.8 bbls
TOV loaded = TOV aboard - OBQ (1,288,713.8 – 1,048.9)
TOV loaded = 1,287,664.9 bbls

GSV aboard = (TOV aboard – water) × VCF for temperature and API
GSV aboard = (1,288,713.8 – 5,827.9) × 0.9960
GSV aboard = 1,277,754.4 bbls

GSV loaded = (TOV loaded – water) × VCF for temperature and API
GSV loaded = (1,287,664.9 – 5,827.9) × 0.9960
GSV loaded = 1,276,709.7 bbls

TCV aboard = GSV aboard + free water
TCV aboard = 1,277,754.4 + 5,827.9
TCV aboard = 1,283,582.3 bbls

TCV loaded = GSV loaded + free water
TCV loaded = 1,276,709.7 + 5,827.9
TCV loaded = 1,282,537.6 bbls

GSV loaded in tons = GSV loaded × conversion factor of long tons per barrel (Table 11)
GSV loaded in tons = 1,276,709.7 × 0.13268
GSV loaded = 169,393.8 LT

PROBLEM:
6. Using the tables found in Appendix B you are scheduled to discharge 1,320,000 bbls of crude oil at Louisiana Offshore Oil Port (LOOP). After you commence discharge operations and have reached the agreed manifold pressure you observe the following ullages at 2000 on March 15.

#1 Center: 8'-01"	#1 Port: 7'-01"	#1 Stbd: 7'-2"
#2 Center: 8'-00"		
#3 Center: 6'-10"		
#4 Center: 6'-8"		
#5 Center: 7'-00"	#5 Port: 7'-05"	#5 Stbd: 7'-05"

At 2100 you take another round of ullages and get the following readings:

#1 Center: 9'-9"	#1 Port: 10'-01"	#1 Stbd: 10'-00"
#2 Center: 9'-06"		
#3 Center: 9'-05"		
#4 Center: 9'-03"		
#5 Center: 9'-00"	#5 Port: 9'-08"	#5 Stbd: 10'-00"

You plan to maintain the rate throughout the discharge. What is your estimated time of completion?

ANSWER:
2000 Totals

#1 Center: 176,148.4	#1 Port: 99,769.0	#1 Stbd: 99,623.8
#2 Center: 193,141.1		
#3 Center: 195,412.3		
#4 Center: 195,949.6		
#5 Center: 182,235.3	#5 Port: 79,881.6	#5 Stbd: 79,869.0

Total On Board at 2000: 1,302,030.1 bbls
Total Discharged: 17,969.9 bbls

2100 Totals:

#1 Center: 172,905.5	#1 Port: 96,304.0	#1 Stbd: 96,351.1
#2 Center: 189,946.2		
#3 Center: 189,910.6		
#4 Center: 190,447.2		
#5 Center: 178,190.1	#5 Port: 77,643.0	#5 Stbd: 77,303.1

Total On Board at 2100:	1,269,000.8 bbls
Total On Board at 2000:	1,302,030.1 bbls
Rate per hour:	33,029.3 bbls/hour

Total on Board at 2100   <u>1,269,000.8 bbls</u>
(divided by) Rate   33,029.3 bbls/hour

Equals hours remaining 38.4 hours

Estimated Time of Completion:

Time/Date of last Total on Board	2100	March 15th
+ Hours remaining	<u>38.4</u>	<u>hours</u>
Estimated Time/Date of Completion:	1124	March 17th

CHAPTER 8

Cargo Pumps

The discharge of cargo on a modern tank vessel is accomplished through the use of a variety of pumping systems. This chapter provides a review of cargo pump theory, operation, and troubleshooting. The person-in-charge of the discharge operation must have a thorough working knowledge of the operating principles and limitations of cargo pumps. Operators should consult the cargo transfer procedures manual on the vessel as well as the pump manufacturers manuals for specific guidance concerning proper operation of the cargo pumping equipment onboard.

The purpose of the cargo pump is to impart energy to the liquid in order to raise its level from the cargo tanks on the vessel to the shore tanks and, at the same time, to overcome friction and flow losses while negotiating the piping system. The type of cargo pumps installed on a particular vessel depends on a number of factors including the following:

1. Degree of segregation desired between cargoes
2. Vessel trade
3. Type of prime mover (drive unit)
4. Characteristics of the cargo (viscosity, specific gravity, corrosivity, and so forth)
5. Reliability and ease of maintenance
6. Pumping capacity (volume of liquid moved per unit of time)

The cargo pumps are located either aft of the cargo tanks in a formal compartment called the pumproom or at the lowest point (well or sump) in each tank. Figure 8-1 illustrates the typical location of the pumproom, which places it in close proximity to the engine room and takes advantage of the tendency of the vessel to be trimmed by the stern. Depending on the trade, some vessels are also equipped with a forward pumproom.

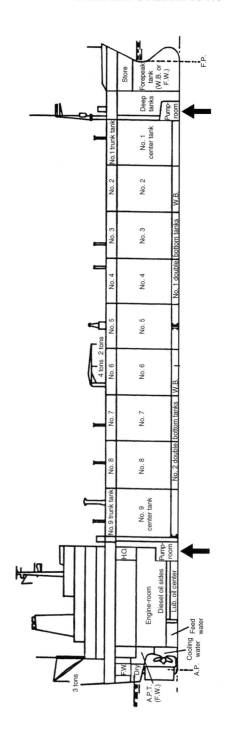

Figure 8-1. The pump room is typically located aft of the cargo tanks. In addition, some vessels are equipped with a forward pump room, as seen above. Copyright International Maritime Organization (IMO), London.

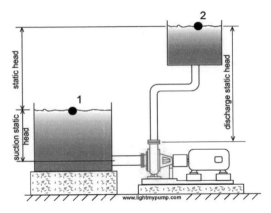

8-1a. Typical cargo pumping arrangement (head condition). Courtesy Lightmypump.com

Figure 8-2 illustrates a submerged pump located at the after end of each cargo tank in a well or sump.

Each of these approaches has distinct advantages; therefore, no one design is suited for all applications.

Pumps are classified as kinetic or positive-displacement. Kinetic pumps are divided into three groups: centrifugal, vertical-turbine (deepwell), and submerged. Types of positive-displacement pumps include reciprocating and rotary (lobe, gear, screw, and vane).

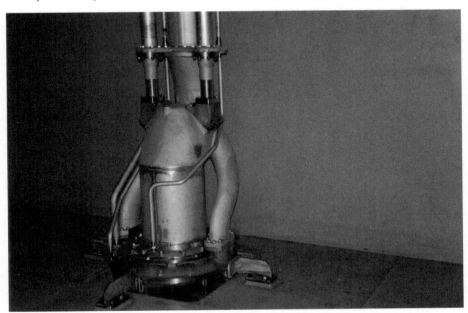

Figure 8-2. Submerged cargo pump located in a sump at the bottom of the cargo tank. Courtesy Scott R. Bergeron.

KINETIC PUMPS

The following section describes the various applications of centrifugal pumps in the cargo system of a tank vessel and explains their operation.

Centrifugal Pumps

Centrifugal pumps are often referred to as the main cargo pumps (MCP), given the fact that they discharge the bulk of the cargo from the vessel. These pumps have a high-volume output which makes them well suited as main cargo pumps. Centrifugal pumps are small in size considering their output and have few moving parts, which generally equates to greater reliability and less maintenance. Another advantage seen with this type of pump is its adaptability to a variety of drive units, leaving the owner with many options at the time of construction. Centrifugal pumps may be driven by a steam turbine, an electric motor, a hydraulic motor, or a diesel engine. On vessels with an after pumproom, the drive units are located in the engine room and are connected to the liquid end of the pump via a drive shaft extending through the bulkhead. A gas-tight gland must be employed in the bulkhead that separates the pumproom from the engine room to prevent the movement of flammable cargo vapors or liquids between the spaces. Where the cargo pumps are located in each tank, the drive unit is placed either on deck or immediately above the pump within the tank. The main drawback in the design of a centrifugal pump is the fact that it is not self-priming. Centrifugal pumps require a continual flow of liquid (prime) to pump, which usually makes them unsuitable for stripping (draining) a cargo tank. This type of pump does not have the ability to reprime itself when suction is lost unless it is outfitted with a special priming feature. To facilitate stripping the cargo tanks and pipelines, many vessels are equipped with either a positive-displacement pump or an eductor.

In a centrifugal pump, the impeller is the main rotating element that imparts energy by increasing the velocity of the liquid and delivering the cargo to the shore facility. Figure 8-3 is a cutaway view of a centrifugal pump showing the location of the impeller within the casing.

Cargo enters the eye of the impeller along the axis of rotation and is thrown outward radially through use of the swept-back vanes. At this point the cargo leaves the impeller vanes at a high velocity. Through the design of the volute, energy in the liquid is converted from high-velocity to a combination of velocity and discharge pressure. To assist in the conversion of energy, some centrifugal pumps are also equipped with diffuser vanes within the casing. The pump impellers operate at a high speed (rpm) capable of generating considerable heat if the pump is operated improperly. Centrifugal pumps require a continual flow of cargo for proper cooling, internal lubrication, and gland sealing. Operating this type of pump in a starved condition (insufficient prime) can result in overheating, leading to damage and the possibility of fire. In a normal discharge operation, these pumps have a smooth delivery and produce little noise or vibration.

Deepwell (Vertical-Turbine) Pumps

Deepwell (vertical-turbine) pumps are utilized on barges and ships carrying a wide array of liquid cargoes. They are typically of centrifugal design, and are located either in each cargo tank or at the after end of a bottom piping system. Figure 8-4 shows the location of the deepwell pumps in the bottom piping system of a barge.

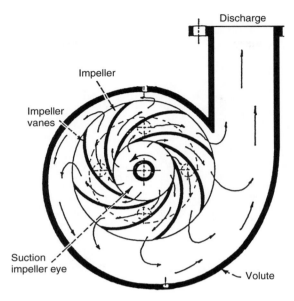

Figure 8-3. Simple view of the internals of a centrifugal pump showing the main rotating element (impeller). Courtesy Ingersoll Dresser Pump.

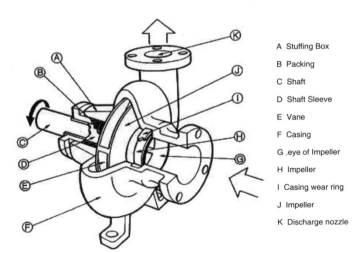

A Stuffing Box

B Packing

C Shaft

D Shaft Sleeve

E Vane

F Casing

G ,eye of Impeller

H Impeller

I Casing wear ring

J Impeller

K Discharge nozzle

Figure 8-3a. Centrifugal pump.
Courtesy Lightmypump.com

Figure 8-4. Deepwell (vertical turbine) pump, relief valve, check valve and discharge valve. Courtesy Stacy DeLoach

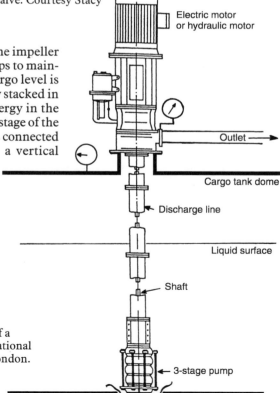

The unique placement of the impeller at the low point in the tanks helps to maintain a useful prime when the cargo level is low. The impellers are generally stacked in such a way as to step up the energy in the liquid as it moves through each stage of the pump. The impeller assembly is connected to the drive unit on deck via a vertical shaft within the discharge pipe. This arrangement takes advantage of the speed and efficiency of a centrifugal pump while minimizing its disadvantages. Figure 8-5 is a cross-sectional view of a deepwell pump showing the main components.

Figure 8-5. Cross-sectional view of a deepwell pump. Copyright International Maritime Organization (IMO), London.

Figure 8-5a. Deepwell pump located in the after corner of the tank.
Courtesy Stacy DeLoach.

Figure 8-5b. Deepwell pump in top view.
Courtesy Stacy DeLoach.

Figure 8-5c. Deepwell cargo pump suction
in sump. Courtesy Stacy DeLoach

The impeller is either enclosed in a barrel (when it is used to pump more than one tank), or it is recessed in a sump at the bottom of each tank. Most deepwell pumps are equipped with an automatic priming feature which reduces the need for a separate stripping system. The advantages of a deepwell-equipped vessel include the following:

1. Enhanced segregation of the cargo
2. Elimination of the cargo pumproom
3. Simplified cargo system
4. Reduced need for bottom piping
5. Pump can be removed for service or replacement from the deck
6. Prime mover on deck is accessible for service
7. Reduced need for separate stripping pumps

Submerged Pumps

Submerged pumps usually refer to hydraulically driven centrifugal cargo pumps located in each tank (Figure 8-6). In modern construction, the shift toward hydraulically powered deck machinery and automated cargo systems has prompted owners to install submerged pumps on their vessels. Figure 8-7 shows a cross-sectional view of a Frank Mohn (FRAMO) submerged pump.

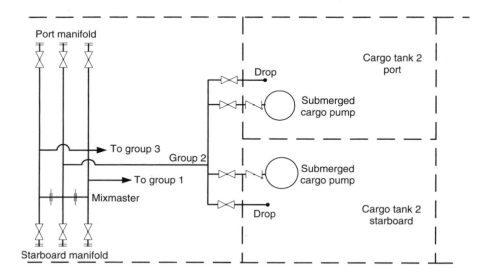

Figure 8-6. A typical piping arrangement on a vessel equipped with submerged pumps in the cargo tanks.

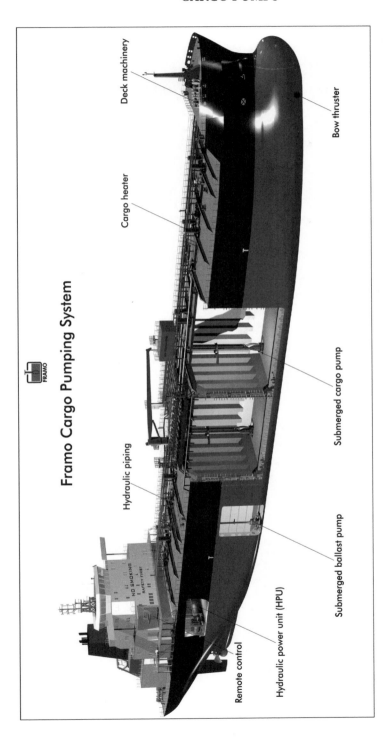

Figure 8-6a. Framo Cargo Pumping System. Courtesy Frank Mohn AS.

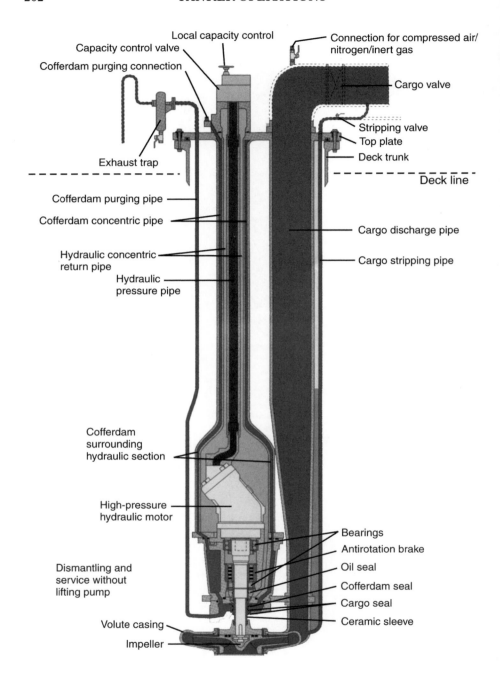

Figure 8-7. Cross-sectional view of a submerged pump showing the major components. Courtesy Frank Mohn AS.

Framo Cargo Pump

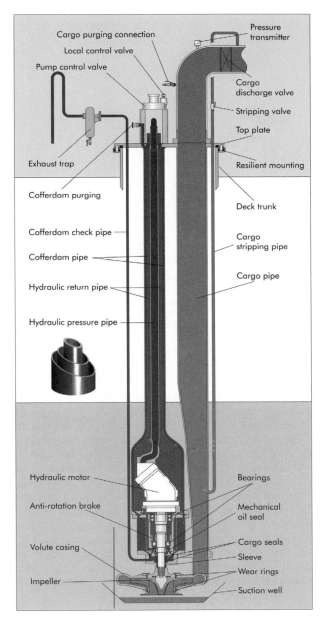

Figure 8-7a. Framo Submerged Cargo Pump - small. Courtesy Frank Mohn AS.

High-pressure hydraulic fluid is delivered to a motor mounted directly above the impeller in the tank. A special concentric pipe directs the high-pressure hydraulic fluid through the center pipe, and the low-pressure fluid returns through the middle line. The outermost pipe serves as a cofferdam which segregates the hydraulic fluid from the cargo. The cargo is delivered via the discharge line to the manifold on deck. When pumping is complete, the discharge line or pipestack (refer to Figure 8-8) can be stripped by idling the pump, closing the discharge valve, and pressurizing the top of the pipestack with air, nitrogen, or inert gas through a connection on deck.

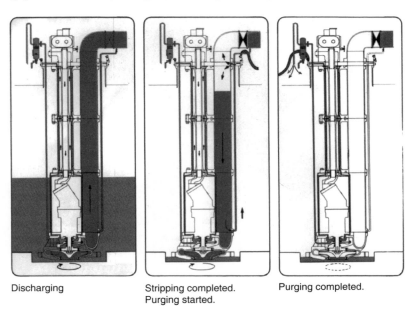

| Discharging | Stripping completed. Purging started. | Purging completed. |

Figure 8-8. The stages of operation of a submerged pump. Courtesy Frank Mohn AS.

The cargo in the discharge pipe is forced through a small-diameter stripping (riser) line directly ashore, thereby minimizing the quantity remaining on board (ROB) at the end of the discharge. Efficient stripping is particularly important when discharging small parcels of high-value cargo. The submerged pump has a single-stage impeller mounted horizontally in a sump at the bottom of the pump assembly. There is no bottom piping in this pumping system; therefore each tank must be equipped with its own pump. The pump speed is controlled through the use of a hydraulic governor on deck or from a cargo control room. The drive shaft has an antirotational brake which allows the discharge piping to also serve as the loading drop. Submerged pumps possess many of the same advantages as the deepwell pump; however, the pump and motor assembly must be repaired in the tank. In the event the pump is inoperative with cargo remaining in the tank, a portable submersible pump is provided to finish the cargo discharge.

PUMP OPERATION

Cargo pumps typically operate in either a head or a lift condition. The optimum pumping situation is the head condition in which the cargo level is some physical height above the inlet of the pump. When operated in the head condition, cargo fills the casing of the pump by gravity, thereby maintaining an adequate prime. On inerted vessels, the positive deck pressure required to be maintained above the cargo in the tanks further assists in keeping the pump primed, particularly when approaching the stripping stage in the discharge. Figure 8-9 illustrates a typical gravity or static suction head condition.

In the lift condition, the cargo level is some physical distance below the inlet of the pump. In a conventional pipeline vessel, the pumps begin to experience this condition when the cargo level falls below the bottom piping in the tank (Figure 8-10). This makes it necessary for the pump to perform work in an effort to draw the cargo up to the inlet.

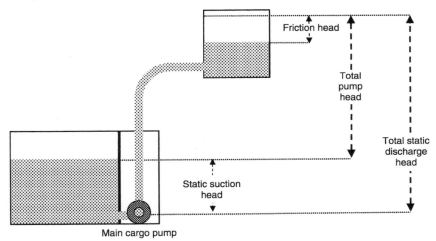

Figure 8-9. Cargo pump operating in a head condition in which gravity maintains a continuous prime. Courtesy Richard Beadon.

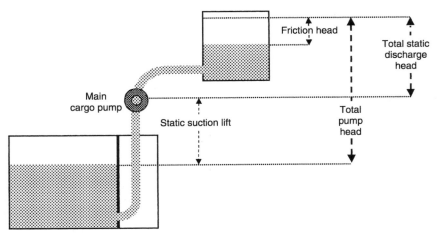

Figure 8-10. Cargo pump operating in a lift condition in which the liquid level is below the inlet of the pump. Courtesy Richard Beadon.

CAVITATION

A centrifugal pump does not operate efficiently in the lift condition and may begin to exhibit signs of cavitation. Cavitation occurs when the pressure in the suction line falls below the vapor pressure of the cargo, resulting in the formation of vapor pockets in the liquid stream; these pockets eventually reach the inlet of the pump. This is frequently referred to as "gassing" up the pump. When these pockets of vapor collapse on the impeller vanes, they create undue noise and vibration in the pump. In the short term, cavitation results in lower pump efficiency as evidenced by a poor discharge rate. In the worst case it can cause physical damage, actually eroding metal surfaces and possibly contributing to bearing or fatigue failure of the pump, resulting in a shorter service life.

Cavitation can be caused by a number of factors including the following:

1. Vortexing action around the bellmouth in the cargo tank. This typically occurs with a low cargo level in the tank. The cargo begins to form an eddy near the bellmouth which can admit IG and cargo vapors (atmosphere) into the suction line of the pump.
2. Cargo characteristics that may contribute to the formation of vapor pockets in the suction piping leading to the pump. For example, high vapor pressure cargoes such as naphtha are susceptible to pumping problems due to vaporization of the liquid in the suction line.
3. High temperature of the cargo, which increases its tendency to vaporize.
4. Leaks or holes in the suction piping allowing air into the line.

The operator of a centrifugal pump must be mindful of the cause and effect of cavitation, particularly when approaching the stripping stage in the discharge of a cargo tank. To minimize the effects of cavitation, it is common practice to partially open a cargo tank with ample gravity head (prime tank) to maintain an adequate flow to the pump. The operator must decide when it is unreasonable to continue discharging the cargo tank with the centrifugal pump or risk losing suction. It is usually necessary to switch over to the stripping pump or eductor to drain the remaining liquid (wedge) in the tank. The point where one must shift to the next tank in the discharge sequence will vary with different pumping systems and operator experience. For example, while discharging a single-hull vessel with a conventional bottom piping system and after pumproom, the operator must carefully monitor the performance of the MCP when the cargo level is getting close to the tank bottom. Even with a prime tank partially opened, the MCP will cavitate or lose suction entirely if a sufficient "slug" of atmosphere from the tank reaches the pump. Consider how different the situation is on a double-hull tank vessel with no bottom piping and submerged pumps located in a sump in each tank. As the cargo level approaches the bottom of the tank, the pump continues to receive an adequate prime until the only cargo remaining in the tank is in the sump. Another benefit of the double-hull vessel is the absence of framing in the bottom of the cargo tank; the smooth inner bottom does not impede the flow of liquid to the pump.

PRIME

Certain telltale signs indicate that a centrifugal pump does not have an adequate prime. The earliest indications that a pump is starved of adequate liquid are slight fluctuations in the pump gauges (tachometer and discharge pressure). If the operating conditions are not changed, the gauge fluctuations become more erratic and the pump generates noise and vibration. Unless a satisfactory flow to the pump is restored, it will lose suction, overspeed, and trip out. To resume pumping it will be necessary to open a tank with sufficient gravity head to reprime the pump.

An operator can employ various techniques to reduce the chance of the main cargo pump losing suction when no prime tank is available.

When a pump first exhibits signs of losing suction, it is common practice to reduce the pump speed (rpm) and partially close (throttle) a discharge valve on the pump. Throttling the discharge valve reduces the velocity of the liquid through the pump, smooths out the flow, and maintains a reasonable delivery. Another recommendation is to shorten the suction pipe by ending the discharge operation close to the pump, thereby minimizing friction and flow losses. Finally, the operator should ensure that the suction line to the pump is isolated by closing block valves and crossovers, thereby minimizing potential air leaks in the system.

CROSSOVERS

To expedite the discharge of the vessel, it is sometimes necessary to cross over cargo pumps. When two centrifugal pumps are drawing from the same suction line, there is a possibility that one pump will receive a greater share of the flow than the other. This inequity is typically caused by the differing lengths of suction piping feeding each pump. Operating the pumps at different speeds can also result in an unequal sharing of the liquid flow. In either case, the operator should keep a close watch on each pump. The distant pump (the one with the longest run of piping) is typically the first to exhibit signs of overheating. Overheating is one indication that a pump does not have adequate flow through the casing for the speed at which it is operating. If the cargo tank level has reached the point where there is insufficient gravity head to properly support the operation of both pumps, the operator should shut down the distant pump and finish with one pump. Figure 8-11 illustrates three pumps connected on the suction side through a crossover.

In the situation where two centrifugal pumps are delivering into the same discharge line, there is a risk of unequal sharing of the load. The pump with the greater delivery tends to put a back pressure on the distant pump discharge. The net effect of this back pressure is equivalent to throttling a discharge valve, resulting in inadequate movement of liquid through the pump and potential overheating. As in the previous case, the net result of this inequity is overheating of the distant pump. Pump manufacturers generally recommend that both pumps be operated at the same speed and closely observed for any difference in temperature.

SHORE BOOSTER PUMP

A shore booster pump is employed when the discharge head at the terminal is too great for the vessel's cargo pump. This situation arises when the shore tanks are located at a considerable elevation above or distance from the vessel. The shore booster pump operates in series with the vessel's cargo pump, adding energy to the system and enabling the cargo delivery to proceed at a reasonable rate. Prior

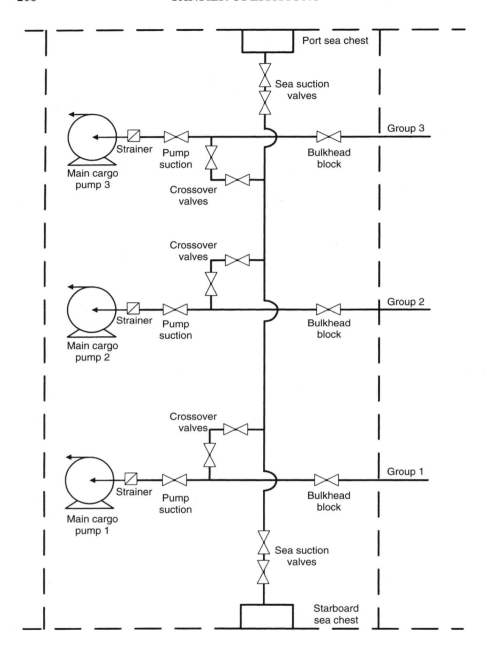

Figure 8-11. Three cargo pumps interconnected on the suction side through a cross-over (top view).

to starting the booster, the vessel cargo pumps typically exhibit a high discharge pressure and a poor delivery rate (velocity). During the pretransfer conference, agreement should be reached concerning advance notification of the booster pump being brought on line. It is common practice for the vessel PIC to stand by the pump controls when the booster is started. The sudden drop in the load on the vessel's cargo pump could cause it to overspeed and trip out. To avoid unnecessary shutdowns or delays, it is essential that good communication exist between the PIC of the vessel and the facility. Once the shore booster pump is operating, there will be a drop in the discharge pressure on the vessel and a corresponding increase in the pumping rate.

AUTOMATIC PRIMING SYSTEMS

The PIC of the cargo discharge typically employs a variety of operational techniques such as those previously discussed to maintain a prime on the centrifugal pump. If performed incorrectly, expensive physical damage or a potential fire hazard can result.

With these concerns in mind, manufacturers developed special automatic priming features for centrifugal pumps to ensure that a minimal flow is always maintained. With certain pumps, the autoprime system virtually eliminates the need for stripping pumps. Figure 8-12 illustrates one type of autopriming system found on cargo pumps in marine service.

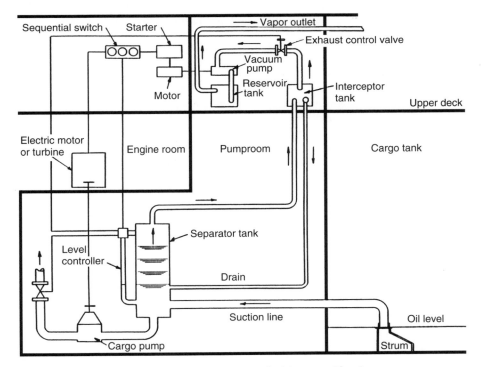

Figure 8-12. One type of autopriming system used with a centrifugal pump.

When the pump begins to lose suction, a venturi activates a recirculation line which returns liquid to the strainer/recirculation tank on the suction side of the pump. The pump is primed by the liquid in the recirculation tank, and the cargo recirculates until a normal pumping condition is restored. At the same time, gases in the recirculation tank and pump casing are vented through an extraction line to the discharge side of the pump. These devices make it possible to discharge the cargo tanks more completely without damaging the pump. They also protect the pump from damage resulting from operator error such as closing a valve on the suction side of an operating pump. There are many designs of self-priming pumps on the market; therefore the operator should consult the manufacturer's manual for a detailed description of the type and method of operation of the one found on the vessel.

Troubleshooting

In any pumping system the problems that typically arise are related to lack of suction or lack of delivery. The operator must carefully monitor cargo pump performance to detect either condition and take corrective action. Cargo pump manuals frequently include a troubleshooting section to assist the operator. If no such guide exists, the operator must locate the source of the problem from experience with the system. The most common source of pumping problems is improper alignment of the valves in the piping system, so this is the simplest place to start. The operator should check the status of the valves on the vessel for correctness, then contact the shore facility and have them verify their alignment. Other less-obvious causes of pumping problems include the following possibilities:

SUCTION SIDE

1. *Air or vapor bound pump.* This condition generally occurs at start-up due to improper priming of the pump. Centrifugal pumps must be initially primed by opening a vent at the top of the casing to bleed off any gas trapped in the pump. The casing and impeller must be filled with liquid at start-up or the pump will not operate.
2. *Clogged strainer.* When a foreign obstruction becomes lodged in the strainer, it restricts the flow of liquid to the pump, which results in a loss of suction. It is then necessary to clean the strainer by closing the appropriate valves to isolate it, draining the cargo from the unit, and removing the access cover. The strainer is generally composed of cylindrically shaped mesh designed to prevent foreign objects in the liquid flow from damaging the internals of the pump.
3. *Cavitation.* This results from air leaks or vaporization of the cargo in the suction line of the pump. Refer to the earlier discussion of cavitation in this chapter.
4. *Insufficient gravity head.* The pump ceases to operate due to a low cargo level in the tank.
5. *Cargo characteristics.* The viscosity or specific gravity of the cargo may be greater than the pump is designed to handle. Pumping problems can also occur when the temperature of a heated cargo is permitted to cool to its pour point. In the worst case scenario, the cargo actually begins to solidify in the tanks.
6. *Mechanical failure.* Mechanical problems including a worn or damaged pump casing, an excessively worn impeller, bearing failure, leaking glands/seals, or a bad drive unit may cause discharge problems.

DISCHARGE SIDE

1. *Excessive discharge pressure.* This may occur at a terminal with an excessive discharge head. It is a common problem when the vessel pumps must deliver the cargo a considerable distance or vertical height to the shore tank. Check for valve malfunctions in the discharge line such as a sheared stem or dropped gate resulting in the pump operating against a closed valve.

2. *Recirculation.* When the drop (loading) valves are left open or an automatic valve in the discharge line is frozen in the open position, recirculation occurs. If the pump is equipped with an automatic priming system, the recirculation valve may be frozen in the open position, permitting the return of cargo to the suction side of the pump. Where fitted, cargo pump relief valves should also be checked to ensure they are in the normally closed position.

3. *Mechanical failure.* A worn or damaged pump, an excessively worn impeller, bearing failure, leaking glands/seals, or a bad drive unit may cause problems.

PUMP CONTROLS AND GAUGES

Most cargo pumps in marine service have variable speed control. This allows the operator to alter the speed of the pump based upon changing operating conditions. The pump speed is changed either by local control of the drive unit or through use of a controller from a remote location. For example, the speed of a deepwell pump on a barge is typically adjusted by changing the speed of the drive unit on deck. On tank vessels equipped with an automated cargo control system, the pump speed is remotely adjusted using an air controller, rheostat, or hydraulic

Figure 8-12a. Manifold pressure gauge. Courtesy Stacy DeLoach.

controller. The performance of the cargo pump is monitored through use of the following gauges:

1. Tachometer (pump rpm)
2. Discharge pressure (discharge side of the pump)
3. Compound gauge (displaying vacuum and pressure on the suction line)
4. Manifold pressure (pressure at the vessel's rail) (Figure 8-12a)

The accuracy of these gauge readouts is vital to an operator when making decisions concerning the pumping operation. The gauges and controls are physically located at the pump or in a convenient central location such as the top of the pumproom or cargo control room. The emergency shutdown control for the cargo pumps is generally located at the midpoint of the vessel in the vicinity of the manifold. All personnel involved in the cargo operation should be aware of the location and operation of the emergency shutdown for the pumps.

POSITIVE-DISPLACEMENT PUMPS AND EDUCTORS

This section describes the different types of positive-displacement pumps as well as the use of eductors to drain the tanks and lines at the end of a cargo discharge.

Reciprocating Pump

One type of positive-displacement pump typically used in stripping service on modern tank vessels is the reciprocating pump. The design generally seen consists of two pistons (duplex pump) that move back and forth in cylinders powered either by steam or air. The flow of liquid to and from the pump is accomplished through spring-loaded valves. On the intake stroke, the movement of the piston creates a vacuum, thus drawing the cargo into the cylinder through an intake valve. Figure 8-13 is a simple view of a reciprocating pump.

During the discharge stroke, the piston forces the cargo out of the cylinder through the discharge valve, creating a delivery pressure on the outlet side of the pump. Additional intake and discharge valves are usually installed to make these pumps double-acting. Simply stated, a double-acting pump is one in which pumping action occurs simultaneously on both sides of the piston for each stroke of the pump. The net result is a smoother delivery and increased pumping capacity. Reciprocating pumps are self-priming, meaning they have the ability to regain suction when suction is lost. These pumps are capable of moving the atmosphere (vapors and inert gas or air) which may be drawn from the cargo tank with the liquid during the final stages of stripping. Unlike centrifugal pumps, they can develop the vacuum necessary to draw liquid into the intake when operating in a lift condition. The main disadvantage of reciprocating pumps is that they are painfully slow. For example, a reciprocating pump on a typical coastal tank vessel generally has a capacity of 800 to 1,000 barrels per hour (bph). One concern when operating a positive-displacement pump is that the discharge line is open. A positive-displacement pump will develop a considerable discharge pressure if the pipeline is blocked or a valve is closed against the pump.

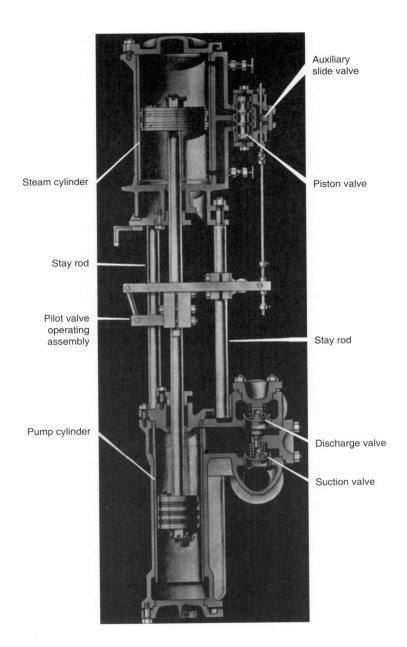

Figure 8-13. Simple view of a reciprocating pump. From *Principles of Naval Engineering*.

Rotary Pump

Another type of positive-displacement pump is the rotary pump, which operates by physically trapping the cargo and carrying it from the inlet to the outlet. This is usually accomplished through the rotation of intermeshing gears, lobes, screws, or vanes. Figures 8-14 through 8-17 illustrate different types of rotary pumps.

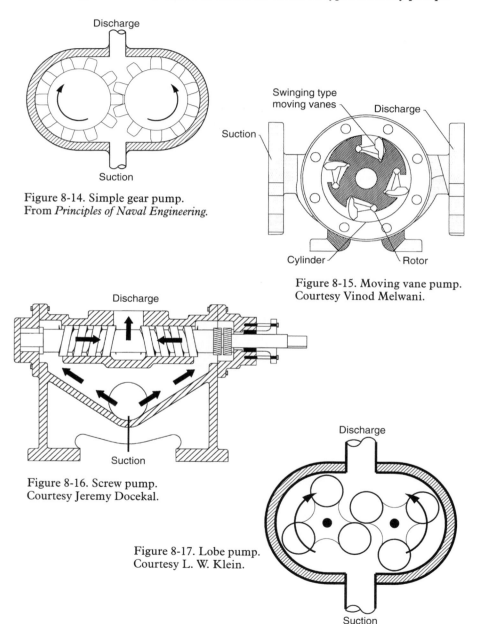

Figure 8-14. Simple gear pump.
From *Principles of Naval Engineering.*

Figure 8-15. Moving vane pump.
Courtesy Vinod Melwani.

Figure 8-16. Screw pump.
Courtesy Jeremy Docekal.

Figure 8-17. Lobe pump.
Courtesy L. W. Klein.

The capacity of these pumps is limited by the volume of the space between the casing and rotating elements. Rotary pumps are self-priming and generally have a better discharge capacity than a reciprocating pump. They depend on liquid flow for lubrication; therefore it is not recommended that they be allowed to operate without liquid for an extended period of time. In marine service they are used primarily as stripping pumps or in situations where a lift condition is encountered, such as with a forward pumproom. This type of pump is also well suited for vessels transporting high-viscosity cargoes such as asphalt.

Eductors

Eductors are typically utilized for stripping on vessels transporting a dedicated cargo. On a crude carrier, for example, eductors are commonly used for stripping during cargo discharge and crude-oil-washing operations. An eductor is a jet-type pump which requires a driving fluid to be delivered to a nozzle to create a low pressure on the suction line. The eductor is typically powered by the cargo taken from the discharge side of one or more of the main cargo pumps or a general purpose pump. Figure 8-18 illustrates the internals of an eductor.

The discharge from the eductor is either sent directly ashore through the discharge main or, more commonly, returned to a designated slop tank. Another application for eductors is in the segregated-ballast system. The eductor is driven

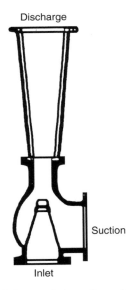

Figure 8-18. Internal view of an eductor.

with seawater from the segregated-ballast pump and used to strip the ballast tanks. There are several advantages associated with the use of eductors in a stripping system:

1. They have no moving parts, which generally translates to less maintenance and greater reliability.
2. Eductors can be upgraded in size to far surpass the stripping capacity of a positive-displacement pump.
3. There is no risk of overheating or inadequate lubrication when the eductor loses suction.

The single disadvantage of an eductor is the fact that a driving fluid (cargo in this case) must be supplied for continued operation. Therefore, an eductor is of little value at the bitter end of the discharge when it is necessary to strip the last cargo tank and drop and strip the pipelines.

REVIEW

1. What type of pump is typically employed for the main cargo pumps on modern tank vessels?
2. What types of pumps are typically employed for stripping service on tank vessels?
3. Define the following terms: gravity head, static suction lift, pump head, discharge head.
4. What is the main rotating element in a centrifugal pump called? In what area of the pump is energy converted from high velocity to a combination of velocity and discharge pressure?
5. During a cargo discharge, what are the telltale signs that a centrifugal pump is being starved of adequate liquid?
6. Define cavitation. What are the short- and long-term effects of cavitating a pump?
7. How does an operator prevent loss of suction in the main cargo pump when the cargo level is low in the tank? (Describe the use of a prime tank.)
8. What techniques are employed to prevent loss of suction in the main cargo pump when discharging the last tank (finishing the cargo discharge)?
9. What is meant by the term "self-priming" when applied to a cargo pump?
10. What are the advantages of stripping with an eductor? When is an eductor of little value in the stripping of the vessel?
11. The person-in-charge must carefully monitor the performance of the cargo pumps when they are crossed over on the suction or discharge side. Why?
12. When is a shore booster pump employed in tandem with the vessel's cargo pumps? What concerns does the shore booster pump create for the operator of the vessel? Describe the performance of the vessel's cargo pumps before and after the booster pump is started.

13. List the possible causes for the main cargo (centrifugal) pump to lose suction during a discharge operation.
14. List the gauges that assist the operator in monitoring the performance of the cargo pumps. Where is the emergency shutdown for the vessel cargo pumps located?
15. What are the primary differences between a deepwell (vertical-turbine) pump and submerged pump? What are the advantages of deepwell/submerged pumping systems over a conventional pumproom?

CHALLENGE QUESTIONS

16. During an apparently normal discharge operation, the discharge pressure on one of the cargo pumps increases abruptly. List the possible causes for this condition.
17. During an apparently normal discharge operation, one of the main cargo pumps suddenly overspeeds and trips out. Describe the possible causes for this condition.
18. Describe the process of replacing a vertical turbine pump in a cargo tank.
19. Describe the process of dropping and stripping the cargo pipelines in various types of pumping systems.
20. What is considered an acceptable ROB in various tanker trades? How is a vessel experience factor determined?

CHAPTER 9

Cargo Transfer Operations

The transfer of cargo between a vessel and the shore facility or between two vessels requires careful planning and execution by the persons-in-charge. The goal of this operation is the safe and efficient transfer of cargo. With this in mind, the PICs should conduct a thorough inspection of the vessel and facility before assuming the cargo watch to minimize the possibility of cargo contamination, spills, fires, and explosions. This chapter addresses the precautions that should be followed by a vessel PIC whenever a cargo transfer is undertaken, and it focuses on the critical points of the loading and discharging operations. Preparation for the cargo transfer begins prior to the vessel's arrival at the berth. It involves the development of a cargo plan outlining such things as the identity of the cargoes, tank layout, sequence of cargo transfer, cargo quantity and calculations, and other details of the upcoming operation (Figure 9-1).

REGULATIONS

The following items, derived from federal regulation as well as company manuals, must be checked prior to commencing a cargo transfer and at the change of watch, before each successive PIC assumes responsibility for the cargo operation. Appendix D contains a copy of the International Ship/Shore Checklist from the International Safety Guide for Oil Tankers and Terminals (ISGOTT).

Title 33 CFR Part 156.120

1. *Vessel moorings:* Moorings shall be strong enough to hold the vessel during all expected conditions of surge, current, and weather. The moorings must be of sufficient length to allow adjustment for changes in draft, drift, and tide during a cargo transfer (Figure 9-2).

MT PETROLAB
January, 2010

VOYAGE # 1

PORT   VALDEZ, AK
BERTH # 4

CARGO
Load approximately 895,000 net bbls ANSCO via 1 – 4 headers
In-tank x-over, deck x-over and both drops ALL OPEN

Start in 4W @ 10,000 BPH and confirm flow. Do a complete check of deck and pumproom. When all is confirmed okay on vessel and shore increase rate slowly doing a good check between rate increases to a maximum of 100,000 BPH. Continue to make rounds of the deck and pumproom. Hold 4W at 8m for last and top off remaining tanks as per diagram in the following order: Slops, 1, 6&5, 2&3, 4. Correct all stops for trim and stagger tanks to advantage for topping off.

Do not exceed 4.0m of trim or 20,000 BPH per tank valve.
Do a comparison with the hermetics prior to topping off.
Check freeboard towards the end of the load and when finished.
Correct list with 4W.

When there is approximately 3 hours of loading remaining:
check salinity alongside, average API and cargo temperature.
Check drafts before topping off 4W and after completion of
loading operation.
Drop arms 4-3-2-1
Finish with IG @ approximately 250 mmwg
Estimated sailing drafts:
F   15.46m
A   16.32m

SEGREGATED BALLAST
Discharge all ballast (2-5, 1, 6) with SBP # 1 and #2 after the boom boats are clear (106 RPM).
Watch stress and adjust cargo accordingly.
When finished with SBP, line up the eductor and strip all ballast tanks.
If time permits, re-strip all J tanks.
Keep accurate entries in Deck Log and Ballast Log.
Call me when discharging 6 J with main pumps.

NOTE:

OBSERVE ALL RULES AND REGULATIONS
MAINTAIN VESSEL SECURITY
MAKE FREQUENT CHECKS OF SURROUNDING WATERS
CHECK PUMPROOM HOURLY
IF IN DOUBT, SHUTDOWN AND CALL ME......ANYTIME

SIGN BELOW ONLY IF READ AND UNDERSTOOD

C/M _____
2/M _____
3/M _____
3/M _____
Cadet _____

	P	S
1	9.00	9.00
2	2.50	2.50
3	2.50	2.50
4	Hold at 8m Last approx 6.35	Hold at 8m Last approx 6.35
5	1.60	1.60
6	1.60	1.60
Slop	2.00	2.00

		MT	
FP			
1		MT	MT
2		MT	MT
3		MT	MT
4		MT	MT
5		MT	MT
6		MT	MT
Wing	MT		MT
Trim	MT		MT

Figure 9-1. Sample Loading Orders.

Figure 9-2. Vessel moorings must be adequate and properly tended.

Figure 9-3. Cargo hoses must have sufficient length and be properly supported to prevent undue strain on the couplings.

2. *Cargo hoses and loading arms, length:* Cargo hoses and loading arms must be long enough to allow the vessel to move to the limits of its moorings without placing strain on the hose, loading arm, or transfer piping system (Figure 9-3).

3. *Cargo hose support:* Each hose must be properly supported to prevent kinking or other damage (such as a bight of the hose getting pinched between the vessel and the dock). The hose should be properly supported to prevent undue strain on its couplings (Figure 9-4).

Figure 9-3a. Steel loading arms. Courtesy Mark Huber.

Figure 9-4. A hose rack used at a terminal.

4. *Transfer system alignment:* The piping system must be properly aligned to permit the scheduled flow of oil or hazardous material. Prior to lining up the piping system, it is recommended that all valves be manually or visually checked to ensure they are initially closed. To minimize the chance of human error, it is standard practice for the PIC to line up the cargo system for the scheduled transfer and then have another individual check it for correctness. On automated vessels, it is also common practice to have personnel on deck verify the opening and closing of valves. With each successive change of the cargo watch, the PIC should check the current status of all cargo system valves. Additionally, the status of the valves in the vent, vapor control and inert gas system should be checked to ensure proper tank venting or inert gas delivery is being conducted based on the operation being undertaken.

5. *Transfer system, unused components:* Any part of the transfer piping not needed for the operation is to be securely blanked or shut off.

6. *Cargo hoses or loading arms, not in use:* The end of each cargo hose or loading arm not connected for the transfer of oil or hazardous material must be blanked-off using proper closure devices such as butterfly valves, wafer-type resilient seated valves, or blank flanges.

7. *Transfer system, fixed piping:* The cargo hose or loading arm is to be connected to fixed piping on the vessel as well as at the facility.

8. *Overboard discharge/sea suction valves:* Each overboard discharge or sea suction valve that is connected to the vessel's cargo piping or tank system must be sealed or lashed in the closed position.

9. *Cargo hoses, condition:* Each cargo hose used in the transfer must be visually inspected for any unrepaired loose covers, kinks, bulges, soft spots, or any other defect that would permit the discharge of oil or hazardous cargo through the hose material. Further, there should be no gouges, cuts, or slashes that penetrate the

first layer of hose reinforcement. The reinforcement layer refers to the strength members in the construction of the hose—fabric, cord, or metal.

10. *Cargo hoses, pressure rating and labeling:* Each cargo hose or loading arm must meet the bursting and working pressure ratings contained in Title 33 CFR Part 154.500. Each cargo hose must have the following information (minimum) marked on the jacket: the name of each product for which the hose may be used or the words "OIL SERVICE." For hazardous materials, the hose must be marked with the words "HAZMAT SERVICE—SEE LIST," followed by a letter, number, or symbol that corresponds to a list or chart with the necessary information in the vessel's transfer procedures manual. In addition to the identity of the products that may be transferred through the hose, the maximum allowable working pressure must be indicated. See chapter 4 for details concerning cargo hose and loading arm inspections and ratings.

11. *Cargo connections:* The cargo hose and loading arm connections must have suitable gasket material to ensure a leak-free seal. When standard flanges are employed, a bolt must be installed at least in every other hole and in no case should there be fewer than four bolts for temporary connections. If the flanges are nonstandard or the connection is permanent, a bolt must be used in every hole of the flange. The details concerning cargo transfer connections can be found in Title 33 CFR Part 156.130.

12. *Monitoring devices:* These devices may be installed at a shore facility to significantly limit the amount of a discharge of oil or hazardous material during a transfer. These devices, when required by local authority, must be installed properly and maintained in proper working order.

13. *Discharge containment equipment:* Each shore facility is required to have ready access to sufficient containment material (i.e., a boom) and equipment to contain any oil or hazardous material discharged onto the water during transfer operations.

Figure 9-4a. Fixed containment under the cargo manifold. Courtesy Stacy DeLoach

14. *Discharge containment area:* This refers to the fixed container (trough) or enclosed deck area located under the vessel manifold (refer to Figure 9-4a) intended to collect any spillage during connection and disconnection of cargo hoses or loading arms. The containment area must have a means of being drained periodically so as to provide the required capacity. The required capacity of the fixed containment (trough) area is dependent on the diameter of the cargo hoses or loading arms connected to the manifold. The details concerning the capacity of the containment area for a vessel can be found in Title 33 CFR Part 155.310. As of this writing, all oil tankers and offshore oil barges with a capacity of 250 barrels or more must have peripheral coamings (athwartships as well as fore-and-aft) that completely enclose the cargo deck area, cargo hatches, manifolds, transfer connections, and other openings where cargo may overflow or leak. The coamings must be at least four inches high except in the after corners on the port and starboard sides of the vessel, where they must be at least eight inches high.

15. *Scuppers or drains:* Means shall be provided to mechanically close each weather-deck drain or scupper in the container or enclosed deck area so that in the event of a tank overfill, ruptured hose, or pipeline leak, the cargo is contained on deck (Figure 9-5). Accumulation of rain water on deck can be problematic therefore it should be periodically stripped to the slop tank. In the event of an oil spill on deck some vessels are equipped with emergency dump valves or covers to drain the spilled oil into a slop tank or empty ballast tank.

16. *Communication system/language fluency:* Two-way voice communications must be provided between the persons-in-charge of the transfer operations. The communications equipment must be suitable for the area of operation, such as the use of intrinsically safe portable radios when transferring flammable or combustible liquids.

Figure 9-5. All-weather deck drains must be plugged during a cargo transfer.

At least one person at the site of the transfer operation must be capable of fluently speaking the language or languages spoken by the PIC on the vessel and the PIC at the facility. Company policy generally prohibits the use of cell phones, pagers, I pods, palm pilots and any other portable electronic devices on deck that are NOT classified as intrinsically safe or approved for use in potentially hazardous environments. Vessel personnel are advised to be vigilant of any visitors such as contractors, vendors or other non crewmembers that may be unaware of these safety requirements.

17. *Emergency shutdown:* Both the vessel and the shore facility must have a means of stopping the flow of oil or hazardous cargo in the event of an emergency during a transfer operation. For vessels, the method may involve a pump control, quick-acting power-actuated valve, or an operating procedure. The emergency shutdown must be operable from the cargo deck, cargo control room, or the usual operating station for the person-in-charge of the cargo transfer on the vessel (Figure 9-6).

The facility must provide an emergency means to enable the PIC of the vessel to stop the flow of oil or hazardous cargo during a transfer. The method of shutting down may be electrical, pneumatic, mechanical linkage to the shore facility, or through electronic voice communication.

For oil transfers, the facility must stop the flow to the vessel within the following limits:

Sixty seconds for facilities that commenced operation on or before November 1, 1980.

Thirty seconds for facilities that commenced operation after November 1, 1980.

Figure 9-6. Emergency Cargo Pump Shutdown. Courtesy Mark Huber.

For hazardous material transfers the flow to the vessel must stop within these limits:

Sixty seconds for a facility that commenced operation before October 4, 1990.
Thirty seconds for a facility that commenced operation on or after October 4, 1990.

18. *Person-in-charge, at site:* There must be a designated person-in-charge on the vessel and at the shore facility. The PICs are required to be at the site of the transfer operation and immediately available to transfer personnel.

19. *Transfer procedures manual:* Each vessel conducting the transfer of oil or hazardous cargo must have a transfer procedures manual. The cargo transfer must be conducted in accordance with the guidelines in the transfer procedures manual.

20. *Sufficient personnel:* The personnel required to conduct the operation as outlined in the transfer procedures manual must be on duty.

21. *Pretransfer conference:* The person-in-charge of the vessel and the person-in-charge of the facility must hold a conference to ensure that each individual understands the following:

Identity of the product to be transferred
Quantity to be transferred
Sequence of the transfer operation
Transfer rate
Name or title and location of each person participating in the transfer
Details of the transferring and receiving systems
Critical stages in the transfer operation
Federal, state, and local rules that apply to the transfer of oil or hazardous cargo
Emergency procedures
Discharge containment procedures
Discharge reporting procedures
Watch or shift arrangements
Transfer shutdown procedures

In addition, proper radio communications must be established.

22. *Agreement to begin transfer:* No cargo transfer is to commence unless the person-in-charge of the vessel and the person-in-charge of the facility agree.

23. *Lighting:* Vessels conducting transfer operations between sunset and sunrise must have deck lighting that adequately illuminates the transfer operations work area and each transfer connection point in use on the vessel.

Title 46 CFR Part 35.35-20

1. *Warning signals displayed:* Tank vessels transferring cargo at a dock must display a red flag (bravo) by day and a red electric lantern at night where visible from all sides (all-round light). When transferring cargo at anchor, only the red flag shall be displayed. A warning sign shall be posted at the gangway or point of access to the vessel with the following wording based on the type of cargo being transferred (Figure 9-7):

If the vessel is transferring a benzene regulated cargo, a warning sign must be posted at each access to the regulated area (Figure 9-7a).

Figure 9-7. Warning Sign. Courtesy Kelly Curtin.

2. *Repair work authorization:* No repair work in way of the cargo spaces may be carried out without permission.

3. *Cargo connections:* (See number 11 above or refer to Title 33 CFR Part 156.130 for details of cargo connections.) Cargo connections must be made to the vessel's piping system and not through an open end of hose led through a hatch. When the cargo connections are supported by the vessel's tackle, the person-in-charge shall determine the weight involved and ensure that adequate support is utilized. Portable drip pans shall be placed under all cargo hose connections where no fixed containment is employed. All cargo valves shall be set for the scheduled transfer and verified by the person-in-charge.

4. *Fires or open flames:* When loading grades A, B, or C cargoes, there shall be no fires or open flames present on deck or in any compartment which is located on, facing, or open and adjacent to that part of the deck on which cargo connections have been made.

5. *Boiler and galley fire safety:* An inspection shall be made to determine that boiler and galley fires can be maintained with reasonable safety.

6. *Safe smoking areas:* Smoking is prohibited on the weather decks of tank vessels when they are not gas free or when they are situated alongside a shore facility or platform. Prior to loading, the person-in-charge shall conduct an inspection to determine whether smoking may be permitted with reasonable safety in areas other than the weather deck of the vessel.

7. *Shore readiness:* The shore terminal or other tank vessel concerned must report its readiness to transfer cargo.

8. *Sea valves:* All sea valves connected to the cargo piping system must be closed.

9. *Inert gas system:* The inert gas system must be operated as necessary to maintain an inert atmosphere in the cargo tanks as required by regulation. Vessels required to operate in the inert condition must verify that all such cargo tanks have an oxygen content of 8 percent or less by volume and a positive pressure (see Title 46 CFR Part 32.53-5 and IMO inert gas requirements contained in SOLAS and TSPP (Figures 9-7b and 9-7c).

Figure 9-7a. Warning sign posted at the access point to the benzene-regulated area on the vessel.

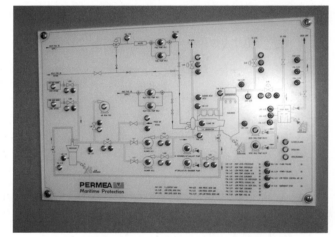

Figure 9-7b. IGG status panel. Courtesy Mark Huber.

Figure 9-7c. IGG status panel. Courtesy George Edenfield.

10. *Vapor control operation:* When conducting a closed loading operation that involves directing the cargo vapors to the shore facility from the vessel's tanks, the person-in-charge shall complete the required additional entries to the declaration of inspection shown in Chapter 5.

The Declaration of Inspection (DOI) is a checklist that contains the items that must be inspected by the PIC. This form (Figure 9-8) must be signed by the PIC prior to taking charge of the cargo transfer on a tank vessel.

Declaration of Inspection - Ship/Shore Safety Checklist; Prior to Transfer of Liquid Cargo in Bulk

Vessel Name	Terminal/Dock	Date and Time of Transfer
Port:	Voyage No:	Type of Transfer: [ ] Vessel to Facility [ ] Facility to Vessel (Check Box) [ ] Vessel to Vessel [ ] Intra Vessel

Cargo Identification

Oil Type	Sulfur	API	Temperature

The following list refers to requirements set forth in 33CFR 155, 120, 150, 156, 158, 160 and 46CFR 35.35-30, U.S. Coast Guard and company policies and regulations. The spaces adjacent to items on the list are provided to indicate that the detailed requirements have been met.

		Deliverer	Receiver
1. Communication system/language fluency	(156.120 (q) (v))		
2. Warning signs and red warning symbols	(35.35-30)		
3. Vessel moorings	(156.120 (a))		
4. Transfer system alignment	(156.120 (d))		
5. Transfer system; unused components	(156.120 (e) (d))		
6. Transfer system; fixed piping	(156.120 (g))		
7. Overboard discharges/sea suction valves	(156.120 (h))		
8. Hoses or loading arms condition	(156.120 (l)) (156.170)		
9. Hoses; length and support/equipment	(156.120 (b) (c) (j))		
10. Connections	(156.130)		
11. Discharge containment system	(156.120 (m)(n))		
12. Scuppers or drains	(156.120 (o))		
13. Emergency shutdown	(156.120 (r))		
14. Repair work authorization	(35.35-30)		
15. Boiler and galley fires safety	(35.35-30)		
16. Fires or open flame	(35.35-30)		
17. Lighting (sunset to sunrise)	(156.120 (y) (z))		
18. Safe smoking spaces	(35.35-30)		
19. Spill and emergency shutdown procedures	(156.120 (w))		
20. Sufficient personnel/designated person in charge at all times	(156.120 (s) (t) (u))		
21. Transfer conference	(156.120 (w))		
22. Inert gas system in operation. (indicate N/A if vessel barge not equipped)			
23. Fire fighting equipment is properly laid out and readily available.			
24. Emergency towing wires rigged, fore and aft. (indicate N/A if vessel/barge not equipped)			
25. Doors and ports are properly fitted and in good condition.			
26. All flames screens properly fitted and in good condition.			
27. Ventilators suitably trimmed with regard to prevailing wind condition			
28. Main Transmitting aerials switched off. (indicate N/A if barge)			
29. Electric cables to portable equipment are disconnected from power.			
30. Window type air conditioning units are disconnected.			
31. No unauthorized craft alongside.			
32. Pumproom bilges are dry and gas free. (indicate N/A if no pumproom).			
33. Agreement to begin transfer.	(156.120 (x))		

I do certify that I have personally inspected this facility or vessel with reference to the requirements set forth in Section 35.35-30 and company policies and regulations and that opposite each of them I have indicated that the regulations have been complied with.

Person in charge of Receiving Unit	Title	Time and Date
Person in charge of Delivering Unit	Title	Time and Date

Time transfer completed

Retain on board for 30 days

Figure 9-8. Declaration of Inspection. A form similar to this one must be signed by the PIC before taking charge of the deck. (see Appendix D on enclosed disc for full form)

Company Recommendations

In addition to the aforementioned regulatory requirements, companies frequently develop standing orders for the cargo and ballast operation on the vessel (Figure 9-8a) and include a number of safety and operational items from their fleet manual in the pretransfer checklist:

1. *Cargo information:* The PIC should be provided with the most current material safety data sheets (MSDS) available, or their equivalent, for the cargoes that are about to be transferred. During the cargo transfer, this information should be displayed in a conspicuous location on the vessel for easy reference by personnel (Figure 9-9).

MT PETROLAB

Standing Orders for Cargo Operations

The ABB VDU's aboard this vessel are an excellent tool for operating and monitoring the pumps and valves. It must not be relied on, however, to show the correct lineup for certain systems. The information available on individual screens is limited and therefore an intimate understanding of the actual piping is required prior to assuming control.

All applicable USCG, State, Local and specific terminal rules must be followed. As an officer aboard this vessel, you are required to have a knowledge and understanding of the Petrol Quality Manuals. These manuals have been written to reflect safety policy and fulfill regulatory requirements and must take precedence at all times. If a discrepancy is observed between these manuals and the cargo orders, the Chief Mate must be notified immediately.

Prior to going on watch you will be expected to make a thorough round of the deck, Pumproom, and Inert Gas Generator room. This round should be recorded in the Deck Logbook as you assume the watch. The DOI must also be signed before the assumption of watch. Once this is done, provided the monitoring systems are functioning correctly, hourly rounds of the Pumproom are not required.

Emphasis should be placed aboard this class of vessel on opening valves slowly and in stages when beginning the flow of oil or water. High pressures will be present even on initial startup and large pressure differentials should be expected when changing pump speeds. Stepping the discharge valve open slowly will help reduce the hydraulic shock both to our pipelines and to the shore's. When washing tanks, the deck crew should also be reminded to open COW valves slowly for the same reason.

Communication is critical throughout all phases of loading or discharging cargo, bunkers and ballast. Keep the deck watch informed so that they may better anticipate operational needs. Remember that the deck watch is responsible for checking the lines, gangway, surrounding water and all deck pipelines in addition to helping with tank sounding as required. CargoMax should be used to project ahead and estimate the effect of all loading or discharging operations prior to starting. Pay particular attention to the stress on the vessel that these actions will cause.

While on cargo/ballast/bunker watch in the CCR, the computer shall only be used for cargo purposes. No personal use of the computer shall be permitted by the Mate on watch at any time.

Never exceed 95% "At Sea" stresses or 5.0m of trim.

Maintain a minimum of 1 meter Under Keel Clearance (UKC)

Keep a record of equipment that is damaged or not functioning properly during your watch. Be mindful of your surroundings, things may sound/look/smell different before problems become readily apparent.

Do not allow valves to continually cycle and damage the solenoids. If you have any valve problems or questions, notify the Chief Mate immediately.

Strictly follow guidelines in the CCR Cargo and Ballast Operations binder should Opacity become an issue.

The deck watch should monitor the position of the vessel relative to the berth and have knowledge of the strength and direction of the current before tending mooring lines. Do not allow the vessel to come away from the dock at any time. If you need extra help tending lines inform the Chief Mate. Your permission should be sought by the watch prior to leaving the deck for a break.

Nothing on this sheet or in the Cargo Orders relieves the mate on watch of her/his duty to understand and be aware of the entire operation. If a question or situation arises that you are unsure of, ask the Chief Mate or Master.

Always remember, *if you are in doubt – Shut Down!* Starting back up will be painless and easy compared to explaining why it was not done.

Oil spill response "Qualified Individual"   #-###-###-####
USCG NRC   #-###-###-####

Chief Mate_____

Figure 9-8a. Standing Orders for Cargo Operations.

CARGO INFORMATION CARD

Enter
Cargo Name _____ **MTBE (METHYL TERTIARY BUTYL ETHER)**

and
Description _____ **FLAMMABLE LIQUID, NOS, CORROSIVE MATERIAL**

CARGO TRANSFER - Use authorized personnel only, properly protected.

HAZARDS

FIRE — Extremely flammable. Ignited by heat, sparks or open flame. NO SMOKING OR OPEN LIGHTS.

EXPOSURE — Breathing vapor in confined areas may cause asphyxiation.

IN CASE OF ACCIDENT

IF THIS HAPPENS

For assistance, phone CHEMTREC
toll free, day or night
800-424-9300
MARINE RADIOTELEPHONES
CALL 202-483-7616

DO THIS

SPILL or LEAK — Shut off all ignition sources. Keep people away. Keep upwind. Shut off leak if without risk. Wear self-contained breathing apparatus. Use water spray to "knock down" vapor. Flush area with water spray. Run-off to waterway creates fire hazard (floats on water); notify fire, health and pollution control agencies.

FIRE — On small fire use dry chemical or carbon dioxide. On large fire use water spray or appropriate foam. Cool exposed tanks with water.

EXPOSURE — Remove to fresh air. If not breathing, apply artificial respiration, oxygen. If breathing is difficult, administer oxygen. Call physician.

Figure 9-9. The cargo information card is usually posted in a conspicuous location on board.

Figure 9-10. Fire-fighting gear is readied prior to a cargo transfer.

2. *Fire-fighting gear and Fire Control Plan:* All fixed fire-fighting gear should be laid out and ready for immediate use in the event of an emergency (Figure 9-10). Vessel personnel should lead out fire hoses in the cargo tank area and position the foam monitors in the ready position. Portable fire extinguishers should be placed in the vicinity of the cargo manifold. The vessel fire control plan should be placed in a sealed container in a location where it is readily available such as at the gangway and cargo control room.

3. *Emergency towing wires* (also referred to as "fire wires"): Towing wires should be rigged fore and aft, with the eye hanging just above the water's edge on the offshore side of the vessel.

4. *Bonding cable/insulating flange:* Although the use of a bonding cable between the vessel and shore facility is discouraged by the maritime community (see *International Safety Guide for Oil Tankers and Terminals*), local rules may still mandate proper bonding of the vessel. Bonding is one method of reducing the possible occurrence of an electrical arc during the connection and disconnection of cargo hoses or loading arms with the vessel. A potential fire risk is possible should there be flammable cargo vapors present in the immediate vicinity of the open pipelines during this operation. The purpose of the bonding cable is to provide an electrical path to earth, thereby preventing the creation of an ignition source locally at the cargo manifold. Where bonding is still required by the shore facility, the person-in-charge should ensure that it is the first item connected to the vessel upon docking and the last item removed from the vessel after completion of the cargo transfer, disconnection of the cargo hoses or arms, and cleanup of any spillage. According to the *International Safety Guide for Oil Tankers and Terminals,* a more effective method of preventing such an occurrence is the use of an insulating flange in the connections between the vessel and shore facility. The insulating flange is typically

installed on the shore end of the hose string or loading arm. It is a permanent fitting that employs a special gasket material, bolt sleeves, and washers to eliminate any metal-to-metal contact across the flange faces. The use of an insulating flange minimizes the chance of an appreciable electrical charge occurring when connecting and disconnecting the cargo hoses or loading arms (Figure 9-11).

   5. *No unauthorized craft alongside:* No vessel may come alongside or remain alongside the cargo area of a tank vessel during loading or discharging of grade A, B, or C cargoes unless granted permission by the person-in-charge.

   6. *Flame screens:* All cargo tank openings must have proper flame screen protection (Figure 9-12). The PIC should check that all flame screens are in good condition (no holes or fouling).

Figure 9-11. Spool-type insulating flange is a one-piece double-sided flange machined from cast nylon. Courtesy Apollo International Corp.

Figure 9-12. Ullage opening with flame screen in position on a tank.

7. *Safety matches:* The use of lighters or matches other than safety matches is forbidden aboard tank vessels at any time.

8. *Doors and ports:* During a cargo transfer or any operation involving possible accumulation of cargo vapors, all doors and ports to the noncargo areas of the vessel (house) must be secured.

9. *Pumproom ventilation and bilge alarm:* On vessels equipped with cargo pumprooms, the PIC shall ensure that power ventilation is operating whenever it is necessary for personnel to enter the space. The pumproom bilge alarm must be tested prior to cargo transfer, and the bilge in the pumproom should be stripped dry.

10. *Cofferdams and voids:* All voids adjacent to the cargo tanks shall be checked to ensure they are empty.

11. *Segregated-ballast system:* When vessels arrive in the ballasted condition, personnel should verify to the best of their ability that the ballast water is "clean" prior to discharge into the harbor. Simultaneous cargo and ballast operations should only be conducted after careful calculation of bending, shear, and stability conditions of the vessel taking free surface into account when operating with numerous tanks in a slack condition.

12. *Portable electric equipment:* Portable electric items must be disconnected from their power source.

13. *Containment boom:* The rigging of a containment boom around the vessel during cargo transfer operations may be required by local authority or some shore facilities (Figure 9-13). In the event of a discharge of cargo into the water, the containment boom limits the movement of the spill with the prevailing winds or currents. The proper use of containment booms not only protects adjoining shorelines and sensitive ecological areas but aids the response and cleanup effort.

14. *Spill response equipment:* Emergency spill response equipment on the vessel shall be readily accessible. The spill response locker typically includes such items as portable (Wilden) pumps, absorbent pads, scoops, buckets, rags, squeegees, mops, boots, gloves, aprons, and goggles.

15. *Draft or freeboard restrictions:* The PIC should confer with the facility concerning the available depth of water alongside the dock and any freeboard restrictions that may be imposed prior to the cargo transfer. If the depth of water at the dock is in question, it is advisable to take soundings around the vessel at the proper stage of the tide. An accurate determination of the depth alongside the dock is necessary to avoid the possibility of the vessel touching bottom during the final stages of loading. While discharging, the change in freeboard, coupled with the range of the tide, could result in undue strain being placed on the connections or mechanical loading arms. The PIC should take corrective action to prevent such an occurrence.

16. *Freshwater allowance (FWA):* When a vessel is being loaded to its applicable limiting load line, the PIC should check the density of the water alongside the vessel to determine any allowance that should be applied if floating in fresh or brackish water. Performing this check will avoid the possibility of overloading or loading the vessel short.

17. *Vessel security:* At no time should the gangway on a tanker be left unattended. Company policy normally requires that visitors have prior clearance to board the vessel, present photo identification such as the Transportation Worker Identification Card (TWIC), sign in at the gangway and have an escort while on-

Figure 9-13. Containment Boom Boat. Courtesy Mark Huber.

Figure 9-13a. Containment boom properly rigged. Courtesy Kevin Dushcenchuk.

board. In the United States, the Department of Homeland Security has established a tiered warning system to advise the maritime community and public of the level of risk to the maritime industry. In summary, Maritime Security (MARSEC as it is called) contains three warning levels starting with Level 1, at which minimum appropriate security measures shall be maintained. Based on a heightened risk of a transportation security incident the levels can be raised to Level 2 or 3, in which specific security measures are implemented when an incident is probable, imminent, or has occurred. At MARSEC level 3 numerous vessel operations will undoubtedly be impacted such as: access to the vessel and cargo operations. The MARSEC details can be found in the vessel and facility security plan and in 33 CFR Part 101.200.

OPERATIONS

After all pretransfer and safety checks have been completed, the cargo operation can begin. When the PICs of the facility and the vessel report their readiness to commence the cargo transfer, the manifold valves are opened. The start of liquid flow is a critical point in the cargo operation, requiring the personal supervision of the PIC. It is standard industry practice to start the cargo transfer at a reduced rate with personnel standing by at the manifold on deck to check the connections, hoses, arms, and piping for leaks. Personnel should make a round of the deck, visually checking the water alongside the vessel and the tank ullages to confirm that cargo is flowing in the correct direction. On vessels equipped with a pumproom, this space should be checked frequently for any leaks. The vessel cargo tanks must be checked for proper venting or inert gas delivery, as the operation warrants. At the start of the cargo transfer, the PIC should keep a close watch on pump, manifold, and deck pressure readings.

When the PIC is satisfied that the cargo transfer is proceeding smoothly, the loading or discharging rate can be increased to the agreed maximum. As the cargo transfer proceeds, it is recommended that vessel personnel make continuous rounds of the deck, checking such items as cargo and ballast tank levels, manifold connections, hoses/arms, valve alignment, pumproom, mooring lines, gangway, vessel traffic, weather, tidal current, and security. Some vessels maintain a status board displaying the current cargo and ballast tanks involved in the transfer with any stop gauges or final ullages noted on the board. The PIC should calculate the cargo transfer rate hourly (to ensure compliance with the agreed value) to determine the anticipated sailing time. In addition, the pump speed and manifold pressures should be recorded by the PIC. The established cargo transfer plan must be followed, and any last-minute change orders or deviations from the plan should be recalculated to avoid excessive trim, list, stress, or stability problems. A number of operations that also require the attention of the PIC at certain points in a cargo transfer include the following:

1. *Sampling:* Tank vessels are subject to sampling and lab analysis at various points in the transport of a cargo. For example, a vessel may be required to perform bottom sampling at the beginning of the loading operation. Bottom sampling is performed for quality assurance and to ensure the vessel was properly prepared to receive the cargo being loaded. The PIC should determine if bottom sampling is required for any of the cargoes during the pretransfer conference. The tanks that must be bottom sampled receive a partial load to a specified level, and then the

operation is shut down. The loading operation can only resume after the PIC receives clearance from the laboratory. Line samples are usually taken off the facility pipeline through which the vessel is being loaded. The PIC of the vessel may request a line sample be taken from the dock manifold, which is then labeled and sealed by a witness. The line sample is usually retained on the vessel for the duration of the voyage and may be consulted should any discrepancies arise concerning the quality of the cargo. The cargo tanks undergo a final sampling toward the end of the loading operation, and a lab analysis of the cargo is typically provided to the vessel prior to leaving the berth (Figure 9-13aa).

Figure 9-13aa. Cargo Sampling. Courtesy Kevin Duschenchuk.

2. *Heated Cargoes:* Tank vessels transport a variety of liquid cargoes that require the application of heat in transit and for pumpability at the discharge terminal. Cargoes such as residual fuel oil, certain crude oils, wax, molten sulphur, vacuum gas oil and asphalt just to name a few have considerably different heating requirements. Therefore when handling heated cargoes, the PIC should consult the operations department or cargo owner for guidance regarding the optimum temperature to be maintained onboard as well as the permissible temperature range (upper and lower limits) that should not be exceeded. It is equally important that the PIC be familiar with the operation of the heating coil system on the vessel and prior to loading, it is recommended that the heating coils in the tanks be checked and any leaks or defective coils repaired. During the loading operation, the temperature of the cargo arriving on the vessel from the shore facility should be determined, and if necessary, heat should be applied to the cargo tanks to attain the desired temperature. Vessel personnel should monitor the cargo temperatures during the loaded passage and make the necessary adjustments to the heating system. When discharging, the PIC should secure the heating coils prior to reaching the stripping stage in the tank to avoid overheating the cargo.

It should be noted that the empty ballast space surrounding the cargo tanks in modern double hull tank vessels provide an insulation factor (air space) from the sea making the job of maintaining cargo temperature easier than with the single hull vessels of the past. On the other hand, caution should be exercised when ballasting in extremely cold water (filling the double hull) when transporting a heated cargo. A thermal shock is possible when the incoming ballast water reaches the cargo tank boundary (inner bottom in this case) resulting in potential damage to coatings or contributing to stress points in the hull.

The type of heating system found on tank vessels usually fall into one of the following categories:

1. Steam coils
2. Thermal oil
3. Hot water

In a traditional steam heating system, low pressure (saturated) steam is supplied to a deck steam main from an auxiliary boiler. The deck steam main branches off to one or more manifolds on deck that contain the valves to control the steam supply to the coils in each cargo tank. The steam supply feeds the coils that are typically installed in a pattern across the bottom of the tank and the closed loop is completed by a return line that carries any condensate back to the steam return line on deck (Figure 9-13aaa).

A thermal oil system normally consists of one or more of the following components (Figure 9-13b):

1. Oil fired heater
2. Economizer
3. Circulating pump
4. Expansion tank
5. Dearator tank
6. Flow valve and control unit

Figure 9-13aaa. Heating coil equipped cargo tank. Image courtesy International Paint Ltd.

Figure 9-13b. Thermal Oil Heater. Courtesy Stacy DeLoach.

The thermal oil used in the system is generally a mineral or synthetic oil that is circulated by pumps through a heat exchanger (heater/economizer) to raise its temperature. The oil then circulates through a closed loop of stainless steel coils mounted in the tanks (Figure 9-13c).

The fluid remains in a liquid state throughout the heating process. Thermal oil heating systems are very efficient, non corrosive and can be used in a variety of vessel applications such as:

1. Cargo tanks/Slop tanks
2. Tank washing heater
3. Bunker tanks

Another option seen today is the installation of deck heaters using a thermal oil system (Figure 9-13d).

In this approach the cargo is circulated through a heat exchanger on deck via a cargo pump and returned to the tank through the drop line.

3. *Static accumulator:* When loading a cargo that is considered a poor conductor, the precautions outlined in the *International Safety Guide for Oil Tankers and Terminals* must be followed. Any cargo capable of accumulating an appreciable electrostatic charge during a loading operation requires special handling. The following precautions apply when loading a static-accumulating cargo into a noninerted tank:

A. *Initial loading rate:* The tanks should be initially loaded at a reduced rate to minimize turbulence and splashing as the cargo first enters the tank. The cargo transfer procedures manual on the vessel specifies the correct initial loading rate.
B. *Cushion the tank:* The reduced loading rate should be maintained until the bottom of the cargo tank is covered to a specified height with cargo.
C. *Maximum loading rate:* The cargo transfer procedures manual of the vessel should be consulted for the maximum allowable loading rate when handling a static-accumulating cargo.
D. *Relaxation period:* After the tank is topped off, a minimum relaxation period of thirty minutes is required before any equipment is introduced into the tank. After the thirty-minute period, any equipment introduced into the tank must be properly bonded to the hull.

4. *Segregation:* On multigrade carriers, proper segregation must be maintained between dissimilar cargoes. There are numerous methods of segregating cargoes; some common approaches include the use of double-block valves, spectacle blanks, and removable spool pieces.

5. *Shutdown:* There are many reasons for the PIC to terminate a cargo transfer. Some are mandated by regulation while others are based on sound tanker practice and common sense. Typical reasons to shut down the cargo transfer include the following:

Oil or hazardous material from any source discharged in the transfer operation work area, into the water, or upon the adjoining shoreline in the transfer area

Severe electrical storms or other extreme weather conditions in the vicinity of the vessel

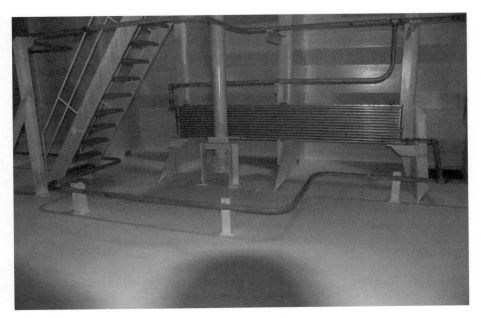

Figure 9-13c. Thermal oil heating coils. Courtesy Stacey DeLoach.

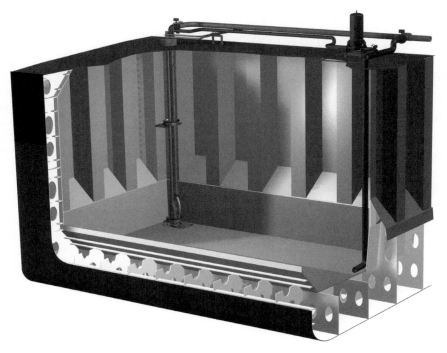

Figure 9-13d. Cargo heating system. Courtesy Frank Mohn AS.

Fire on or in the vicinity of the vessel or on the wharf

Cargo tank overfill alarm activated

Pumproom bilge alarm sounding

Cargo pump overheating

Mechanical failure in the cargo system (i.e., loss of hydraulic system, inoperative valve, inoperative gauging system, inoperative high-level alarm, overfill alarm system, etc.)

Vessel traffic passing at unsafe speed

Excessive motion of the vessel (surging/drifting)

Mooring lines parting

Excessive cargo pump discharge pressure

Inoperative inert gas system

Extreme deck pressure condition (high/low)

PIC experiencing fatigue or loss of orientation with respect to the transfer operation

The person-in-charge of a cargo transfer should not hesitate to shut down the operation if a problem arises or if in doubt—always remember *shut down!*

TOPPING-OFF OPERATION

Another critical point in the loading of a tank vessel is the topping-off of cargo tanks. Topping-off is that stage of the loading operation when a cargo tank is filled to the maximum permissible level consistent with safe operation and regulatory requirements. A number of factors are taken into account when determining the topping-off point in a tank, including potential expansion of the liquid cargo due to temperature variations, ability of the facility to shut down, experience of vessel personnel, and the quantity of cargo to be loaded. As the liquid level in a cargo tank approaches the topping-off point, a number of precautions are routinely observed, including the following:

1. *Control the loading rates:* If necessary, the loading rate to the tank being topped-off can be reduced by opening the next tank(s) in the loading sequence. This is one way to control the rate at which the tank fills and to ensure that proper flow is established to the next tank. Another approach is to have the shore facility reduce the cargo loading rate.

2. *Stagger the tanks:* To prevent too many cargo tanks approaching the topping-off point simultaneously, it is considered sound practice, when possible, to stagger the cargo levels in the tanks.

3. *Valve function:* Any valves involved in the topping-off operation should be checked to ensure they are functional prior to reaching the final ullage in the tank. Both manual and automated valves should be operated and verified visually by vessel personnel.

4. *Tank levels:* On vessels equipped with a closed gauging system, it is standard practice to check the liquid level using a second (independent) gauging method. Manually operated sonic tapes are commonly used to compare readings with the closed gauging figures. Vessels typically carry a number of these handheld tapes that are inserted through a standpipe equipped with a vapor valve arrangement on each tank for the purpose of taking ullage readings.

5. *Backup tank:* If a problem arises while topping-off (for example, a jammed valve), have a backup tank into which the cargo can be directed. Vessel personnel

are advised never to close the valves against the flow from the shore pumps due to the risk of damage from the resultant pressure surge.

6. *Switching tanks:* When the tank in question reaches the prescribed ullage, the next tank to be loaded should be opened before closing the valve on the topped-off tank. The operator should never shut down entirely against the shore pumps as the resultant surge pressure could cause serious damage to components in the cargo transfer system.

7. *Request for help:* During the topping-off operation, the PIC should call for additional help if necessary. If no one is available and too many tanks need to be topped off at the same time, the PIC may selectively shut down the individual grades being loaded, with the exception of the sailing cargo. The sailing cargo is generally the grade that will finish last, thereby determining the sailing time of the vessel. It is therefore possible to shut down the other cargoes without adversely affecting the sailing time of the vessel, and to resume loading each grade as the tanks are topped off. Another problem that frequently arises during the topping-off operation is the condition of the mooring lines. If the PIC is confronted with slack mooring lines due to freeboard change during the topping-off operation, additional help should be called to go fore and aft to tend the lines. If no one is available, it may be necessary to shut down the loading operation to take up the mooring lines.

8. *Shutdown:* If there is any doubt about the topping off operation, the PIC should shut down.

FINISHING CARGO LOADING

When approaching the end of the loading operation or the completion of one of the grades, the PIC of the vessel is dependent on a shore shutdown. At this point, the PIC has few options in the event of a problem, given the absence of a backup tank. With this in mind, the PIC should give the shore facility sufficient advance notice (standby) of the shutdown and communicate the following:

A. *Vessel identification:* Clearly identify the vessel, particularly at facilities that handle the loading of more than one vessel simultaneously.

B. *Cargo identification:* Clearly specify which cargo is being completed.

C. *Cargo loading rate:* Reduce the cargo loading rate if necessary and whenever possible finish the loading operation by gravity.

D. *Cargo tank status:* Advise the facility PIC if the loading operation is ending in a slack or topped-off tank on the vessel.

E. *Dock manifold standby:* When possible, have the facility PIC stand by the dock manifold valve in the event of any problems stopping the flow to the vessel.

F. *Shutdown:* Once the order to shut down the cargo is given, the PIC should verify that flow to the cargo tank has ceased.

G. *Manifold-valve position:* Secure the vessel manifold valve(s) for the cargo(s) in question.

When possible, it is advisable to end the loading operation in a slack tank, which is particularly helpful when trimming the vessel to a certain draft. When the loading operation is complete, all manifold valves should be secured. The cargo hoses or loading arms are then drained, disconnected, and blanked. The hoses are typically drained by either gravity or vacuum pump, or they are blown clear with air or nitrogen. The cargo manifold valves on the vessel may need to be open depending on the method employed to drain the hoses or arms. Once the hoses or loading arms are clear of the vessel, the manifold should be blanked and the entire cargo system secured for sea. At this point, the deck pressure of inerted vessels should be checked and the inert gas system started if it is necessary to top up the pressure.

Line Displacement

During the final stages of loading, the facility may want to change products in anticipation of the next operation. This involves displacing the contents of the shore pipelines with the next cargo to be handled. The shore facility will calculate the total quantity necessary to perform the line displacement and advise the PIC of the vessel. The PIC must then calculate the proper stop gauge in the last tank to be loaded on the vessel to accommodate the line displacement from the dock.

Pigging

Pigging is a process in which the contents of a facility (shore) pipeline can be cleared toward the vessel's cargo tanks at the conclusion of a loading operation. It is accomplished through the use of a bullet or cylindrically shaped polyurethane foam plug that is forced through the piping system under pressure. Clearing the shore pipeline in this fashion ensures more complete delivery of small batches (parcels). It also minimizes the risk of contamination or downgrade that is possible when a line displacement is performed with the next product. Pigging toward a vessel's cargo tanks is not a common practice, therefore care should be exercized by both the facility and vessel PIC to ensure that the correct operating procedure is understood and properly executed.

Switch Loading

A term used in connection with loading a low vapor pressure (high flash point) liquid into a cargo tank that previously contained a high vapor pressure (low flash point) product is "switch loading." There is a two-fold risk assciated with performing this operation:

(1) Contamination of the cargo being loaded

(2) In the case of non-inerted vessels, a potential fire/explosion risk resulting from loading a static accumulating cargo into a tank in which the atmosphere may be within the flammable range. This is due to the presence of bottoms (residuals) from the previous cargo as well as the contents of pipelines and pumps.

Companies employ different approaches to deal with the foregoing problem. For example, in the case of non-inerted cargo tanks, cargo pipelines and pumps should be flushed, cleaned and gas freed prior to reloading. Another approach may be to drop and thoroughly strip the contents of the cargo tanks, pipelines and pumps followed by purging with inert gas. Regardless of the method employed to prepare the cargo tanks, it is normally recommended that bottom sampling be performed for quality assurance at the beginning of the load.

Commingling

At the beginning of the loading operation some vessels are required to perform a blending operation. This involves loading a specified quantity of one grade into each cargo tank, then stopping the operation. The partial load in each cargo tank is then blended with the next grade in the loading sequence to meet the customer's specifications.

Final Draft and Trim

When the vessel is loaded to its allowable load line, vessel personnel must ensure that the cargo calculations are correct, and they must visually check the drafts during the final stages of the loading operation. At this point the draft and trim are closely monitored, and any allowance for fresh or brackish water at the dock should be applied. To determine the specific gravity of the water alongside the vessel, a water sample is taken and measured using a hydrometer. Reference to the vessel tables will reveal the permissible allowance when floating in other than salt water. The final trim is also an important consideration for some vessels. The PIC has a number of options to adjust the trim of the vessel including the following:

1. *Trim tanks:* Toward the end of the loading operation, the PIC can reach the desired drafts by selectively filling cargo tanks located fore and aft of the tipping center.
2. *Shifting cargo:* After loading operations are complete, the correct trim can be achieved by shifting cargo within the vessel.
3. *Shifting bunkers:* When it is not possible to shift cargo, the trim of the vessel can be adjusted by shifting bunkers.
4. *Ballasting:* The trim can be adjusted by taking on ballast into the segregated ballast system provided this does not result in overloading or excessive stresses on the hull.

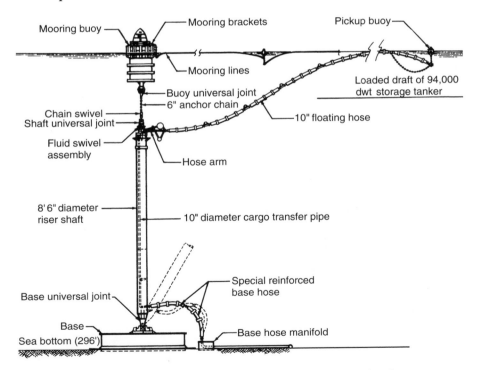

Figure 9-14. Diagram of a single-point mooring used in the South China Sea.

OFFSHORE TERMINALS

As tank vessels have increased in size, the number of ports that can safely accommodate these vessels has declined. This has led to the development of deepwater terminals located some distance offshore. The use of offshore mooring buoys, platforms, and lightering zones has reduced the need for larger tank vessels to negotiate busy harbors and restricted navigable waterways.

Single-Point and Multipoint Moorings

With single-point or multipoint moorings, a tankship can tie up to one or more buoys in deep water well offshore. Figure 9-14 illustrates a single-point mooring and special hose assembly through which cargo is loaded or discharged.

The hose is attached to a submerged pipeline that carries the cargo along the seabed to or from the onshore terminal. At a single-point mooring the vessel makes fast to a buoy at the bow. This permits the vessel to pivot freely around the mooring buoy with changes in wind and current. A floating hose is hoisted aboard the vessel and made fast to the manifold. One example of a successful single-point mooring arrangement is the Louisiana Offshore Oil Port (LOOP), located eighteen miles offshore in the Gulf of Mexico (Figure 9-15).

Good communications, training, and prior planning are essential for the safety of all involved in the use of offshore moorings. In multi-point moorings, the vessel is carefully maneuvered into position, typically by letting go of each anchor and backing to a position within a cluster of five or more mooring buoys. The mooring lines are run to the buoys by line boats, and the vessel is held securely in position by the combination of mooring lines and anchor chains. An example of a multi-point mooring arrangement can be seen in Figure 9-15a.

Figure 9-15. The Louisiana Offshore Oil Port (LOOP). Courtesy LOOP LLC.

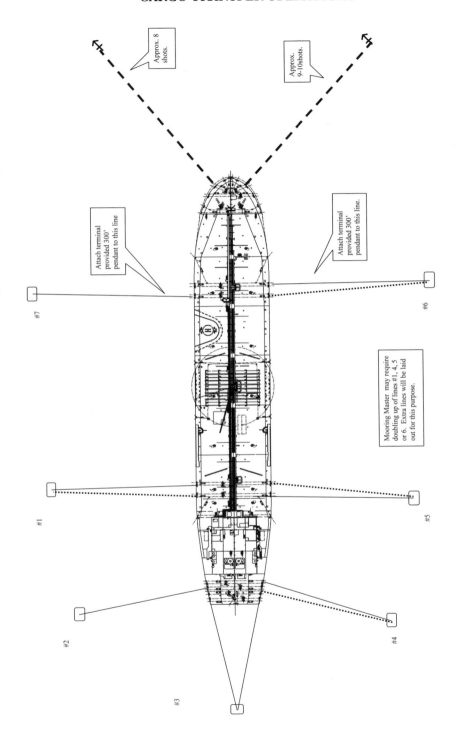

Approx. 8 shots.

Approx. 9-10shots.

Attach terminal provided 300' pendant to this line

Attach terminal provided 300' pendant to this line.

Mooring Master may require doubling up of lines #1, 4, 5 or 6. Extra lines will be laid out for this purpose.

#7

#6

#1

#5

#2

#4

#3

Figure 9-15a. Chevron mooring diagram. Courtesy Chevron Shipping Company LLC.

Platforms
Another approach at some facilities is the use of offshore platforms. Figure 9-16 illustrates such a platform located in deep water offshore.

Lightering Zones
Lightering zones represent another option when dealing with larger vessels, as seen in Figure 9-16a. In this instance, the VLCC is discharging part of her cargo into a smaller lightering vessel for delivery to the tank farm or refinery.

DISCHARGING OPERATION

Upon completion of the pretransfer procedures mentioned earlier and after rechecking the lineup of the vessel, the discharge operation can commence. It is advisable to start the cargo pumps at a reduced speed, while personnel check connections, tank ullages, and proper flow to the facility. (Consult Chapter 8 for a detailed discussion regarding the operation of cargo pumps).

Discharge Procedure
As in the case of loading, there are many approaches to the discharge of a vessel. The discharge plan that is developed should endeavor to maintain a reasonable distribution of weight over the length of the vessel, thereby minimizing stress on the hull and ensuring proper stability (Figure 9-17).

During the discharge of the vessel, the PIC follows a number of sound tanker practices:

1. *Trim:* The discharge of the cargo tanks should be arranged to get the bow up at an early stage in the operation. Discharging the forward tanks increases the trim by the stern, which aids the draining and stripping of the cargo tanks. On the other hand, the vessel should not be allowed to develop an excessive trim, as this can contribute to false tank level alarms, unacceptable hull stresses, engine room problems, and underkeel clearance problems.

2. *Pumproom:* Frequent inspections of the pumproom should be made during the discharge operation, checking for leaks. The PIC should periodically check the cargo pump seals, bearings, and casings for overheating. When cargo pumps are operating properly, there should be minimal noise or vibration. No one is permitted to enter the pumproom unless the ventilation system is operating and a standby is present or the PIC is notified.

As the number of double-hull tank vessels entering service increases, many owners are installing fixed gas detection systems to the boundary spaces around the cargo tanks. When one considers the amount of surface area shared between the cargo and ballast areas in a double hull vessel, it is apparent that early detection of cargo leaks (migration) into the normally gas-free areas of the vessel is imperative. In response to this concern, fixed gas detection systems (Figure 9-17b) are being installed to monitor such areas as the pumproom, double hull ballast tanks, cofferdams/voids, and any other normally gas-free spaces connected to the cargo area.

3. *Cargo pump and manifold pressure:* During the cargo discharge, the cargo pump and manifold pressures should be closely monitored. Any abnormal fluctuations or sudden changes in the discharge pressure warrant a shutdown of

Figure 9-16. Tank barge moored at an offshore platform. Courtesy TOSCO Riverhead.

Figure 9-16a. Lightering operation. Courtesy Abigail Robson.

MT PETROLAB   VOYAGE # 1   PORT   ANACORTES, WA
January, 2010   BERTH   TOSCO REFINERY

CARGO
Discharge approximately 280,000 net bbls ANSCO via 1 – 4 headers using MCP's 2 and 4
In-tank x-over, pumproom x-over and deck x-over OPEN.

Start discharge from 2W at reduced rate until shore booster pump is online. Do a complete check of the deck and
pumproom. When all is confirmed okay on the vessel and shore, increase rate to maximum shore pressure (125 psi)

Discharge 3W for 1 hour and then take 2 and 5 out at the same rate for 2 hours. Hold 5 (at approximately 5.0m ullage) and
continue in 2 while taking slop tanks to the stop gauge. Then use 2 to take 6W to 23m for cowing and start the wash at that
time. After wash is complete continue in 2W for approximately 1 ½ hours. Then use 5W to strip 2W and take 5 to the stop.

Maintain 500 mmwg IG pressure and oxygen content of 5% or less.
Compare pressure and rate with the dock hourly or after changes
to maintain maximum pressure.
NOTE: DO NOT EXCEED 4.0m TRIM OR 12.80m DRAFT
Estimated sailing drafts:
         F   9.10
         A   11.50

COW
Do a complete bottom cycle and top cycle of 150-0-40 B crank to 40W-0
Crank machine with supply open
Strip washings ashore with MCP's maintaining a 0.3 to 0.8m innage.
Start wash at 0.8m innage.
Check and log oxygen in each tank 1m below deck and at the mid-point
of the ullage space before washing.
Strip final amount to the slop tank
Complete COW operations checklist as applicable and notify the dock
when the COW operation commences and is completed

SEGREGATED BALLAST
Gravitate 6J for trim when taking 6 COT to 23m before COW.
During wash, gravitate 1J then 3J for draft/trim control.
After wash and 3J, gravitate 2J (keep at least 3m trim for stripping 2 COT)

NOTE:

OBSERVE ALL RULES AND REGULATIONS
MAINTAIN VESSEL SECURITY
CHECK PUMPROOM FREQUENTLY
UTILIZE CARGOMAX TO CHECK VESSEL STRESSES
MAINTAIN ACCURATE DECK LOG AND BALLAST HANDLING LOG
DO NOT HESITATE TO SHUTDOWN IF IN DOUBT
CALL ME ANYTIME IF NEEDED OR IF IN DOUBT
C/M _____
2/M _____
3/M _____
Cadet _____

	P	S
1	Hold	Hold
2	MT Strip	MT Strip
3	Hold	Hold
4	Hold	Hold
5	11.75	11.75
6	Cow Strip	Cow Strip
Slop	Hold at 10M	Hold at 10M

FP	MT	
1	5.00	5.00
2	8.00	8.00
3	8.00	8.00
4	MT	MT
5	MT	MT
6	5.00	5.00
Wing	MT	MT
Trim	MT	MT

Figure 9-17. Cargo discharge orders and ballast plan.

VESSEL CONDITION WHILE DISCHARGING VOY 094 US Oil Tacoma, WA

Tank	0	2	4	6	8	10	12	14	16	18	20	22	24	26
CARGO														
1P	7.0	8.0	8.0	8.0	8.5	15.8	18.3	MT	MT	MT	MT	MT	MT	MT
1S	7.0	8.0	8.0	8.0	8.5	15.8	18.3	MT	MT	MT	MT	MT	MT	MT
2P	18.5	21.6	MT	MT	MT	MT	MT	MT	MT	MT	MT	MT	MT	MT
2S	18.5	21.6	MT	MT	MT	MT	MT	MT	MT	MT	MT	MT	MT	MT
3P	2.0	3.0	5.6	9.2	9.2	9.2	9.2	13.7	13.7	13.7	13.8	18.4	23.2	MT
3S	2.0	3.0	5.6	9.2	9.2	9.2	9.2	13.7	13.7	13.7	13.8	18.4	23.2	MT
4P	MT	MT	MT	MT	MT	MT	MT	MT	MT	MT	MT	MT	MT	MT
4S	MT	MT	MT	MT	MT	MT	MT	MT	MT	MT	MT	MT	MT	MT
5P	9.0	9.0	9.0	10.0	10.0	10.0	13.0	14.5	14.5	19.0	MT	MT	MT	MT
5S	9.0	9.0	9.0	10.0	10.0	10.0	13.0	14.5	14.5	19.0	MT	MT	MT	MT
6P	18.0	18.0	18.0	18.0	MT	MT	MT	MT	MT	MT	MT	MT	MT	MT
6S	18.0	18.0	18.0	18.0	MT	MT	MT	MT	MT	MT	MT	MT	MT	MT
P Slop	15.0	15.0	15.0	15.0	15.0	15.0	15.0	15.0	15.0	15.0	15.0	15.0	15.0	MT
S Slop	15.0	15.0	15.0	15.0	15.0	15.0	15.0	15.0	15.0	15.0	15.0	15.0	15.0	MT
GSV	449520	413,520	377,520	341,520	305,520	269,520	233,520	197,520	161,520	125,520	89,520	53,520	17,520	0
COW			2W's	2W's			6W's	6W's						
BALLAST														
FP													5.0	11.5
1P						8.0	8.0	8.0	25.0	25.0	25.0	25.0	25.0	25.0
1S						8.0	8.0	8.0	25.0	25.0	25.0	25.0	25.0	25.0
2P					8.0	8.0	8.0	25.0	25.0	25.0	25.0	25.0	25.0	25.0
2S					8.0	8.0	8.0	25.0	25.0	25.0	25.0	25.0	25.0	25.0
3P						8.0	8.0	8.0	25.0	25.0	25.0	25.0	25.0	25.0
3S						8.0	8.0	8.0	25.0	25.0	25.0	25.0	25.0	25.0
4P		8.0	8.0	8.0	8.0	8.0	8.0	8.0	25.0	25.0	25.0	25.0	25.0	25.0
4S		8.0	8.0	8.0	8.0	8.0	8.0	8.0	25.0	25.0	25.0	25.0	25.0	25.0
5P							8.0	8.0	8.0	8.0	25.0	25.0	25.0	25.0
5S							8.0	8.0	8.0	8.0	25.0	25.0	25.0	25.0
6P					8.0	8.0	8.0	8.0	8.0	8.0	25.0	25.0	25.0	25.0
6S					8.0	8.0	8.0	8.0	8.0	8.0	25.0	25.0	25.0	25.0
E/R W Port													5.0	8.3
E/R W Stbd													5.0	8.3
Port Trim													5.0	5.0
Stbd Trim													5.0	5.0
BBls	0	35,777	35,777	35,777	97,510	155,823	189,705	230,828	280,068	300,188	321,186	321,186	335,288	346,694
Bunkers														
Draft FWD	10.31	9.55	8.43	7.66	9.07	9.76	9.18	9.56	10.73	10.70	10.36	9.41	8.85	9.54
Draft MID	10.28	10.40	9.96	9.48	9.94	10.39	10.47	10.64	10.90	10.74	10.60	10.12	9.87	9.79
Draft AFT	10.40	11.27	11.48	11.31	10.88	11.05	11.71	11.64	11.13	10.85	10.94	10.95	11.08	10.31
TRIM	0.09	1.72	3.05	3.65	1.81	1.29	2.53	2.08	0.40	0.15	0.58	1.54	2.23	0.77
Bend %	49%	30%	30%	32%	39%	32%	27%	35%	31%	33%	45%	49%	70%	78%
Shear %	66%	50%	45%	51%	35%	44%	41%	57%	31%	34%	42%	35%	43%	80%

Cargo Rate: **36,000** bbls/2 hr

NOTE:

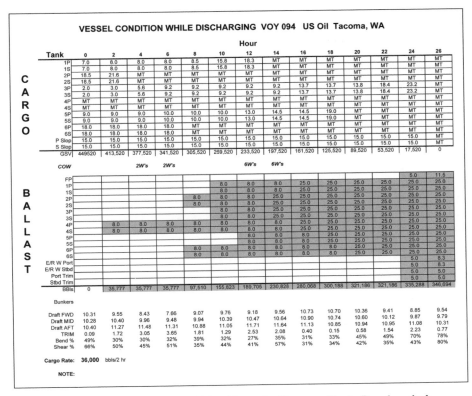

Figure 9-17a. Vessel Discharge Condition Example. Courtesy Kevin Duschenchuk.

Figure 9-17b. Gas Detection panel.

the operation. The cargo pump speed (RPM's), discharge pressure and manifold pressure should be periodically recorded to ensure compliance with company and charter obligations. The discharge pressure of the vessel can be expected to change gradually as the shore tanks are filled or when shifting tanks.

4. *Cargo system crossover:* On vessels equipped with submerged pumps in each tank, the PIC should exercise caution when two or more cargo pumps are discharging into the same pipeline. There is a risk of overflowing a tank if the delivery from one pump overpowers the other. The absence of a check valve on the discharge side of the pump increases the possibility of return flow to the tank from either another pump or the shore facility. When discharging into a common line, check the tank ullages to ensure that cargo is flowing in the correct direction. On vessels equipped with loading drops for each tank, also verify that these valves are closed prior to the start of the cargo discharge. If the drop valves are open, there is a distinct possibility of overflowing a full cargo tank that is not currently being discharged. Vessel personnel should never ignore tank high-level or overfill alarms that activate during a cargo discharge, as there may be a real problem.

5. *Stability and stress:* While discharging cargo and ballasting the vessel, the PIC should follow the prescribed transfer plan to avoid stability and stress problems (see Figure 9-18). Any last-minute changes to the cargo plan should be recalculated to ensure that no such problems arise during the operation.

6. *Emergency shutdown:* All personnel involved in the cargo discharge must know the location and operation of the emergency shutdown controls for the

Figure 9-18. The effect of free surface on a vessel conducting simultaneous cargo and ballast operations.

cargo pumps. Personnel should be made aware of the conditions or situations that warrant an emergency shutdown of the cargo pumps.

7. *Inert gas system:* On vessels that are obligated to maintain an inert condition in the cargo tanks, the inert gas system must be operating during the discharge operation. Vessel personnel must monitor the performance of the inert gas system throughout the discharge to ensure that it is delivering the required quantity (deck pressure) and quality (oxygen content) of gas to the protected tanks. (Refer to Chapter 16 for a more detailed discussion of inert gas systems).

8. *Slop tank level:* When stripping cargo to a designated slop tank on the vessel, the operator must guard against overflow of the tank. On some vessels it is a common practice to consolidate the strippings from each tank into a designated slop tank. This expedites the operation as the contents of the slop tank can then be discharged ashore with the main cargo pump at the end. For this reason, one or more of the after tanks near the pumproom usually serve as a slop tank. It is advisable to initially pump the slop tank down to a level that can accommodate the strippings from the other tanks. The PIC should closely monitor the slop tank level during the discharge operation to ensure that it does not overflow.

9. *Dropping and stripping the pipelines:* At the completion of the discharge operation, the contents of the pipelines and pumps on the vessel should be delivered to the facility. The pipelines are typically drained and the remaining cargo is stripped to the facility. In the case of crude-oil vessels, a special small-diameter line is provided for this purpose. The small-diameter pipeline is installed from the stripping (discharge) pump to the outboard side of the manifold valve on deck. The MARPOL line (as it is frequently called) permits the PIC to bypass the large discharge mains on deck when performing the final stripping of the cargo system. This final stripping is critical to minimize the cargo remaining on board (ROB) and contamination of any ballast water that may be subsequently introduced into the cargo system.

10. *Mooring lines:* During the discharge operation, personnel must closely monitor the tension on the mooring lines, gangway, cargo hoses, and arms. Necessary adjustments should be made during the transfer to avoid parting the mooring lines or connections.

11. *Heating coils:* The heating coils should be secured when approaching the stripping stage in the discharge of a cargo tank to avoid overheating the remaining cargo.

12. *Line flush:* When the sequence of discharge of different cargoes may result in the contamination of a high-grade product, it is often necessary to perform a line flush. This is a fairly common practice when multiple grades are carried in a single cargo system on the vessel. In this situation, there is a risk of contamination from the previous cargo that remains in the piping and pumps of the vessel. The contents of the vessel pipelines are carefully flushed (cleared) by pumping a small quantity of the next cargo into a slop line or into a tank with lower grade product on the dock. To ensure the piping is sufficiently flushed, the facility PIC usually checks for the color change (dyed cargoes) at the dock manifold.

13. *SHUTDOWN:* If any question or problem arises during discharge operations, the PIC should shut down.

REVIEW

1. What document identifies the "person-in-charge" of the cargo transfer on the vessel and facility?
2. When conducting a visual inspection of a cargo hose, list the possible causes for rejection of the hose.
3. What warning signals/signs must be displayed when a tank vessel is transferring cargo?
4. List the information that should be exchanged between the PICs during a pretransfer conference.
5. What is the requirement concerning language fluency when conducting a cargo transfer on a tank vessel in the United States?
6. What is the difference between a line sample and a bottom sample?
7. List some of the typical reasons for terminating a cargo transfer on a tank vessel.
8. What is the function of a bonding cable? Describe the correct sequence of connection if a bonding cable is employed.
9. What is the function of an insulating flange? Where is the insulating flange typically installed?
10. Describe the precautions commonly followed when topping off a cargo tank.
11. Describe the process of finishing cargo loading on a tank vessel.
12. What are the typical methods of draining the cargo hoses or loading arms at the completion of a cargo transfer?
13. What is the purpose of dropping and stripping the pipelines at the end of a cargo discharge?
14. When stripping the cargo tanks to a designated slop tank on the vessel, what precaution should be followed?
15. When discharging a heated cargo, when should the heating coils in the tank be secured?

CHALLENGE QUESTIONS

16. What safety concerns does the use of portable electronic devices pose on the deck of a tank vessel? Explain what is meant by the term "intrinsically safe" device.
17. Describe the precautions that should be followed when loading a static accumulating cargo into a non-inerted cargo tank.
18. When conducting cargo transfers at a facility using steel arms (chicksans) describe some of the concerns and precautions that should be followed by the person in charge of the vessel.
19. When crossing over cargo systems during a cargo transfer, describe the precautions that should be exercised by the person in charge of the operation?
20. What concerns can arise for the vessel when commingling cargoes during loading?

CHAPTER 10

Chartering and Operations

SCOTT R BERGERON

With the exception of the use of a pipeline, the most cost-efficient method of transporting bulk quantities of liquids is ocean transportation. It is estimated that, worldwide, tankers carry approximately 2 billion tons of oil annually. To put the cost of transportation into perspective, approximately 2 to 3 cents per gallon of gasoline at the pump can be attributed to transport of the oil by sea.

The United States is the largest oil consumer in the world. To feed this demand, the U.S. imports over 500 million gallons of oil per day or approximately 185 billion gallons per year. (Source: Energy Information Administration)

An operator or Person-in-Charge (PIC) of a tankship or barge should be familiar with the commercial side of the tanker transportation industry. Decisions made onboard the vessel by a PIC can have serious commercial implications affecting such areas as profitability, customer satisfaction, maintenance of cargo quality, and environmental impact.

This chapter provides an overview of the interface between commercial operations and onboard operations. The better both sides understand each other, the more successful each operation will be.

VESSEL OWNERSHIP

Reference will be made in this chapter to the owner, shipowner, operator, manager, and charterer. While many different organizational arrangements have been established to operate ships, we refer to a basic structure in this chapter. To eliminate any confusion, the following vessel ownership and operating structure is assumed.

Owner/shipowner: The vessel's registered owner is not always the party that is responsible for running the vessel. It is common that a ship is owned by a single-purpose company for financial, accounting, and liability reasons. When using the term owner or shipowner, we refer to the managing owner or the party who takes the responsibility to put the vessel into service.

Operator: The operator is the party who is responsible for the day-to-day commercial operation of a vessel. This includes finding spot and voyage charterers for the vessel and coordinating the necessary voyage details. The operator is normally responsible for providing bunkers to the vessel. In the case of a ship under time charter, the time charterer normally becomes the operator of a vessel. When operating a vessel, the operator does so "on behalf of" the shipowner.

Manager: The manager is the party who is responsible for the technical management and maintenance of the vessel. Normally the technical manager is the one who coordinates the crew, purchases the spares, and handles the day-to-day maintenance and periodic drydocking of the ship. In a traditional shipowning company, the same organization may be responsible for the operation and maintenance of the ship. However, it is becoming more common that the shipowner, the operator, and the technical manager are three different parties working together to make a profit.

Charterer: The charterer is the party who has taken the vessel on a voyage charter. Cargo owners and cargo traders are commonly charterers. The charterer is involved with a cargo that must be moved from one place to another. The charter party is the contract between the charterer and the operator of the vessel.

TYPES OF CHARTER AGREEMENTS

There are four ways for a charterer to purchase ocean transportation. The first is a voyage charter. In this case, the charterer will charter or "rent" a vessel to carry cargo from the loading port(s) to the discharge port(s). The second method is through a period or time charter. During a time charter, the vessel is "rented" for a specified period of time—six months or two years, for example. The vessel will perform multiple voyages at the direction of the charterer during a time charter. In a voyage charter, the operator maintains responsibility for the daily operation of the vessel, while with a time charter, the charterer normally assumes this responsibility. In both voyage and time charters, the technical manager retains the responsibility of running the vessel (maintenance and crew).

A third method for a charterer to procure transportation is through a bareboat charter. When taking a vessel on a bareboat charter, the charterer is not only responsible for the voyage charter, but must also assume responsibility for onboard operation or running the vessel, including providing the crew and maintenance. In such a case, the charterer actually becomes a ship operator and a manager.

Finally, cargo can be transported under a contract of affreightment or COA. Commonly, a supplier or producer of a product will make a COA with an operator of vessels. The COA will specify a total volume of cargo to be carried through the period of the contract in certain size shipments. An example of this arrangement is a chemical company that produces caustic soda. The chemical company needs to ship caustic soda in regular shipments to various locations in the world. As these shipments are fairly predictable, the chemical company could conclude a COA with a ship operator to provide transportation for 1 million tons of cargo per year. The COA specifies where the cargo is to be loaded and discharged and also establishes the size of the shipments (for instance, 1 million tons in 20,000-ton shipments). A COA normally involves multiple options and clauses for flexibility to suit the needs of both the charterer and the ship operator.

There are numerous benefits with a COA. First, the charterer has purchasing power due to the large size and number of shipments. Second, charterers can insulate themselves from the price fluctuations in the transportation market. Finally, the charterer can predict fixed costs for transportation for better overall cost control when selling the cargo. The vessel operator also benefits as the COA provides a certain amount of guaranteed cargo to be carried over the period of the contract.

Once the charter is negotiated, the terms of the agreement are specified in a contract known as the charter party. In order to identify suitable vessels and to subsequently facilitate negotiations with the operators of these vessels, a charterer normally uses a broker. The "fixture" of a charter party is the conclusion of the negotiation.

THE CHARTER PARTY

As the requirements involved with bulk liquid transportation do not change, the basic terms are made on a standard contract form (charter party). There are several versions of the standard contract; each is named after the author and the type of charter (voyage or time charter). COAs are normally contracted under the terms of a voyage charter party with many clauses attached. Examples of common charter party contract templates include ASBATANKVOY (Association of Ship Brokers and Agents), SHELLVOY (Shell Oil Company), EXXONVOY (Exxon-Mobil), BEEPEEVOY (British Petroleum) and TANKERVOY (International Association of Independent Tanker Owners).

The type of charter party form that is used will depend on the charterer, the nature of the transportation needs, and, in many cases, the state of the market. Each standard charter party has different strengths, weaknesses, and emphasis. Also, clauses can be added to the standard charter party form to allow modification to suit the charterer or shipowner. As with the freight rate, the clauses are subject to negotiation. In a weak market, where the charterer has a strong position, many clauses can be added to the charter party to improve the charterer's position. In a strong market, the shipowner will have the leverage to eliminate standard clauses and perhaps add some that favor the vessel.

The fixed charter party with clauses will cover all aspects of the carriage contract. Some of the issues negotiated and agreed to in the charter party include the identity and description of the vessel, cargo(es) and the amounts to be loaded, the freight rate, charterer's expenses, port fees, port agents, cleaning procedures and costs, laytime, demurrage, loading and discharge rates, loading and discharge ports, oil pollution liabilities, voyage liabilities, cancellations, and even weather and force-majeure (act of God) issues. The potential list of issues covered in the charter party is endless.

The Role of the Broker

A charterer will provide a broker with information regarding transportation needs and will commission the broker to negotiate a contract. The broker then "comes out on the market," looking for indications of freight rates. During this period, the broker will identify suitable vessels from position lists of vessel operators and from replies to direct solicitations. At that point, the broker starts to "work" the cargo, often playing one shipowner against another to get better rates. Depending on the type of cargo, charter parties can be negotiated on a prompt basis (voyage commencing within 24–48 hours of fixture) or for very large cargoes. With contracts of affreightment for highly specialized chemical parcels, the charter party could be fixed months in advance of the commencement of the voyage.

The role of the broker is important in the negotiation of bulk transportation. The risks and the amount of money involved require a certain level of trust among the charterers, vessel operators, and owners. The broker is the middleman who can build the personal relationships and can also serve as the intermediary in solving problems between the parties.

Sometimes a shipowner or operator will commission a broker also. In such a case, both brokers work to fix the charter party. Due to depressed markets in the late 1990s, the use of owners' brokers has decreased; in some cases, charterers have begun working charters directly with ship operators and owners. The normal fee charged by a broker is 1.25 percent of the gross transportation fee, referred to as the gross freight. If more brokers are involved, more commissions are added to the fixture and the charter becomes less attractive. Increased transparency and efficiencies brought by the Internet continues to influence the role of the broker.

TERMS OF THE CHARTER PARTY

To ensure a profitable voyage, the tankerman should be familiar with the terms in a charter party. Careful negotiations are made on many particulars of each charter with the assumption that vessel personnel will act properly and within the terms of the agreement. Ideally, a copy of the concluded charter party should be placed on board the vessel for the crew to review, but this seldom happens because the actual signed document may not arrive until after the voyage has been completed. In any case, copies of standard charter parties should be available to the master.

Practical details and items not addressed or specified in sufficient detail in the charter party should be explained in the voyage instructions, which are normally sent to the operator and the vessel via the broker involved in fixing the cargo. The following points addressed in each charter party should be reviewed and understood.

Laytime
Laytime is the amount of time allotted to load and discharge the cargo. There is normally a window of time known as the "lay/can" (laytime commencing/canceling), when the vessel must present the notice of readiness (NOR) to indicate that the vessel has arrived and is ready in all respects to load the cargo. If the vessel does not arrive and present a valid NOR in this window, the charter is subject to being canceled. The times involved with the NOR and the disconnection of cargo hoses normally determine when laytime begins and ends.

Typically, a vessel is allowed thirty-six hours to load and thirty-six hours to discharge a single cargo at one loading and one discharge port. Further allowances are made for multiple cargoes or grades of cargo and multiple berths. The total of seventy-two hours is referred to as laytime. While the rule of thumb is that a vessel should physically load or discharge her cargo within twenty-four hours, the thirty-six hours of laytime on each side anticipates the additional time required in port. In some cases, the laytime is reversible, meaning that if only twelve hours are used to load, up to sixty hours can be allowed for the discharge. When a vessel is prevented from completing the load and discharge within the agreed laytime, demurrage ensues. Demurrage is defined as the compensation from the charterer paid to the shipowner/operator for time exceeding laytime. For example, laytime may be exceeded due to slow loading or receiving caused by the terminal's operations. In such a case, the ship operator will lodge a claim against the charterer for demurrage expenses. (The charterer may then claim the terminal, if appropriate.) If the delays are due to a fault of the vessel (for example, malfunctioning cargo pumps), then the charterer will institute a claim against the shipowner/operator for the lost time. Demurrage rates are normally negotiated at a fixed price per day (a negotiated time-charter equivalent rate) as part of the charter party.

Cargo Quantity

It is very difficult to load an exact amount of liquid cargo, due to variations in the temperature and density of the cargo as well as the problems encountered when measuring liquid volumes in a floating vessel. However, the charter party specifies the amount of cargo to load, with some variables. Common terms referring to quantity include "min/max" (minimum and maximum amount), percentage moloo (a percentage of the nominated cargo quantity which can be more or less at the shipowner's option. Example: 1,000 tons at 5 percent Moloo means the owner can load anywhere between 950 and 1,050 tons), and CHOPT (charterer's option, normally referring to the fact that the charterer can load as much as he wants, up to the full capacity of the vessel).

Notice of Readiness

As per the terms of a charter party, the master of a vessel must present the notice of readiness (NOR) when the vessel arrives and is ready in all respects to conduct cargo operations (Figure 10-1). The significance of this notice is that, under normal charter parties, it establishes when laytime will begin.

This simple act can raise serious questions and conflicts. While the NOR may be tendered, it may not always be accepted. The charter party clearly specifies under what conditions—when, where, and how—the NOR is to be tendered. Common clauses indicate how it must be sent (verbally to the agent or via telex or fax) and whether it must be tendered during the workday or the workweek. The following example of the NOR clause is derived from the standard ASBA-TANKVOY charter party. This clause is subject to negotiated changes. NOR clauses will also vary among other standard charter parties.

The ASBATANKVOY notice of readiness clause is as follows:

> Upon arrival at customary anchorage at each port of loading and discharge, the Master or his agent shall give the Charterer or his agent notice by letter, telegraph, wireless or telephone that the Vessel is ready to load or discharge cargo, berth or no berth, and laytime, as hereinafter provided, shall commence upon the expiration of six (6) hours after receipt of such notice, or upon the Vessel's arrival berth (i.e. finished mooring when at sealoading or discharging terminal and when all fast when loading or discharging alongside a wharf), whichever first occurs. However, where delay is caused to Vessel getting into berth after giving notice of readiness for any reason over which the Charterer has no control, such delay shall not count as used laytime.

It is important for the master to know when to present the NOR. For example, if the charter party clearly specifies that the NOR must be tendered between

LAURIN MARITIME

Notice of Readiness

M/T   PORT

VOY NO.   Date:

TO:

To whom it may concern

This is to inform you that the above vessel has arrived at the Port of   at
hours LT on the   in every respect ready to commence load of her cargo of
               in accordance with the terms and conditions of the charter party governing this
shipment.

                                        Yours Faithfully,

                                        Signature

                                        Master

For and on behalf of shippers/receivers

Notice of Readiness Accepted/Received

At   Hours on the

By

         Signature

         Name

Figure 10-1. Notice of readiness document. Courtesy Laurin Maritime.

0800 and 1800 Monday through Friday, then there is no need to waste fuel and expenses to try to meet an ETA of 1900 on Friday. Under these conditions, at the direction of the commercial operator, it is probably more prudent to run at an economical speed the last few days to plan an early Monday arrival. The same theory applies to tides. Under most charter parties, laytime will begin either six hours after a valid NOR is tendered (in the event that the vessel must anchor due to berth unavailability) or once the vessel is moored. Time waiting for tides after the NOR is tendered is normally not considered as laytime and, therefore, is not paid for by the charterers. Time used during transit from the anchorage is also not counted as laytime.

It is important to note that the NOR should be tendered only when the vessel has arrived at the location specified by the charter party (normally the customary anchorage) and when the vessel is ready to load or discharge. Under most conditions, the vessel is ready to load or discharge when she is ready to proceed to berth. While delays caused by such things as port clearance formalities have to be accepted, there are no excuses if the cargo tanks are not properly cleaned or inerted. A falsely presented NOR is not valid and could cause significant delay and monetary loss in the resulting disputes.

Cargo Hoses

While the NOR normally sets the time for laytime to begin, the disconnection of hoses normally determines the end of laytime. It is in the best interest of the shipowner/operator to have the hoses connected as long as possible after loading or discharging has stopped.

Under most charter parties, the cargo hoses shall be furnished, connected, and disconnected by the charterer at the charterer's risk and expense. In practical terms, the terminal hoses will be used. If it becomes necessary to use vessel hoses, it should be done only if the charterer assumes liability for the condition of the hoses. For a shipowner/operator, responsibility for the cargo ends at the manifold rail. Anything that happens to the cargo once it has passed the rail should remain the charterer's responsibility.

Pumping Clause

The following is a reprint of the BEEPEEVOY 3 Pumping Clause and is an example of normally agreed upon pumping performance.

Owners undertake that the Vessel shall discharge full cargo, as defined hereunder, within 24 hours, or prorata thereof in respect of part cargo, from the commencement of pumping or that the Vessel shall maintain a minimum discharge pressure of 100 psig at the Vessel's manifold throughout the period of discharge provided that the shore receiving facilities are capable of accepting discharge of the cargo within such time or at such pressure.

As a rule of thumb, the vessel should always provide enough pumping pressure to maintain the 100 psig. Normally, the shipowner is financially responsible for the discharging time exceeding twenty-four hours when 100 psig is not maintained. Exceptions to this include discharging of multiple grades or terminal restrictions. A pumping log must be maintained by the vessel's crew in order to defend any possible claims in this respect.

With regard to loading, the normal rule is that the vessel will accept cargo at the rates requested by the charterer, with due regard for safety. The loading rate of the vessel is normally determined by the venting capacity of either the common vent line, the pressure relief valve, or the vapor collection (emission) control system, depending upon which venting system is in use while loading. Suitable reductions of the loading rate are allowed to safely top off the vessel's tanks.

Protests

It is incumbent for the vessel's master to protest against violations of the charter party. This is done by a letter of protest. The vessel operator normally specifies the format of the letter of protest (see Figures 10-2 and 10-3).

Common reasons for protest include the following:

Slow loading rates (anything less than the vessel can safely accept for loading)

Slow discharge rates (any reduced discharge rates ordered by the terminal or when the terminal will not allow the vessel to maintain 100 psig at the manifold)

Deadfreight (when the terminal does not load the minimum nominated cargo quantity)

Cargo quantity discrepancies (differences between the charterer's cargo surveyors' totals and the vessel's own cargo quantities determined by gauging)

Shore connections (insufficient number of shore connections provided, or the diameter of the hoses is less than that of the manifold)

Multiple grades of cargo (multiple grades of cargo are not loaded or discharged simultaneously)

Safe Berth

In most if not all charter parties, it is the charterer's obligation to nominate a safe berth for both loading and discharge. The safe navigation of the vessel remains the responsibility of the master. However, if the master feels that the nominated berth is unsafe or if it does not have enough water available for the draft of the vessel, then the master must advise the operator as soon as possible.

For further study, it is recommended that one consult the Intertanko publication: "Tanker Voyage Charters."

OTHER CONSIDERATIONS

The following items, while not specifically addressed in a charter party, should be familiar to vessel personnel:

CLEAN TANK CERTIFICATE

Prior to loading cargo, the terminal representative or cargo surveyor should sign a certificate indicating that the tanks are of suitable cleanliness. This is particularly important when the charterer requires visual tank inspection prior to loading (Figure 10-4).

DRY TANK CERTIFICATE

In order to defend a claim against shortlanding of cargo, a certificate should be signed by an independent surveyor or the cargo receiver, acknowledging that the cargo tanks have indeed been stripped of all pumpable quantities (Figure 10-4).

LAURIN MARITIME

LETTER OF PROTEST

TO:

FROM: **The Master of M/T Bolero I**

DATE:

I, the Master of M/T Bolero I , hereby, on behalf of the Owners and/or charterers, protest

against the slow loading rate alongside your terminal,   during the loading of

on the

Actual loading rate was only: m^3 / h

M/T Bolero I can receive m^3 / h in the tanks loaded.

I hereby also declare, on behalf of the Owners and/or charterers, the right to revert in this

matter at a later date.

Yours faithfully,

the Master of M/T Bolero I

Kindly sign the copy for receipt and return the same to the Master

Figure 10-2. Sample letter of protest—difference between ship and shore figures. Courtesy Laurin Maritime.

LAURIN MARITIME

PROTEST OF DIFFERENCE
BETWEEN SHIP & SHORE FIGURES

M/T   BOLERO I   16-06-00

To:

Please be advised that there is a discrepancy between ship and shore figures covering the

loaded at your terminal this   day of

Shore figures	<u>Metric Tonnes</u>
Ship figures	<u>Metric Tonnes</u>
Difference	<u>Metric Tonnes</u>

On behalf of the Owners and/or charterers, I hereby protest this difference holding you responsible for any and/or all claims which may occur due to this difference.

Yours faithfully,

Master

Received:

Figure 10-3. Sample letter of protest. Courtesy Laurin Maritime.

LAURIN MARITIME

TANK INSPECTION CERTIFICATE

M/T

PORT

TERMINAL

We, the undersigned, hereby declare that the cargo tanks of M/T   were jointly

inspected at   hours LT on the   and found well drained after previous cargo

suitable to load the cargo(es) of   as follows:

Cargo tanks Nos.	for	
Cargo tanks Nos.	for	
Cargo tanks Nos.	for	
Cargo tanks Nos.	for	

The ship's pumps, lines, manifolds, deck heaters were also inspected by opening them and found dry and clean as far as visible, so acceptable for loading the above cargo(es).

Remarks:

Date:

Time:

Surveyor: Chief Officer:

Signature   Signature

Name   Name

Figure 10-4. Sample tank inspection certificate. Courtesy Laurin Maritime.

CARGO SAMPLES

It is a normal procedure that samples of the cargo are taken during the loading operation and prior to discharge. The samples should be taken and sealed by a cargo surveyor. The vessel should maintain a complete set of all samples taken. These samples should be kept in the sample locker for at least one year following the discharge of the cargo. In the event of a cargo quality issue, the vessel's sample may be required to defend the shipowner.

BILL OF LADING

The bill of lading is the charterer's receipt that the cargo has been delivered to the vessel. It is used as a financial tool, sometimes considered as the title of ownership document of the cargo used to trade the cargo once afloat, and it is also the document required to discharge the cargo. While an in-depth discussion of this legal instrument is beyond the scope of this book, the following should be understood.

Great care must be taken when dealing with bills of lading. There are normally three original copies issued which the master should sign after the cargo has been loaded. If the cargo type and quantity do not match the vessel's calculations, then a protest must be issued.

After loading, the bills of lading are sent to the receiver. The cargo should not be discharged until the original bills of lading are presented. It is common that the vessel will arrive at the discharge port before the bills of lading. This is often the case when the cargo is sold or traded during the voyage as the bills of lading follow each transaction. Upon instruction from the vessel operator, it is usually acceptable to discharge the cargo if the charterer has issued a letter of indemnity (LOI) which holds the vessel's owner and operators harmless for discharging the cargo without the original bills of lading.

A ship's master should further study the issues concerning bills of lading. The vessel operator will provide specific instructions about handling bills of lading.

PREPARING FOR THE NEXT VOYAGE

Close communication should always be maintained with the chartering department of the vessel operator. In order to save time and expense, the vessel's crew should be aware of the anticipated voyage in advance of the discharge. Issues of tank cleaning, inerting, and other requirements such as port restriction, regulation, and carriage requirements will help the vessel's crew to efficiently prepare for the next voyage.

PRICING

Before cargo is loaded onboard a vessel, supply and demand must be determined. The transportation of the product is one part of the larger picture of fulfilling demand. In some cases, the purchaser coordinates transportation; in other cases the seller will include delivery costs in the selling price. In either situation, the people wishing to purchase transportation are commonly referred to as the shippers or charterers.

The price for transportation may fluctuate on a daily and even an hourly basis. These changes are due to vessel supply and cargo demand: the greater the number of suitable vessels at a given location, the lower the price to transport cargo from that location. Likewise, when a ship is the only suitable one at the location, the rate will increase.

Many factors affect the suitability of a vessel. A potential charterer must consider the following questions:

1. *Cargo authorization:* Is the vessel approved to carry a certain cargo?
2. *Vessel size:* Will the size of the vessel allow it to reach the load and discharge ports? Can it load the desired amount of cargo within the required draft limits?
3. *Charterer approval:* Does the charterer accept the vessel, or has the vessel been rejected due to poor operational history and/or technical condition? For example, has the vessel been rejected due to unsatisfactory cargo tank conditions, a history of pollution, accidents, or poor vetting inspections?
4. *Cabotage issues:* Are there any restrictions from using the vessel because it is not eligible to carry the cargo? For example, in order to load cargo from one port in the United States and discharge it in another port within the United States, the Jones Act requires the vessel to be built, owned, operated, and crewed by U.S. citizens.
5. *Prior cargo:* Are the prior cargoes compatible with that to be carried? Many cargoes require that the previous three cargoes do not contain lead. Edible cargoes require that prior cargoes do not violate certain restrictions. Chemical cargoes must be compatible with prior cargoes in order to avoid dangerous chemical reactions or cargo contamination.

FREIGHT RATE

The price charged for transportation is normally referred to as the freight rate. In most cases, the freight rate is made inclusive of all trading costs. Along with the operation of the ship, there are costs for bunkers, pilots, tugs, line handlers, custom fees, agency fees, clearance costs, and other tolls. There are different means of calculating and offering the freight rate: Worldscale, lumpsum, time-charter equivalent, and rate per ton.

Worldscale

The new Worldwide Tanker Nominal Freight Scale, commonly referred to as Worldscale, is a standard freight rate system established and governed jointly by the Worldscale Associations of London and New York. The original concept of the Worldscale was to determine uniform voyage costs and revenues after the variable port and bunkers costs were taken into account. Today, the Worldscale rate system defines a common reference and procedure to assist charterers, brokers, and vessel operators in calculating freight rates for the carriage of oil from all loading points in the world to all discharge locations in the world.

Driven by the needs of the British and American governments to establish a consistent and unbiased charter calculation method after World War II, the Worldscale system was born. The American and British systems were combined in 1969 (old Worldscale) and later revised in 1989 (new Worldscale). Together, the Worldscale Associations maintain a schedule of freight scales based on a standard calculation method. The schedule is normally revised annually, with new rates becoming effective on January 1 of the revision year.

In addition to the calculation procedures, the Worldscale schedule lists standard rates per metric ton and mileage for voyages between all conceivable ports in the world. Variable costs including bunker prices, port costs, canal fees, and unique port cost differentials are addressed and adjusted during each revision of the schedule.

When freight rates are booked on a Worldscale basis, then the actual fixture rate is specified in percentage reference to the Worldscale calculation. The baseline or "flat rate" is specified as WS 100. A fixture concluded at WS 125 is 125 percent of the cataloged Worldscale calculation whereas WS 80 is 80 percent of the calculation for the load and discharge ports specified in the charter party.

The basis of the Worldscale calculation is best explained in the following Worldscale Preamble, copied directly from the 2009 edition (www.worldscale-usa.com):

> All rate calculations, which are made in USD, are per tonne for a full cargo for the standard vessel based upon a round voyage from loading port or ports to discharging port or ports and return to first loading port using the undermentioned factors.
>
> All of the factors shown are purely nominal and for rate calculation purposes only. In particular, the fixed hire element of USD 12,000 per day is not intended to represent an actual level of operating costs, nor to produce rates providing a certain level of income or margin of profit, either for the standard vessel or for any other vessel under any flag.

(a) Standard Vessel

Total Capacity (i.e. the vessel's capacity for cargo plus stores, water, and bunkers, both voyage and reserve)	75,000 tons
Average service speed	14.5 knots
Bunker Consumption Steaming	55 tons per day
Purposes other than steaming	100 tons per round voyage
In port	5 tons for each port involved in the voyage.
Grade of fuel oil	380 cst

(b) Port Time

4 days for a voyage from one loading port to one discharging port; an additional 12 hours being allowed for each extra port on a voyage.

(c) Fixed Hire Element

USD 12,000 per day

(d) Bunker Price*

USD 554.05 per ton

*This price represents the average worldwide bunker price for fueloil (380 cst) during the period 1st October 2007 to 30th September 2008 as assessed by LQM Petroleum Services, Inc.

(e) Port Costs

Port costs used are those assessed by the Associations in the light of information available to them up to the end of September 2008, the rate of exchange used for converting costs in a local currency to USD being the average applicable during September 2008.

(f) Canal Transit Time
24 hours is allowed for each transit of the Panama Canal
30 hours is allowed for each transit of the Suez Canal
Mileage is not taken into account in either case.

The following sample calculation illustrates how the Worldscale rate per ton provides a consistent method to calculate the freight rate, regardless of how the actual vessel characteristics differ from the standard vessel defined by the Worldscale Preamble. The prevailing Worldscale percentage compared against WS 100 for any particular cargo size and route identifies the price fluctuations in the market. (Due to the higher costs of operating U.S.-flag tankships, a similar system to Worldscale has been established for Jones Act fixtures. This system is referred to as the American Tanker Rate Schedule or ATRS.)

Simplified Worldscale Calculation
EXAMPLE: Cargo—50,000 tons of oil from Bullen Bay, Curacao, to New York, N.Y. (3,540 nautical miles)

Rate per Worldscale schedule: $3.75/metric ton:
WS 100 or flat rate: 50,000mt @ $3.75/mt = $187,500
WS 45 rate: 45% of $187,500 = $ 84,375
WS 200 rate: 200% of $187,500 = $375,000

Voyage calculation based on WS 200:
(*Note:* Fixed rate differentials have been omitted for clarification.)

Days:

Ballast—1,770 nm @ 14.5 knots	5.1 days		
Loaded—1,770 nm @ 14.0 knots	5.3 days		
Loading	2 days		
Discharging	2 days		
WS 200 freight rate			375,000
Commission (1.25%)		(4,688)	
Port expenses (Curacao)		(12,000)	
Port expenses (New York)		(25,000)	
Bunkers:			
Fuel oil at sea—10.4 days @ 35mt/day @ $110/mt		(40,040)	
Fuel oil in port—4 days @ 10mt/day @ $110/mt		(4,400)	
Total			288,872
Daily running costs 14.4 @ $5,000/day		(72,000)	
(*Note:* excluding owner's capital costs for vessel)			
Results			$216,872
Result per day			$ 15,061

Lump Sum
As the name implies, a lump-sum rate is a fixed price for the delivery of the nominated cargo and voyage. This fixed price is normally inclusive of all voyage-related costs and probably will not be affected by the amount of cargo ultimately loaded. When offering a lump-sum rate, the shipowner/operator must have precise figures for port costs and other expenses or the profit margin will be directly reduced.

Rate per Ton

While Worldscale rates are generally used for the movement of petroleum oils, the rate per ton is more common for the fixture of chemicals. In this case, a certain price is mutually established for each ton of cargo loaded. Unless agreed otherwise, this rate will not include the voyage and port costs. A fixture made on a rate-per-ton basis usually involves an option involving the quantity. An example would be the nominated quantity of 20,000 tons, 5 percent moloo (more or less, owner's option). The shipowner/operator normally wants to exercise the option to load the additional 5 percent quantity in order to improve his earnings for the voyage.

Time-Charter Equivalent

For practical reasons, it is common to determine the time-charter equivalent rate for a cargo fixture. When a vessel is hired under a time charter, the charterer agrees to pay a daily rate for the vessel. This rate includes all of the technical costs of the ship, such as maintenance and crew wages, but it excludes voyage-related costs such as bunkers and port fees. This daily rate makes it very easy for a shipowner to compare his daily costs against the daily earnings. When working with freight rates on a spot market, it is common for the shipowner to convert the freight rate calculation into a day rate. This day rate is known as the time-charter equivalent.

REVIEW

1. What are some of the factors taken into consideration by a charterer when selecting a vessel for hire?
2. List four ways that ocean transportation can be contracted.
3. Describe the role of the broker in arranging a charter.
4. What is Worldscale?
5. Define laytime.
6. What is the purpose of the notice of readiness?
7. List some of the common reasons to write a protest for violations of the charter party.
8. What is the significance of a dry certificate?
9. Why is sampling an important function in the transport of bulk liquid cargoes?
10. Describe the purpose of a bill of lading.

CHAPTER 11

Vetting Inspections
Scott R Bergeron

M arine disasters in modern times have created increased scrutiny of those involved with the ownership and operation of tankships. The public opinion has become less and less tolerant of oil pollution caused by accidents involving ships.

Recent times have seen reduced participation of vessel ownership by oil companies. Therefore, oil companies make up the largest sector of tanker charterers. In order to minimize the risk of being exposed to a marine disaster involving their cargo, most oil companies have established risk management or vetting departments within their chartering organization.

INSPECTIONS

The term vetting is analogous to screening or reviewing a vessel prior to chartering that vessel. While many oil companies undertake their risk management responsibilities using different methods or procedures, there are many common principles involved in the vetting process.

A favorable review of the vessel will result in an approval from the vetting/risk management department. The respective company may then charter the vessel. On the other hand, a vessel with a poor record and unfavorable screening normally prevents that vessel from being chartered. Prior to chartering a vessel, a review of the following records is normally made.

1. Inspection

Most major oil companies employ vessel inspectors to physically inspect vessels being considered for charter. These inspectors are normally former masters or chief engineers. They will attend a vessel for several hours and complete a thorough checklist and questionnaire after observing the operation and materiel condition of the ship. Areas of inspection include test of navigation, safety and cargo related equipment. Also included is the close up examination of the engine room, cargo tanks and their coating condition, cargo pumps and piping. Finally, reviews of onboard safety procedures and maintenance routines are also included in the inspection as well as a check of validity of all certifications for the vessel and her crew. Depending on the charterer, vetting inspection intervals vary. Some require annual inspection, while others re-inspect every second year.

2. Vessel's History

Most vetting departments maintain databases of vessels that they may be likely to charter. Any publicized reports of accidents, such as pollution incidents, groundings or collisions are recorded and studied. It is also common for these companies to maintain details of a vessel's performance while on charter. This record is then analyzed before the vessel is chartered again.

3. Owner/Manager's Record

In a similar fashion, vetting groups will also keep records of the companies involved with the operation of tankships. By doing this, they are able to analyze the performance and accident records of a fleet. Before being chartered, a vessel run by a company with frequent mishaps or continuous records of poor performance may be determined to be an unacceptable risk to the chartering company.

SIRE System

The Oil Companies' International Marine Forum (OCIMF) recognized that their industry would be well served if they could share the results of vessel inspections. By doing so, not only could they reduce burden and demand of financial and personnel resources in conducting the many vessel inspections, but they could also learn from the actions of their counterparts who perhaps had already inspected a particular ship.

The result of the desire to share inspection information became what is known as the Ship Inspection Reporting Exchange (or SIRE) system. The SIRE system is a database operated by OCIMF. A member company uploads inspection reports into this database. The report then becomes available for review by other member companies. The SIRE system facilitates the ability of a company to make use of an inspection report in the SIRE system in lieu of conducting their own inspection.

Shipowners are allowed to make comments to inspection reports. These comments are also uploaded into the database and appended to the inspection report. This allows vetting department to review inspections of ships and also consider the ship owner's comments to any deficiencies noted during the inspection. The report remains on the SIRE database for 2 years.

The SIRE system is continually improving and being made more and more standardized. All member companies of the OCIMF have agreed upon a common Vessel Particular Questionnaire (VPQ). The VPQ is normally completed by the shipowner and identifies the most common particulars of a vessel that interests charterers. These particulars include all types of information such as LOA, beam, tonnages, certificate expiry dates, cargo pump details and mooring equipment.

Likewise, a standard Vessel Inspection Questionnaire or VIQ has been established. The VIQ contains hundreds of standard questions related to the vessel's operation, maintenance, safety procedures and pollution prevention techniques. A vetting inspector verifies each of these inspection questions during the inspection.

Beyond Inspections and the ISM Code – The TMSA

In 2004, the OCIMF introduced the Tanker Management and Self Assessment (TMSA) program. TMSA is a framework designed to help shipowners and charterers document and evaluate the management system employed in the operation of the shipowner's vessels. The International Ship Management (ISM) Code requires shipowners to operate in accordance with a safety management system to formally foster a continuously improving culture of safety and environmental awareness and practices. The TMSA is designed to bridge implementation gaps of the ISM

Code and help distinguish between shipowners that merely attempt to meet the minimum requirements of the ISM Code and those that have embraced the full intent of the ISM with aims of achieving incident-free operations.

To this end, the TMSA focuses on four key elements of continuous improvement for the safe and environmentally conscious vessel operations: Plan, Act, Measure and Improve. Through their self assessment and reporting criteria, the TMSA has developed so-called Key Performance Indicators and require shipowners to benchmark their systems, procedures and performance against industry best practices.

Through the comprehensive evaluation of this reporting system, OCIMF intends to provide shipowners with a means to demonstrate their commitment to safety and environmental excellence.

CDI

Chemical company charterers established a comparable system to SIRE. This system is known as the CDI (Chemical Distribution Institute) Inspection Scheme. In a similar fashion, the chemical companies maintain a common database for the inspection reports made on chemical and liquefied petroleum and liquefied natural gas carriers.

There are some differences in the way these two report-sharing systems operate, however, the principle is the same. There is an effort for closer cooperation between the two organizations. For example both CDI and OCIMF have agreed to make use of the common VPQ, as all charterers need this standard information.

While not all charterers are members of OCIMF or CDI, these two systems are the most common systems for vetting inspections. Charterers who are not members normally have their own questionnaires and procedures for risk management. Needless to say, the crew's role remains the same, regardless of the charterer or the vetting method. It should be kept in mind that a poor inspection report remains in the SIRE database for other perspective charterers to see and ultimately to reject a vessel on that basis.

THE CREW'S ROLE IN VETTING

A typical vetting inspection starts with a meeting between the inspector and the master and chief engineer. The inspector should introduce him/herself, present his credentials and explain the scope of inspection. The inspection then normally proceeds with a review of vessel documentation and records, observation of cargo operations, inspection of all areas of the vessel including engine room, accommodation, safety appliances and cargo and mooring areas on deck. Interviews with key vessel staff are also usually conducted.

As a crew member, you should always be prepared for a vetting inspection. While most inspections are made with prior notice, this is not always the case. Many terminals will conduct an inspection during loading or discharge. Other times a vetting inspection could be made without any forewarning.

The role of the vetting inspector is to take a snapshot look at the operation of the vessel. He or she will report on the status and condition of all areas of the vessel back to the vetting/risk management department of the chartering company or potential charterer for evaluation. The inspector should be treated with respect, but he should not be allowed to interfere with the vessel's operation. By properly maintaining the areas of the ship you are responsible for, you will help to ensure a successful vetting inspection. Clear violations of regulations or standard industry accepted practices will have a negative influence on the inspection and the possibility of vessel employment.

A positive vetting inspection will help to ensure acceptance by perspective charterers and thereby allow the vessel to trade without restrictions. A poor vetting inspection will probably result in the charterer rejecting or disapproving of the vessel. A vessel that is not approved by different charters cannot be readily employed.

It is very important to correct any deficiencies noted by a vetting inspector as soon as possible. The vetting inspector's comments and the corrective actions taken by the crew should be reported back to the vessel's superintendent or port captain by the master as soon as possible. The superintendent or port captain should then indicate that corrective actions have been taken when submitting the owner comments to the SIRE database. These comments can then be taken into consideration by the vetting departments that access the inspection report for determining the suitability of the vessel for possible service.

If all goes well and the vessel is found to be of a good standard, then the vessel can be chartered. If there are many deficiencies, or some significant violations identified, the vessel may be required to undergo a subsequent inspection to ensure that corrective action has been taken.

To better understand how to be prepared for a vetting inspection, the following information has been reprinted with permission from Intertanko's "A Guide for Vetting Inspections" 7th edition.

A GUIDE TO VETTING INSPECTIONS

The inspection result establishes whether the tanker is operated in a safe way in accordance with valid rules, regulations and best industry practices.

The onboard inspection can only be successful if the tanker is prepared for the inspection. The inspector who is to carry out the inspection will start to collect impressions from even before the time he takes his first step onto the gangway and will continue to do so until he takes the last step off the gangway when leaving the tanker after completing the inspection.

Almost all accredited vetting inspectors are former seafarers who have either deck or engine room experience. Most likely the first impression formed from the time the tanker is sighted until the inspector's arrival at the Master's cabin will be the strongest. It will be subjective at this point. Inspectors will undertake the inspection of the tanker looking for objective criteria by which to judge the tanker. It is a fact of life that, however subconscious the urge may be, the inspector will look for objective evidence to support his initial subjective opinion. Thus the importance of the route from ship side to Master's cabin should not be underestimated. Remember you do not get a second chance to make a first impression.

Preparation for the Inspection

Make sure that the inspection is scheduled at a convenient time for the vessel, so it does not conflict with other inspections or similar matters. This could easily be arranged through the port agent.

Make sure that each head of department has completed his own inspection before arrival at port and that any deficiencies have been reported/corrected. This should be incorporated into the normal routine guidelines.

An effective way of administering this is to introduce a Self-Assessment form covering the following areas. The allocation of tasks for the specific areas is a suggestion and will depend on individual, company defined areas of responsibility.

• Tanker Particulars	Master
• Certification/Documentation	Master
• Crew Management	Master
• Safety Management	Master / Chief Engineer
• Lifesaving Equipment	Second / Third Mate
• Fire Fighting Equipment	Chief Engineer
• Pollution Prevention	Chief Officer
• Cargo / Ballast System	Chief Officer
• Inert Gas System	First Engineer
• COW Installation	Chief Officer
• Mooring Equipment	Chief Officer
• Bridge Equipment	Second Officer
• Radio Equipment	Radio Officer / Master
• Engine Room and Steering	Chief Engineer
• Load Lines Items	Chief Officer
• Chemical Supplement	Chief Officer

This is meant as an example. The next layer in this table is the delegation given to petty officers and in turn, to the rest of the crew. It is important to have a working organization that delegates. This will achieve an understanding all the way down through the ranks. Prior to the inspection preparations can be made in certain areas.

The Inspector may need to have a copy of the following:
- Classification Document
- IMO Certificate of Fitness
- Certificate of Registry
- IOPP Certificate & Supplement
- Cargo Ship Safety Construction Certificate
- Certificate of Financial Responsibility
- Cargo Ship Safety Equipment Certificate
- A Crew List
- Safety Radiotelegraphy Certificate
- Vessel's Safe Manning Document
- Load Line Certificate
- A Drawing of the vessel's cargo tank arrangement

The following should be available for Inspection (some are not applicable to all vessels):
- Officers Licenses
- Settings for vessel's P/V valves
- Health Certificates
- Shipping document and cargo manifest
- P&A manual
- Proof of life raft servicing
- Approved COW manual

- Safety Manual
- Approved Ballast manual
- Inert Gas Manual
- Vessel Response Plan/SOPEP/SMPEP, etc
- Oil/Cargo record book
- Safety Management System Manuals
- Vessel Operation Manual
- Oil transfer procedures
- Waiver Letters, if any
- Proof of cargo hose/piping testing
- Garbage log for compliance with MARPOL Annex V
- Proof of fixed and portable fire fighting equipment servicing
- Proof of professional servicing of breathing apparatus
- Company's policy for upgrading and training.
- Certificate of inhabitation or stabilization of cargo
- Declaration of inspection if transferring bunkers
- Cargo information cards and/or Material Safety Data Sheets for the cargo on board

Be prepared to calibrate and/or demonstrate the proper operation of:

- Combustible gas detectors or fixed gas detection system
- Oxygen analyzer
- Toxic gas detector
- Oil Discharge Monitoring Equipment (O.D.M.E.)
- Cargo Pump Emergency Shutdown and bearing alarms
- High Level alarms
- Overfill alarms
- Quick closing valves.

Be prepared to demonstrate the proper operation of the following systems/ alarms:

- Inert Gas system alarms
- Emergency Generator
- Oily Water Separator
- Engine room ventilation shutdowns
- Fire fighting systems
- Lifeboat engines
- Steering Gear

In addition, the following items may be checked and should be ready:

- Fireman's outfits
- EPIRB, pyrotechnics and hydrostatic releases
- International ship/shore connection
- Flame screens on bunker and ballast tanks
- Navigation equipment
- Paint locker smothering system
- Charts, publications, and corrections
- Marine sanitation device

Reference should also be made to the particular requirements of the oil major inspecting the vessel.

The following items are of vital importance, as these provide an overall impression of the vessel, and will play an essential part in how the inspection will be conducted.
1) Gangway: Correctly arranged - is the gangway net rigged? Is there a life ring near by?
2) Signs: All warning signs posted
3) Crew: All crew working on deck should have hard hats and the necessary protection gear.
4) Deck Watch: Is he present in the area? Hard-hat, emergency equipment handy, necessary for cargo loading/discharging; walkie-talkie; ask the inspector who he is and who he wants to see; confirm with duty Officer. One crew member should follow the inspector to the ship office.
5) Fire Equipment at the Manifold: Correctly rigged and present.
6) Deck: Clean, free of oil/water and obstructions.
7) Scuppers: Plugged and oil spill pumps in position and discharge connected.
8) Cargo Information: Make sure that all personnel involved in the cargo operation are briefed regarding what cargoes are being loaded/discharged, particularly the deck watch. All MSDS' to be posted and easily readable.
9) Emergency Equipment: Working, readily available and clearly marked.
10) Moorings: In good order and a good mooring watch maintained.
11) Accommodation: All doors and ports closed, clean and tidy and ventilation with a slightly positive pressure.

THE INSPECTION

Make sure that the inspector is accompanied on the vessel during the inspection. The best people to do this would be the Master, Chief Engineer, Chief Officer and the First Assistant Engineer (Second Engineer) who can divide the areas of inspection amongst themselves.

Normally, the inspector will start by checking all certificates and documentation with the Master. He would then move into the areas listed in the inspection protocol. However, it must be remembered that the order and schedule of the inspection can be changed to achieve less disturbance to the normal operations onboard. The inspector will have a pre-planned inspection format, which he will wish to follow, though there is the possibility to inspect areas in a different order. With the OCIMF VPQ, much of the data referring to the tanker will have been completed in advance. Make sure that you have a completed up to date copy available for the inspector as this will save much time.

Below are easily avoided, yet common, deficiencies found during vetting inspections:

Bridge and Radio Room
The most common deficiencies encountered in the Bridge / Radio room area are related to publications.
• Passage plan only pilot to pilot. Ensure that the filed passage plan covers berth to berth navigation

- Missing publications or old editions onboard when new publications have been issued
- Missing Masters standing orders and night order book
- No logs for gyro error
- No entry of position on the navigation chart during transit of pilotage to berth

Cargo Control Room and Tank Deck
- No cargo/ballast plan available
- Hydraulic leaks on deck
- Officers and ratings not wearing hard hats on deck
- No screens inside the vents for the ballast tanks
- No calibration gas for gas detection instruments
- Crew not wearing personal protection gear
- No policy for entering tanks

Engine Room and Steering Gear
- No procedures or instructions posted for foam system
- Emergency steering procedures not posted properly in steering gear room
- Hot work/procedures not used or not present in the manuals
- No safety guidelines available for engine room/workshop welding equipment
- No eye protection warning notices posted for engine workshop machinery
- No clean goggles by grinders and lathes

Accommodation/Galley
- Untidy
- Overhead ventilation greasy - fire hazard
- Accommodation ventilators with no identification labels

After the Inspection
All inspectors should sit down and discuss observations and comments after the inspection is completed. If not the Master should record a written objection that this has not taken place, and inform his company immediately. Additionally, the inspector should give the Master a written list of the deficiencies found. This is a very important part in the inspection process. It is also the last chance for input before the inspector will file his report with the oil major. Be sure to discuss everything thoroughly. Misunderstandings can be the cause of deficiencies.

- Correct all deficiencies as soon as possible
- Send the report to the head office or department in charge

REVIEW
1. What is vetting? Why is it necessary in the present-day transportation of cargo?
2. Describe the SIRE system established by the Oil Companies International Marine Forum.
3. List the major areas of a vessel that are examined in an inspection.
4. Why is a successful vetting inspection crucial to the vessel? What are some common deficiencies identified by vetting inspectors?
5. Describe the role of vessel personnel in a vetting inspection.

CHAPTER 12

Ballasting and Deballasting Operations

Tank vessels generally carry cargo in only one direction, hence the need to carry seawater ballast on the return leg of the voyage. There are exceptions to this rule such as parcel carriers that often backload as they are discharging cargo and older tank barges that lack the necessary equipment to ballast. Properly ballasting a vessel is extremely important; it represents a large percentage of the operating life of a tanker (see Figure 12-1). Careful planning and execution is also essential when conducting simultaneous cargo discharge and ballasting operations (Figure 12-3).

Figure 12-1. The ballast leg of a voyage constitutes a large percentage of the operating life of a vessel. Courtesy Abigail Robson

BALLASTING OPERATIONS

A proper ballast plan for the vessel should take into account a number of factors:

1. Meeting minimum draft and trim requirements consistent with safe operation
2. Minimizing hull stresses
3. Minimizing vibration
4. Reducing sail area (freeboard)
5. Maintaining maneuverability and handling
6. Ensuring fuel economy and speed
7. Stability of the vessel

Tank vessels are extremely vulnerable to structural fatigue (stresses) and mis-handling when operated in a light condition. Consistently operating the vessel in an underballasted condition can lead to damage and ultimately shorten the life expectancy of the vessel. Therefore, it is essential that operators carefully devise a suitable ballast plan for the vessel, taking into account the anticipated weather en route to the next loading port. Modern tankers are equipped with segregated-ballast systems consisting of tanks, piping, pumps, and sea chests used exclusively for ballast service during the life of the vessel. These systems were mandated as a means of reducing the routine operational discharges of oil that typically occurred during the ballast trip. Operational discharges at sea included the disposal of dirty ballast, tank washings, pipeline flush-ings, and pumproom bilges. The installation of segregated-ballast systems on tankers has dramatically reduced these discharges by simply eliminating contact between oil and water. The minimum quantity of ballast carried is dictated by the draft and trim requirements that are contained in Regulation 13-2 (a), (b), and (c) of MARPOL 73/78 and Title 33 CFR Part 157.10 which stipulate the following:

Minimum draft amidships = 2.0 + .02 (length of vessel) in meters
Maximum trim = .015 (length of vessel) in meters
Full propeller immersion

The rules further state that a vessel must comply with these requirements during *all phases* of the ballast passage. The segregated-ballast tanks must also be *protectively located* around the cargo area to minimize pollution in the event of collision or grounding. The total amount of ballast needed to meet the minimum draft and trim requirements is generally considered a fair-weather ballast condition. Adverse or deteriorating weather conditions offshore during the ballast trip often necessitate taking on storm ballast. Storm ballast is additional ballast carried in one or more cargo tanks which then require special handling or disposal at reception facilities ashore. In a storm condition, the vessel may carry 40 percent or more of the deadweight tonnage in ballast.

Ballast System Configurations

The segregated-ballast system on a tanker generally falls under one of three possible designs. The most common segregated ballast configuration today uti-lizes the space between the inner and outer hulls of a double hull tank vessel.

(Figure 12-2c) In the past, single hull tankers carried the segregated-ballast in a series of wing tanks (Figure 12-2a) that either partially or fully complied with the aforementioned draft and trim requirements. In the case of a double-bottom tanker (Figure 12-2b), the space between the inner and outer bottom is used for the carriage of seawater ballast.

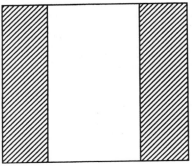

(a) Wing tank ballast arrangement on a single-hull vessel

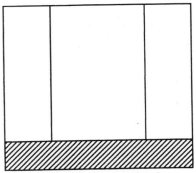

(b) The segregated-ballast space in a double-bottom tanker.

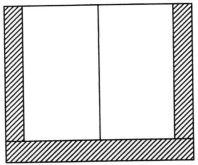

(c) The ballast space in a double-hull tanker.

Figure 12-2. Ballast system configurations.

| MT PETROLAB | VOYAGE # 1 | PORT | ANACORTES, WA |
| January, 2010 | | BERTH | SHELL NW |

CARGO

Discharge approximately 630,000 GSV bbls ANSCO via 2 and 3 headers using MCP's 2 and 3. Estimated discharge rate to be approximately 27,000 BPH (23.3 hours).

Start line displacement from #4W with #2 and #3 MCP's on low speed and discharge valves open 25% until flow is confirmed. After verifying that the ship and shore are ready to increase rate switch the pumps to high speed and gradually open the discharge valves until wide open. After the line displacement resume discharge with #2 and #3 MCP's from 1W. Once the discharge valves are wide open we will run the engine RPM at 127 or maximum dock pressure of 125 psi (whichever comes first). Once at the maximum rate, make a complete check of the deck and pumproom. Bring 1W, 2W and 6W down 1m before continuing discharge. Concentrate on discharging 3W and 5W in order to COW. Plan on finishing in the starboard slop tank.

IGS

Start and operate the inert gas system in the (IG generator mode) using the auto control feature with the deck pressure set point at 500 mmwg. Follow the checklist attached to the IG panel for start-up procedure. Notify the engine room prior to starting and stopping the IG system. Ensure system is delivering IG with an oxygen content of 5% or less.

COW

Utilize the COW pump to power the wash. Prime the pump prior to use. Start the wash in 3W and 5W at 1m innage and with at least 3m trim. Open COW supply valve slowly until the line is packed. Wash program is 0-150-0-50B/50W-0-50B/50W-0. Wash program is not complete until all Machines make a full cycle and at least 120 minute run time. Maintain a minimum of 7.6 bar (110 psi) on the tank cleaning main while washing. Prior to COW verify and log oxygen content of designated tank in the Deck Log and COW checklist. Complete COW checklist prior to and during the wash.

SEGREGATED BALLAST

Once 3W are at 15m ullage begin gravitating ballast in following this order: 3J, 5J, 1J, 2J, 4J and 6J. Gravitate one set of tanks at a time. Once finished gravitating use the SBP to fill to the stops shown in order: 1J, 2J, 3J, 5J, 4J and 6J. Prior to finishing begin to fill the forepeak and ER wings. Monitor stresses before and during ballasting operation and adjust accordingly.

DRAFTS

NOTE:

	Docking	Sail
	F 12.78m	F 9.55m
DEPTH ALONGSIDE DOCK = 13.7m	M 12.72m	M 9.77m
MAXIMUM DRAFT = 12.8m	A 12.80m	A 10.25m

OBSERVE ALL RULES AND REGULATIONS
MAINTAIN VESSEL SECURITY
CHECK PUMPROOM FREQUENTLY
UTILIZE CARGOMAX TO CHECK VESSEL STRESSES
MAINTAIN ACCURATE DECK LOG AND BALLAST HANDLING LOG
DO NOT HESITATE TO SHUTDOWN IF IN DOUBT
CALL ME ANYTIME IF NEEDED OR IF IN DOUBT
C/M _____
2/M _____
3/M _____
Cadet _____

	P	S
1	MT	MT
2	MT	MT
3	MT / COW	MT / COW
4	MT	MT
5	MT / COW	MT / COW
6	MT	MT
Slop	MT	MT

Segregated Ballast

	FPK 11M	
1	25M	25M
2	25M	25M
3	25M	25M
4	25M	25M
5	25M	25M
6	25M	25M
ERW	6M	6M
Trim T.	MT	MT

Figure 12-3. Ballasting Plan.

Ballast Terminology

A degree of confusion often arises concerning the terms used to describe the ballast on a vessel. The most commonly used terms are defined as follows:

Dirty ballast refers to seawater introduced into cargo tanks upon completion of cargo discharge. Residual (unpumped) cargo, clingage adhering to the sides of a tank, and cargo remaining in the pumps and pipelines all become part of the incoming ballast water, contaminating it. This ballast contains significant quantities of oil; it requires special handling at sea, or it must be retained on board for disposal at a reception facility at the loading port.

Departure ballast is a term used on crude carriers to describe seawater introduced into a cargo tank that has been crude-oil-washed during the cargo discharge. Crude-oil-washing of a cargo tank prior to taking on seawater ballast reduces the quantity of oil remaining on board to a fraction of what it would be otherwise.

Clean/arrival ballast is defined by the USCG as ballast "if discharged from a vessel that is stationary into clean, calm water on a clear day would not produce visible traces of oil on the surface of the water or on adjoining shore lines; or cause a sludge or emulsion to be deposited beneath the surface of the water or upon adjoining shore lines." Clean ballast discharged overboard through an approved oil discharge monitoring and control system has an oil content that does not exceed 15 parts per million.

A *segregated-ballast system* consists of tanks, piping, pumps, and sea chest (opening in the hull). The system is designed exclusively for ballast service and is completely separate from the cargo system. In the ideal sense, segregated ballast should always be "clean." However, structural failures such as bulkhead or inner-bottom fractures and pipeline leaks can result in contamination of the ballast water. It is, therefore, considered prudent to inspect and verify the condition of the segregated ballast prior to discharge overboard.

Ballasting Procedure

With a segregated-ballast system, seawater is introduced into the vessel by first lining up the ballast tanks, piping, and sea suctions. The ballast water enters the vessel either by gravity or by using dedicated pumps to bring water through a sea chest. As with any operation, the lineup should be checked to verify that ballast is flowing to the correct tanks.

DIRTY (STORM) BALLASTING

Today there are operational concerns that may still make it necessary to take on dirty ballast into the cargo tanks of the vessel. Taking seawater into the cargo system is a critical operation as it involves opening up the cargo system of the vessel to the sea. For this reason, established procedures must be followed when commencing "dirty ballasting" of the vessel. The cargo tanks to be ballasted, the piping, and the pumps should be correctly lined up. With the double sea-suction valves *closed*, the cargo pump should be started and run at an idle speed, thereby creating a vacuum in the sea lane between the sea-suction valves and the pump. At this point, the operator should first open the inboard sea-suction valve and then open the outboard sea-suction valve. When it is verified that seawater is flowing to the correct tanks, the speed of the cargo pump can be increased. The object of this procedure is to prevent oil in the cargo system from escaping (gravitating) from the vessel to the sea when the sea-suction valves are opened. Failure to adhere to this procedure could result in a serious pollution incident. In any event, whether ballasting with the pumps or by gravity, the sea-suction valves must be the last valves opened and the first closed.

VAPOR EMISSION CONTROL

Another issue presently confronting vessel operators is the need to ballast cargo tanks without venting any cargo vapors to the atmosphere. A number of states and localities have imposed strict limits on the emission of vapors during operations such as the ballasting of cargo tanks. During the ballasting of cargo tanks, all deck openings and vents should remain closed for the duration of the operation. The atmosphere being displaced by the ballast water coming into the cargo tanks can be handled in different ways:

With a *vapor control system*, the vapors from the cargo tanks being ballasted are directed ashore via collection piping on deck for processing or destruction.

Internal transfer is a method that involves simultaneous cargo discharge and ballasting operations within the vessel. The atmosphere from the cargo tanks being ballasted is directed to the cargo tanks being discharged, as shown in Figure 12-4.

Compression ballasting involves reducing the deck pressure of the vessel to a minimum at the end of cargo discharge. Then, with all cargo tanks common via the inert gas piping, the atmosphere from the tanks being ballasted is directed to the rest of the empty cargo tanks (Figure 12-5).

During this operation all vents and deck openings are secured. The atmosphere in the vessel's cargo tanks is compressed up to a safe margin below the pressure-relief-valve settings for the tanks.

All three operations should be conducted under the close supervision of senior personnel, as damage to the vessel, cargo, and environment can result.

DEBALLASTING OPERATIONS

Upon arrival at the loading berth, the ballast must be disposed of to permit the vessel to load a full cargo. Clean ballast, from either properly prepared cargo tanks or the segregated-ballast tanks, may be pumped into the harbor without risk of pollution. Dirty ballast is pumped to a designated reception facility at the terminal or to a barge.

Load-on-Top

Over the years, various ballast handling techniques have been utilized to minimize pollution of the sea. One such technique developed in the 1960s is called "load-on-top" (LOT). This is an operational method of reducing the quantity of dirty ballast and slops discharged into the sea. The technique (Figure 12-6) is still practiced by a significant number of vessels worldwide.

To understand the process, envision a VLCC at a terminal conducting a full discharge of cargo. The ship is fitted with a crude-oil-washing system; therefore, ballast preparations begin with a crude-oil-wash of all cargo tanks to be ballasted. The cargo tanks to be ballasted prior to leaving the dock at the discharge terminal must be washed as well as those cargo tanks to be used for arrival ballast. Upon completion of cargo discharge, the *departure ballast* tanks are filled with seawater

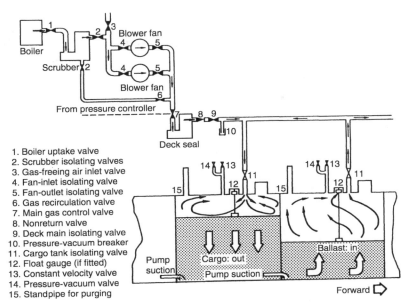

1. Boiler uptake valve
2. Scrubber isolating valves
3. Gas-freeing air inlet valve
4. Fan-inlet isolating valve
5. Fan-outlet isolating valve
6. Gas recirculation valve
7. Main gas control valve
8. Nonreturn valve
9. Deck main isolating valve
10. Pressure-vacuum breaker
11. Cargo tank isolating valve
12. Float gauge (if fitted)
13. Constant velocity valve
14. Pressure-vacuum valve
15. Standpipe for purging

Figure 12-4. Vapor emissions can be controlled by simultaneous cargo discharge and ballasting operations. The vapors displaced by the incoming ballast are transferred to the tanks being discharged. Reprinted with permission from *Controlling Hydrocarbon Emissions from Tank Vessel Loading*, 1987. Published by National Academy Press, Washington, D.C

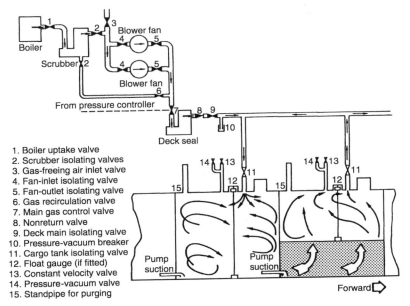

1. Boiler uptake valve
2. Scrubber isolating valves
3. Gas-freeing air inlet valve
4. Fan-inlet isolating valve
5. Fan-outlet isolating valve
6. Gas recirculation valve
7. Main gas control valve
8. Nonreturn valve
9. Deck main isolating valve
10. Pressure-vacuum breaker
11. Cargo tank isolating valve
12. Float gauge (if fitted)
13. Constant velocity valve
14. Pressure-vacuum valve
15. Standpipe for purging

Figure 12-5. When taking on ballast in a locality where vapor emissions are prohibited, the atmosphere in the ballast tank can be handled in several ways. One method, known as "compression ballasting" (shown here), involves transferring the vapors from the ballast tank to the available empty cargo tanks in the vessel. Reprinted with permission from *Controlling Hydrocarbon Emissions from Tank Vessel Loading*, 1987. Published by National Academy Press, Washington, D.C.

'Load on top' system of controlling pollution at sea

After discharging cargo, a tanker requires quantities of sea-water in some of its tanks to serve as ballast. When the water is loaded it mixes with oil residues in the tanks and becomes 'dirty'. During the voyage this dirty ballast water has to be replaced by clean ballast which can be pumped back to the sea without risk of pollution when the tanker reaches the loading port. Some empty tanks must therefore be cleaned at sea to ensure that the sea-water pumped into them as ballast remains clean and free of oil.

1 During the voyage tanks to be filled with clean ballast water are washed and the oily washings are collected into one slop tank.
The oil in the neighbouring 'dirty ballast' tanks floats to the top.

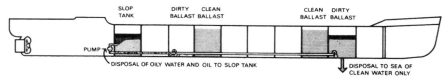

2 The now clean tanks are filled with ballast water which will remain clean and suitable for discharge at the loading port.
In the 'dirty ballast' tanks, the clean water under the oil is discharged to the sea and the oily layer on top is transferred to the slop tank.

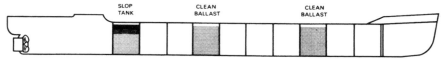

3 In the slop tank, the dirty washings and the oil from the dirty ballast settle into a layer of oil floating on clean sea-water.

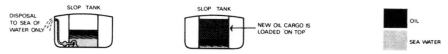

4 This clean water under the oil is carefully pumped back into the sea and the oily waste left on board. The next cargo is loaded on top of the remaining oil and all of it is discharged when the tanker berths at the refinery.

Figure 12-6. A summary of the load-on-top method of pollution control. Courtesy Shell International Petroleum.

to meet the minimum draft and trim requirements discussed earlier. The vessel departs the discharge port and begins the ballast leg of the voyage en route to the next loading port. The vessel is informed that it must arrive at the next loading port in a clean/arrival ballast condition. While at sea, the cargo tanks that were crude-oil-washed and left empty at the discharge terminal are now water-washed and stripped. The washings and slops are transferred within the vessel and retained in a designated slop tank. The washed tanks are then filled with clean seawater, considered *arrival ballast.* In the meantime, the oil in the departure ballast tanks has separated into a defined layer on the surface of the ballast. Using an interface meter, the depth of the oil layer can be determined and the water underneath this layer can be carefully pumped overboard through the oil discharge monitoring and control equipment (ODME). The pollution regulations that presently govern the disposal of cargo residues in this fashion can be found in Chapter 14. The water in the departure ballast is decanted overboard until the oil-water interface reaches a "safe" height above the bottom of the tank, thus reducing turbulence or eddying that could disturb the oil layer and cause unacceptable quantities of oil to be discharged to the sea. The small quantity of oil still in suspension in the ballast water being pumped overboard is dissipated in the wake of the vessel. The remaining oil and water in the departure ballast tanks is transferred to the slop tank where the process of separation and decanting is repeated. The vessel arrives at the loading port with clean ballast and a small quantity of oil and water in the slop tank. Before the clean ballast is disposed of, it may be necessary to flush the pumps and bottom piping into the slop tank to avoid any possible contamination. The clean ballast is pumped overboard at the loading port, and, depending on the trade of the vessel, the small quantity of oil and water in the slop tank is measured and the next cargo is either loaded on top of it or pumped ashore to a slop tank.

The effectiveness of the load-on-top procedure is influenced by a number of factors including these:

1. Length of the voyage
2. Type of cargo carried
3. Motion of the vessel
4. Cleaning procedures and effectiveness
5. Experience factor

Reception Facilities

The load-on-top procedure works well for most crude carriers with a sufficiently long ballast passage which permits ample separation time to process the ballast and slops. Product carriers and vessels on short hauls along the coast must often handle the dirty ballast and slops in other ways. These vessels utilize reception facilities at the loading port for the disposal of their ballast and slops. Reception facilities are shore tanks used exclusively for the receipt of dirty ballast and slops from vessels. The water and oil is processed ashore using separators, and the water is returned to the harbor. Today, many terminals are outfitted with sufficient reception facility capacity to accept whatever a vessel has on arrival. In the instance where a terminal cannot take a vessel's dirty ballast or slops, it may be necessary to transfer it to a barge or retain it on board.

Ballast-Water Exchange

In the past, efforts to reduce sea pollution have primarily focused on the oil content of the ballast water. In recent years, increasing concern over other forms of pollution have prompted regulators to impose restrictions on ballast-water disposal. The problem stems from the worldwide transport of various types of bacteria, viruses, plants, and marine organisms in the ballast water of vessels, which poses a serious threat to the marine environment. Following is a notice from one fleet operator concerning the need for ballast water exchange:

Ballast Water Exchange

During the past few years there has been increasing concern over the introduction of unwanted aquatic organisms and pathogens through the discharge of ships' ballast water. In many cases the organisms (plant, animal and bacterial species) have been able to flourish in their new surroundings, often to the detriment of indigenous marine life.

In 1993 the IMO Assembly adopted a resolution which contains guidelines for combating the problems, and the Marine Environment Protection Committee (MEPC) is currently drafting a possible new annex to MARPOL 73/78, which will introduce mandatory regulations.

However, IMO is anxious to ensure that any measures which are introduced or recommended to protect the environment do not threaten safety. One suggestion for reducing contamination from ballast water has been to exchange ballast water taken near the coast for water loaded at sea, where there is very little marine life. However, some concern has also been expressed about the safety of this practice, especially if the ship were to be caught by bad weather before the ballast exchange could be completed.

The draft guidelines provide two methods of carrying out exchange of ballast water at sea. One is the sequential method, in which ballast tanks are pumped out, bottom flushed, and refilled with clean water; and the other is the flow-through method, in which ballast tanks are simultaneously filled and discharged by pumping in clean water.

The draft guidelines go on to identify the precautions which need to be taken to ensure the ship's safety—such as taking into account weather conditions, ensuring the ship's stability, being aware of the possible effects on the ship's structure and other factors. They also cover crew training and familiarization and the long-term evaluation of safety aspects.

Reprinted with permission from Chevron Shipping Company, LLC.

Invasive species can cause damage in sensitive areas to the tune of billions of dollars. In the United States, the National Invasive Species Act of 1996 requires vessel operators to implement a ballast-water management scheme which includes a voluntary regime for ballast-water exchange. Certain areas such as the Great Lakes and California are protected by a mandatory ballast-water exchange program whereby prior to arrival , vessels must conduct a full exchange of the ballast water at sea beyond the exclusive economic zone (EEZ) in depths greater than 2,000 meters. Upon conclusion of the exchange, the ballast water that will be discharged must have a minimum salinity level of thirty parts per thousand. Alternatively, vessels entering the Great Lakes can retain the ballast on board during the time in the lakes or dispose of the ballast at a shore reception facility.

Tankers currently employ two basic approaches when exchanging the ballast at sea:

- Flow through method-ocean water is continuously pumped through the ballast tank in an effort to flush out or displace the original coastal ballast water. This typically involves volume changes of 300% or more to ensure effective removal of unwanted organisms.
- Empty and Refill method-the ballast tanks are emptied of coastal ballast water and refilled with open ocean water. The effectiveness of this method can be influenced by a number of factors including the ability to thoroughly strip the ballast tanks as well as the presence of mud, silt and other sediments that can remain trapped in the emptied tanks. Another concern associated with this method is the potential development of unacceptable hull stresses during the process at sea. Tank vessels that must conduct ballast water exchange need sufficient time (sea passage) in transit between ports as well as distance offshore to properly conduct this operation. Consequently, ballast water exchange may not be performed by vessels in near coastal trades that have an adverse impact in limiting the spread of non-indigenous species (NIS).

Whatever approach is ultimately selected to deal with this problem should not compromise the operational safety of the vessel or its crew.

To date, numerous studies have been conducted to quantify the effectiveness of ballast water exchange in certain bodies of water such as the Great Lakes and Chesapeake Bay. While research continues, it is generally accepted that ballast water exchange is a highly effective method in reducing the discharge of nonindigenous species (NIS) when conducted in accordance with the guidelines already mentioned.

Despite the obvious benefits of ballast water exchange in reducing the risks from NIS it continues to be viewed as an interim pollution control measure until a more effective shipboard control technology is developed, approved and implemented. Currently several approaches being tested to treat the ballast water onboard include:

1. microwave
2. oxygen depletion
3. infrared
4. ultraviolet
5. filtering

Ballast Water Treatment

As of this writing, OceanSaver AS is one of the first companies to receive IMO approval for installation of its ballast water treatment system on three new VLCCs being built by Hyundai. The ballast water treatment system is based on a three-pronged approach, namely:

initial filtration of ballast water entering the sea chest

exposure of the ballast water to a cavitation chamber

injection of a mixture of nitrogen and activated water into the ballast

The OceanSaver system has proven compliant with the IMO Ballast Water Performance Standard.

Another alternative being investigated involves a design change in the vessel essentially eliminating the need to conventionally ballast. The ballast free vessel design employs slow flow ballast tubes or trunks open to the sea when there is no cargo onboard thereby reducing the global movement of NIS.

The tanker industry continues to research methods of ballast-water management, treatment, and control to reduce the introduction of marine organisms and plant life not indigenous to the coastal areas and inland waterways of the United States.

REVIEW

1. What are the IMO draft and trim requirements for a 300-meter tanker in ballast?
2. Clean ballast is considered water with an oil content of less than _____ ppm.
3. List four factors that should be taken into account when devising the vessel ballast plan.
4. Define departure ballast.
5. Define arrival ballast.
6. A segregated-ballast system consists of what components?
7. Describe the procedure for commencing ballasting operations using the cargo pumps and piping.
8. Describe the load-on-top method of pollution reduction.
9. Why is ballast-water exchange necessary on certain vessels today, particularly those trading overseas?
10. List four factors that could adversely affect the performance of the LOT method of pollution reduction.
11. Why must the pipelines and pumps be dropped and stripped prior to introducing ballast water into the cargo system?
12. When ballasting cargo tanks in a state that limits vapor emissions, list the various ways the tank atmosphere can be handled.
13. What is a reception facility?
14. How can the water in a segregated-ballast system become contaminated?
15. If your vessel is required to exchange ballast prior to arrival, according to current guidelines where should it be conducted?

CHALLENGE QUESTIONS

16. Do the segregated ballast tanks of newly constructed double hull tankers meet the MARPOL draft and trim requirements for a tanker in ballast?
17. In the event that cargo has migrated into one of the double hull tanks contaminating the ballast describe how you would legally handle the disposal of the contaminated ballast?
18. In the previous question, explain the process of finding the location of the leak on the inner hull.
19. If the vessel trade involves repeatedly ballasting in rivers with heavy silt content what methods are employed to reduce the accumulation of solids and mud in the ballast tanks?
20. Discuss the importance of ongoing visual inspection of the condition of coatings by vessel personnel in the ballast spaces of a double hull tank vessel.

CHAPTER 13

Tank Cleaning Operations

In the operation of any tank vessel, it is necessary at some point to clean tanks. There are numerous reasons for cleaning:

1. Change of cargo or vessel trade
2. Preparation for clean ballast
3. Sludge control
4. Preparation for gas-freeing and tank entry
5. Preparation for shipyard

A successful tank cleaning operation involves careful planning and execution to avoid wasting time and energy. The tank cleaning plan should include such considerations as the sequence of tanks to be washed, the method to be employed, the number of machines, line pressure, temperature, stripping method, slop tank use, and atmosphere requirements. Fortunately, oil and chemical transporters provide extensive guidance in the form of cleaning charts or manuals that suggest the optimum method to clean a tank for a particular cargo or operation. Figure 13-1 is a cleaning chart from one operator for use within their fleet.

As an example, if the last cargo carried was commercial gasoline, special preparation would be required to load jet fuel (JP-4) on the next trip. Reference to the cleaning chart reveals the following recommendations:

1. Machine wash the tank with cold water
2. Flush all associated piping, heating coils, and pumps
3. Thoroughly strip the tank of washings
4. Dry the tank through ventilation and mopping
5. Remove loose sediment, sludge, and scale

TANK CLEANING OPERATIONS.

MSTS CHART 4020

LAST CARGO CARRIED		NEXT CARGO TO BE LOADED										
Military Products	Commercial Products	AvGas MIL-G-5572	White Gas VV-G-109	MoGas MIL-G-3056 VV-G-76	Jet Fuel MIL-J-5624 JP-3 & 4	Kerosene VV-K-211	Diesel Fuels MIL-F-16884 VV-F-800	Jet Fuel MIL-J-5624 JP-5	Burner Fuel Oil VV-F-815 Gr. 1&2	Lubricating Oils	Boiler Fuel Oil MIL-F-859	Burner Fuel Oil VV-F-815 Gr. 4,5&6
AvGas MIL-G-5572	Leaded Aviation Gasoline	A-B-E	A-C-D-E	A-B-E	A-C-D-E (3)	A-C-D-E	A-C-D-E	A-C-D-E (3)	A-C-D-E	(2)	A-C-D-E	A-C-D-E
White Gas VV-G-109	White Gasoline	A-B-E	A-B-E	A-B-E	A-B-E (3)	A-C-D-E	A-C-D-E	A-C-D-E (3)	A-C-D-E	(2)	A-C-D-E	A-C-D-E
MoGas MIL-G-3056 VV-G-76	Motor Gasoline	A-B-E	A-C-D-E	A-B-E	A-C-D-E (3)	A-C-D-E	A-C-D-E	A-C-D-E (3)	A-C-D-E	(2)	A-C-D-E	A-C-D-E
Jet Fuel MIL-J-5624	JP-5	A-C-D-E	A-B-E	A-B-E	A-B-E (3)	A-B-E	A-B-E	A-B-E (3)	A-B-E	(2)	A-B-E	A-B-E
	JP-3 & 4	A-C-D-E	A-B-E	A-C-D-E	A-B-E (3)	A-C-D-E	A-C-D-E	A-C-D-E (3)	A-C-D-E	(2)	A-C-D-E	A-C-D-E
Lubricating Oils		1	A-C-D-E	A-C-D-E	1	A-C-D-E	A-C-D-E	1	A-C-D-E	(2)	A-B-E	A-B-E
Kerosene VV-K-211	Water White or Standard White Kerosene	A-C-D-E	A-C-D-E	A-C-D-E	A-B-E (3)	A-B-E	A-C-D-E	A-B-E (3)	A-B-E	(2)	A-B-E	A-B-E
Burner Fuel Oil R-F-815 Gr. 1 & 2	Fuel Oil Gr. 1 & 2	A-C-D-E	A-C-D-E	A-C-D-E	A-C-D-E (3)	A-C-D-E	A-B-E	A-C-D-E (3)	A-B-E	(2)	A-B-E	A-B-E
	Dyed Kerosene	A-C-D-E	A-C-D-E	A-C-D-E	A-C-D-E (3)	A-C-D-E	A-C-D-E	A-C-D-E (3)	A-B-E	(2)	A-B-E	A-B-E
Diesel Fuels MIL-F-16884 VV-F-800	Diesel Gas Oil	A-C-D-E	A-C-D-E	A-C-D-E	A-C-D-E (3)	A-C-D-E	A-B-E	A-C-D-E (3)	A-B-E	(2)	A-B-E	A-B-E
	Diesel Oil Commercial Diesel	F	F	F	F (3)	F	F	F (3)	F	(2)	A-B-E	A-B-E
Boiler Fuel Oil MIL-F-859		F	F	F	F (3)	F	F	F (3)	F	(2)	B-G	A-E
Burner Fuel Oil VV-F-815 Gr. 4,5&6	Bunker Oil or Bunker C Fuel Oil	F	F	F	F (3)	F	F	F (3)	F	(2)	A-C-D-E	B-G
	Crude Oil	F	F	F	F (3)	F	F	F (3)	F	(2)	A-C-D-E	A-C-D-E
	Molasses, Linseed Oil, Waxes, Cotton Seed Oil & Tar (1)	H-I (1)	H-I (1)	H-I (1)	H-I (1)	H-I (1)	H-I (1)	H-I (1)	H-I (1)	H-I (2)	H	H

Numbers listed as (1), (2), and (3) above apply to NOTES on page 15

Figure 13-1. Cleaning charts provide guidance to the operator concerning proper cargotank preparation.

Figure 13-2. Typical portable tank cleaning machine used on smaller vessels. Courtesy Gamlen Chemical.

If no written cleaning procedures exist aboard the vessel, personnel often depend on an experience factor with the operation, or they turn to the cargo owner for advice on how to prepare for a certain cargo. Preparation for loading sensitive chemical cargoes often involves additional work such as freshwater rinsing, hand hosing, chemical washdown, and drying.

EQUIPMENT

A typical tank cleaning system consists of a sea chest, tank cleaning pump, heater, and fixed piping on deck.

Machines

The cleaning operation is accomplished through fixed machines permanently mounted in each tank or portable machines supplied by hydrants and flexible tank cleaning hoses suspended in each tank. Cleaning equipment and technology have advanced dramatically over the past several decades. Figure 13-2 illustrates a modern tank cleaning machine that is smaller, weighs less, and operates at a lower pressure than its predecessors.

Tank cleaning machines have smooth bore nozzles that deliver the washing fluid at high pressure, cleaning the surfaces of a tank by direct impingement (scouring action) of the jet as well as splashback. Figure 13-3 illustrates a typical permanently installed (fixed) tank cleaning machine in operation in a cargo tank.

The machines are commonly driven by the washing fluid and rotate in the vertical and horizontal planes to achieve the necessary coverage of the tank surfaces for effective cleaning. Nearly all surfaces of the tank are hit by the cleaning jet with the exception of "shadow areas," surfaces which may be shielded from the machine by framing members or other obstructions within the tank. If a significant portion of the tank surface is within the shadow, additional submerged machines (side- or bottom-mounted) or the spotting of a portable machine may be necessary to reach the areas in the shadow.

Figure 13-3. A permanently installed (fixed) washing machine operating in a cargo tank. Courtesy Hudson Engineering.

Portable Water-Washing

Although permanently mounted tank cleaning machines are becoming more prevalent in new construction, portable machines still predominate in smaller vessels. A portable tank cleaning machine (as shown in Figure 13-4) is used in conjunction with standard 2½-inch tank cleaning hose connected to a hydrant on the supply main.

Seawater is delivered to the washing machines from a tank cleaning pump. The temperature of the wash water can range from that of the ambient seawater to as high as 180°F if a heater is employed. The supply pressure to be maintained at the machines on deck is a critical parameter that varies considerably from one machine to another; therefore, the manufacturer's manual should be consulted for the correct value. Modern tank cleaning machines operate at lower pressures than their predecessors, so a standard value is difficult to assign; it can range from 100 to 180 psi (7 to 13 kg/cm^2). The correct supply pressure is critical to the success of the wash operation, as it affects the jet length and the cycle time of the machine. The cycle time represents the time necessary for the machine to rotate through all the angles at a particular location in a tank. The cycle time for portable machines varies depending on the manufacturer; however, cycle time on most machines averages approximately thirty minutes. Since most portable machines are driven (powered) by the washing medium, insufficient supply pressure results in slower machine rotation which has an adverse affect on the wash.

The machines should be electrically bonded via the hose to the tank cleaning supply main. Safe industry practice dictates that the tank cleaning hose be tested for continuity in a dry condition and visually inspected for damage prior to use. A measured resistance exceeding 6 ohms per meter of hose is cause for rejection of the hose. The tank cleaning hoses should be clearly marked and a record of the test results maintained on the vessel. All connections should be made up prior to lowering the machine into the tank—the machine is connected to the hose and the hose is connected to the hydrant prior to lowering the machine to the first drop (level) in the tank. These connections should *not* be broken until the operation is complete and the machine is removed from the tank. To facilitate draining the hose after the cleaning operation is complete, a bleeder valve should be opened or the coupling loosened to break the vacuum, then resecured.

The machine and hose assembly is lowered into the tank (Figure 13-5) through openings in the deck; it is typically supported on a fairlead called a saddle.

Figure 13-4. A portable tank cleaning machine connected to a standard tank cleaning hose. Courtesy Apollo International Corp.

Figure 13-5. Tank cleaning hose and machine suspended in a cargo tank on a fairlead called a saddle.

The machines generally have a fitting to which a tag line is connected. This facilitates changing the position of the machine as well as securing it at a particular drop. Some vessels are equipped with a special hose reel (Figure 13-6) assembly for portable washing operations.

The tanks are washed in a series of drops with the machine remaining at each level for the specified cycle time of the machine or in some cases longer. The number of drops and the amount of time spent at each drop can vary based on a number of factors including the following:

1. Experience factor with the cleaning operation
2. Tank size and internal configuration (complexity)
3. Time elapsed since the last cleaning operation
4. Previous cargoes carried
5. Coated or mild steel tanks
6. Hot- or cold-water wash
7. Rinsing or full wash

The number of tank cleaning machines that can be operated simultaneously is usually dependent on the dimension of the supply main and the ability to maintain the design pressure at the machine. A typical tank cleaning error occurs when personnel attempt to operate more than the recommended number of machines simultaneously. As mentioned earlier, this can result in low supply pressure leading to slower machine rotation and inadequate jet length. Operators should be mindful of the limitations of the cleaning system that can influence the effectiveness of the wash. An equally important consideration that can affect the success of the cleaning operation is the stripping capacity of the vessel. Effective stripping is a vital part of any tank cleaning operation. Wash water and slops must be continuously removed while the cleaning machines are in operation. The washings are stripped to a designated slop tank for processing and disposal. Stripping is accomplished through the use of a positive-displacement pump, an eductor, or a self-priming centrifugal pump. When conducting a bottom wash, stripping system capacity must exceed the total throughput of the tank cleaning machines in use. The object is to prevent the buildup of washing fluid on the tank bottom, which would shield the dirtiest part of the tank from the cleaning jets. It is therefore important to monitor the tank to ensure the machines are washing over a "dry" tank bottom. Should a buildup occur, the operator should stop the wash, strip the accumulated wash water, and reduce the number of machines in use.

The use of coatings on the tank surfaces (for example, epoxy) is very effective in reducing the time and effort necessary to clean a tank. The use of such coatings also reduces the quantity of scale and cargo impregnated in the steel of a tank. Coating manufacturers should be consulted for any cleaning limitations that might contribute to coating failure. Typical restrictions involve the temperature of the wash water, supply pressures, and the use of chemical additives. In addition, contact with certain types of cargoes, heating requirements, excessive stress, and vibration can also contribute to the premature failure of the coatings in a tank. It is recommended that tank coatings be regularly inspected for evidence of failure and maintained in accordance with manufacturers' guidelines.

Figure 13-6. Portable tank cleaning operation using a hose reel on deck. Courtesy Gamlen Chemical.

Fixed Machines

The use of fixed washing systems with permanently mounted machines in the cargo tanks paralleled the development of the modern supertanker. With the advent of larger ships came the daunting task of how to effectively clean the immense cargo tanks. It soon became apparent that the portable cleaning methods of the past were inadequate for the job.

A number of equipment manufacturers developed fixed tank washing machines connected to fixed supply piping on deck. Figure 13-7 illustrates a typical deck-mounted tank cleaning machine that can be powered with seawater or crude oil.

When compared to portable machines, fixed machines have the advantages of greater throughput and jet length, which are more effective when cleaning larger tanks. The higher output fixed washing machines are typically classified according to their throughput. A high-capacity washing machine (HCWM) is defined as any fixed machine with a throughput exceeding 60 cubic meters per hour. Fixed machines in this category were linked to the development of a significant static charge in the tank atmosphere which ultimately led to the inerting requirements (see Chapter 16). Other claimed advantages of fixed washing systems are the reduction of labor and the fact that it is easier to control the tank atmosphere since all deck openings remain closed during the operation. In new construction, some owners are now installing fixed washing systems in their smaller product carri-

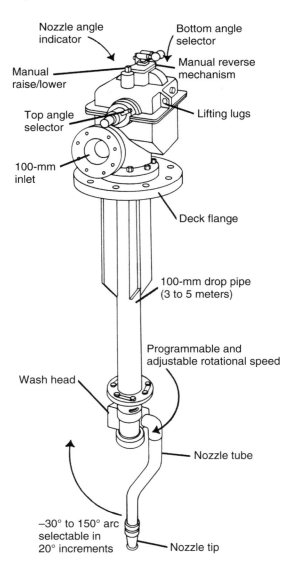

Figure 13-7. A deck-mounted washing machine that can be powered with crude oil or water. Courtesy Butterworth.

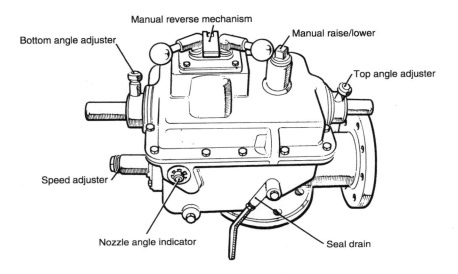

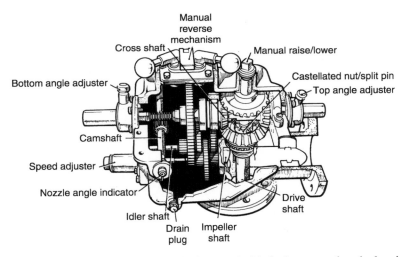

Figure 13-8. Gearbox and controls for the lavomatic SA deck-mounted tank cleaning machine. This model contains (1) the in-line washing fluid turbine that powers the machine and (2) the system of gears controlling the speed and direction of nozzle rotation. Courtesy Butterworth.

ers, a departure from the traditional use of portable washing methods. Fixed tank washing machines can be divided into two groups: programmable deck-mounted machines and nonprogrammable submerged machines. Deck-mounted machines are usually the single-nozzle variety and can be programmed to perform the wash in two or more distinct stages in a tank. The operator sets the machine to wash between prescribed angles using a control unit on deck (Figure 13-8).

Figure 13-8a. COW machine gear box on deck. Courtesy Mark Huber.

Twin-nozzle nonprogrammable machines are strategically positioned (side- or bottom-mounted) in the tank to reach the shadow areas which are missed by the deck-mounted machines. The shadow areas represent the tank surfaces that are shielded from the cleaning jets by large primary structural members such as main girders, stringers, transverses, web frames, and so on (Figure 13-9). These machines are factory set to perform a complete washing pattern every time, cycling through all the angles. The number and location of tank cleaning machines is determined by the coverage of the tank surfaces. The IMO rules for crude-oil-washing systems state that "all horizontal and vertical areas are washed by direct impingement or effectively by deflection or splashing of the impinging jet."

During the washing operation, the machine settings, cycle times, and proper rotation should be checked by personnel on deck. With double-hull vessels, the smooth internal surfaces of the cargo tanks and the absence of shadow areas should make the job of cleaning tanks much simpler. The smooth inner bottom of the cargo tank combined with the use of suction wells or sumps should enhance stripping effectiveness and reduce the quantity of muck and sediments left behind in the tank. With correct positioning of the tank cleaning machines, operators should find the job of cleaning the cargo tanks on a double-hull vessel far easier and more effective than on the single-hull vessels of the past. On the other hand, should it become necessary to machine wash the space between the hulls (segregated-ballast tanks), the structural complexity of this area will make the job of cleaning and gas-freeing difficult at best.

Figure 13-9. This internal view of a cargo tank illustrates the difficulty of reaching all the areas of the space during a cleaning operation.

Pipelines and Pumps

When cleaning tanks it is important to remember that it may be necessary to flush all associated cargo piping and pumps. For example, before water-washing tanks in preparation for clean ballast, it is imperative that all bottom piping, branch lines, and pumps be thoroughly flushed with water. Failure to do so may result in the contamination of the clean ballast by cargo remaining behind, trapped in dead-end sections of the piping system. The contents of cargo piping and pumps should also be considered when preparing for a change of cargo, repair work, and the shipyard.

CRUDE-OIL-WASHING (COW)

Growing world concern over pollution of the seas prompted the development of a new cleaning technology called crude-oil-washing. Crude-oil tankers have always been viewed as a major contributor to the worsening condition of the seas. This conclusion was based on the fact that they constituted the majority of world tanker tonnage and crude oil was considered a "persistent" oil. A persistent oil is one that is not fully processed by nature through weathering, a combination of wave and wind action. When crude oil is discharged or spilled into the sea, the light fractions weather away through vaporization, typically leaving behind the heavier asphalt and waxy substances in the form of "tar balls." The problem originated from what was termed "routine operational discharges" of crude carriers, namely the disposal of dirty ballast, tank washings, slops, pipeline flushings, and pumproom bilges at sea during the ballast leg of the voyage. The estimated quantity of oil being dumped into the seas annually was staggering. Over the past forty years the industry developed several approaches to combat the problem including load-on-top procedures, installation of shore reception facilities, segregated-ballast systems, and crude-oil-washing systems.

A crude-oil-washed tanker has a significantly reduced quantity of oil remaining on board at the completion of a discharge. This minimizes contamination of the departure ballast water and eliminates the need for extensive water-washing of cargo tanks at sea. The net result is a reduction in the generation of oily-water mixtures (slops), the basis of the pollution problem. Tanks are washed with the cargo (crude oil) during the normal discharge of the vessel. The first meter of any cargo tank to be used as a source of crude-oil-washing fluid must be pumped out. This reduces excessive electrostatic generation due to the presence of water in the crude-oil-washing fluid. Cleaning the tanks in this fashion takes advantage of the solvent properties of the crude oil to degrease the cargo adhering to the surfaces of the tank (clingage) as well as to dissolve the solid residues (muck) that accumulate on tank bottoms. The need for manual removal of sludge from cargo tanks (mucking) has been greatly reduced as the majority of the solids become part of the delivery being pumped ashore during the discharge of the vessel. This saves a substantial amount of time when it is necessary to clean the entire vessel in preparation for drydock. There is also an economic benefit to owners as crude-oil-washing increases the quantity of cargo received in the shore tank. Comparative studies have revealed that a crude-oil-washed VLCC can deliver an additional 1,000 to 1,500 tons of cargo which would otherwise have remained in the vessel's tanks and pipelines.

During a typical discharge, crude oil is routed to a special line on deck referred to as the COW main, which branches off to the individual fixed cleaning machines in each tank. According to regulation, the crude-oil-wash piping and machines must be permanently installed and constructed of steel or an equivalent material. The COW machines must be supported in a manner that will enable them to withstand the vibration and pressure surges associated with the operation. Each COW machine must be equipped with stop valves or equivalent in the branch line. The nozzles are driven by the washing fluid acting on impellers and gearing within the body of the machine or through the use of portable drive units mounted on deck. Where portable drive units are employed, there must be a sufficient number carried on board so that no more than two moves from the original position are necessary to carry out the COW program. Figure 13-10 depicts the COW piping connected to a fixed washing machine mounted under the deck.

The crude oil is supplied to the COW main from either the discharge side of the main cargo pumps or a special general purpose pump. The COW machines operate at pressures ranging from approximately 110 to 150 psi, the details of which can be found in the crude-oil-washing operations and equipment manual for the vessel. While conducting the wash, if the cargo pump discharge pressure is inadequate to drive the machines, it may be necessary to throttle a discharge valve to reach the recommended operating pressure. Doing so has a negative impact on the discharge (turnaround) time of the vessel. Some owners opted to install an additional "general purpose" pump for this operation rather than punish the delivery from the main cargo pumps. As in the case of water-washing, maintenance of the design pressure to the crude-oil-washing machines is critical

Figure 13-10. The crude-oil-wash-supply piping and stop valve at the machine on deck.

to the success of the operation. Crude-oil-washing generally takes place while the cargo tank is being discharged. As the cargo level drops, the exposed surfaces of the tank are washed in a series of stages (i.e., top, middle, bottom). Cleaning the tank in this fashion has several advantages:

1. It saves time, as crude-oil-washing can begin earlier in the discharge operation.
2. Freshly exposed crude oil is removed before it has time to cool and solidify.
3. Washing in stages reduces the use of the stripping system.

Figure 13-11 illustrates the typical settings for a multistage wash.

Reference to the COW operations and equipment manual for the vessel will give the prescribed ullages and machine angles for the cleaning operation. It is imperative that operators follow these guidelines, as improperly set machines can result in wasting time and energy "washing" the surface of the cargo instead of the tank surfaces. With most crude oils, effective cleaning can be accomplished with the machine performing one and a half cycles per arc. In the case of poor cleaning crudes or an excessively dirty tank, two to three cycles of the machine may be necessary. Overwashing a tank should be avoided as it is a waste of time. The bottom wash of a tank usually begins when the tank is nearly empty (approximately 1 meter left). As the bottom of a cargo tank accumulates the greatest quantity of oil, sediments, and waxy residues, one must ensure the machine jets are washing over a "dry" bottom. With this in mind, the COW rules stipulate that the stripping system must be capable of removing oil at a rate of 1.25 times the throughput of all the tank cleaning machines operated simultaneously. At this point, the vessel must have an adequate trim by the stern to aid in stripping the tanks. In order to verify the performance of the stripping system, suitable arrangements must be provided to confirm that the bottom of the cargo tank is dry at the end of the wash. For a single-hull vessel, a tank is considered dry, according to IMO, "if there is no more than a small quantity of oil near the stripping suction with no accumulation of oil elsewhere in the tank." Upon completion of crude-oil-washing and cargo discharge, the contents of all pumps and pipelines must be drained and stripped ashore through a special small-diameter line that terminates outboard of the vessel's manifold valve. This final stripping is critical, as unacceptable quantities of oil remaining in the lines will contaminate the incoming ballast water. It is important to remember the goal of the crude-oil-washing operation is to reduce the possibility of accidental or intentional sea pollution by ensuring the vessel leaves the discharge port with minimal quantities of oil and residues on board. Vessels performing crude-oil-washing are subject to inspection by port state authorities at any point in the discharge operation to ensure continued compliance with the regulatory and operational guidelines.

Atmosphere Control

Vessels that engage in crude-oil-washing operations must have an operational inert gas system. Prior to crude-oil-washing a tank, the oxygen levels must be determined at a point 1 meter below the deck and at half the ullage space. Additional oxygen readings may be necessary dependent on the internal

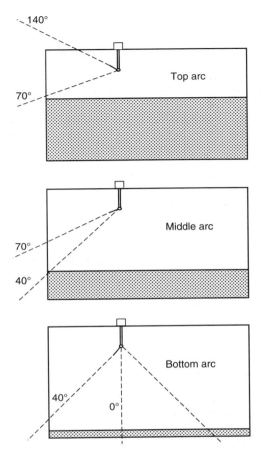

Figure 13-11. A typical multistage crude-oil-wash operation. Courtesy John Hanus and Mark Huber.

configuration of the tank. The oxygen readings must not exceed 8 percent by volume and a positive deck pressure is required throughout the wash. The object is to maintain the cargo tanks in a nonflammable condition throughout the cargo discharge and crude-oil-wash. The person-in-charge should carefully monitor the quality and quantity of the inert gas being delivered to the tanks during the washing operation.

For a more detailed discussion concerning the use of the inert gas system see chapter 16.

Tankage to be Crude Oil Washed

In the ideal situation, every cargo tank would be washed each time the vessel discharges. However, time constraints and charter obligations are frequently limiting factors. According to IMO, during the discharge of the vessel, a sufficient number of cargo tanks should be crude-oil-washed to enable compliance with the following criteria:

1. MARPOL 73/78 draft and trim requirements must be met throughout the ballast leg of the voyage (chapter 12).
2. Heavy weather ballast, should sea conditions necessitate taking on additional ballast.
3. After all tanks to be ballasted have been washed, 25 percent of the remaining tanks must be washed on a rotational basis to control the buildup of solid residues. Cargo tanks need not be washed more frequently than once every four months.
4. Ballast water may not be placed into a tank that has not been crude-oil-washed.
5. All crude-oil-washing must be completed before the vessel leaves its final port of discharge.

In the case of double hull crude oil tankers the number of cargo tanks that need to be washed during a typical discharge has been greatly reduced as the MARPOL draft and trim requirements are likely to be met by the segregated ballast tanks. Therefore the cargo tanks that are required to be washed are those designated for the receipt of "storm" ballast as well as the tanks being washed for sludge control. Additionally, the smooth internal surfaces and reduction of "shadow areas" in the cargo tanks has resulted in fewer fixed washing machines and less involved washing programs, a benefit to the operators of these vessels. Despite the reduction in the number of cargo tanks that must be washed during a discharge, the workload in port is hectic to say the least. Vessel personnel are expected to safely and efficiently carry out the simultaneous discharge of cargo, crude-oil-washing of tanks, operation of the inert gas plant, and ballasting of the vessel. The complexity of the operation has reached the point where everyone involved must be particularly vigilant to prevent any possible mishaps.

Slop Tank

The quantity of oil that can be legally discharged at sea from the cargo system is limited to a tiny fraction of the total cargo transported. It is therefore necessary to utilize a slop tank arrangement to retain and process the oily-water mixtures generated during the ballast trip. Existing vessels and new tank vessels under 70,000 dwt must have at least one slop tank. Vessels above 70,000 dwt must have at least two slop tanks. These tanks usually have sufficient capacity to retain the slops generated from tank washing, oil residues, pipeline flushings, and dirty ballast. Any other contaminated water within the cargo system is usually stripped to the slop tank. On smaller vessels, it is common to use one of the aftermost cargo tanks or a small independent tank on deck for receipt of the slops. There are a number of vessels equipped with multiple (wing) slop tanks that can be used in stages to assist in the processing of the oily-water mixtures. On these vessels, the separation process is more complete, thereby permitting the disposal of the water from the slop tank at sea. Where two slop tanks are used (Figure 13-12), the first tank to receive the oily-water mixture is the dirty tank.

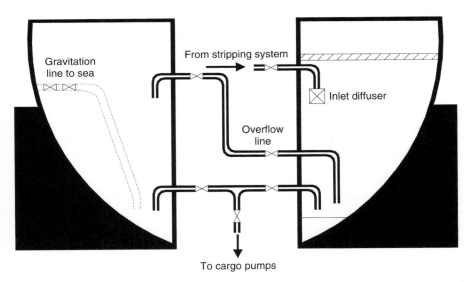

Figure 13-12. The piping arrangement on vessels eqipped with dual slop tanks ensures efficient separation of oil-water mixtures. Courtesy Richard Beadon and Eric Ma.

Both slop tanks are initially filled partway with clean seawater. Slops are introduced into the dirty tank about midheight in the tank. The oil forms a defined layer on the surface of the dirty slop tank while the clean (bottom) portion of the tank continuously gravitates (through a special line) to about midheight in the clean slop tank. The water at the bottom of the clean slop tank is either recirculated to the tank cleaning machines or pumped overboard through an oil content monitor (see Chapter 14).

Upon completion of the washing operation, the slop tanks are allowed to settle and the water under the oil layer is pumped overboard to a certain level above the suction point. The remaining oil and water mixture can be discharged to a reception facility at the loading port or commingled with the next cargo. The processing of the slops in this manner is called load-on-top, an operational technique that has been used on tankers to reduce sea pollution for many years (see chapter 12 for a detailed discussion of the load-on-top procedure). Unfortunately, this technique is labor-intensive and its effectiveness can be influenced by a number of factors including the following:

1. Motion of the vessel
2. Length of the ballast trip
3. Characteristics of the cargo
4. Seawater temperature
5. Human error

For this reason, many operators retain the slops on board for disposal at a reception facility at the next port.

SUPPLEMENTAL CLEANING

After a tank has been gas-freed, it is sometimes necessary to perform additional washing.

Hand hosing is one way to spot wash the areas of a tank not reached by the fixed or portable machines. It can also be used as an aid in the mucking process, sweeping the mud, scale, and sediment toward one location in a tank. Generally, a 1½-inch hose with a smooth bore nozzle is used, charged with water at 100 psi. Personnel must be cautious of slick surfaces while working in the tank in order to ensure good footing.

Mucking is an operation that involves the removal of scale, sludge, and other physical residues that have accumulated on tank bottoms, shelves, and platforms over an extended period of time. When preparing a space for "hot work" (see chapter 15), the removal of these residues is particularly important to minimize the regeneration of flammable vapors that could create a fire hazard. Personnel involved in the mucking operation should be suitably protected from contact with the cargo residues through the use of personal protective equipment. (Chapter 15 discusses the precautions to be taken when working in enclosed spaces.) The periodic mucking of tanks helps to control sludge buildup, reduces possible cargo contamination, and permits more cargo to be loaded in subsequent voyages.

Steaming is sometimes necessary to clean tanks that have transported certain cargoes. When steaming a tank, the following precautions should be observed. Steaming should only be conducted in cargo tanks that are properly inerted or gas-freed. The concentration of flammable gases in the space should be measured prior to steaming. The measured value should not exceed 10 percent of the lower explosive limit.

If *cleaning chemicals* must be used to clean a cargo tank, the operation should be conducted following the most stringent safety precautions. Operators should check the compatibility of cargoes and cleaning agents, particularly when stripping to a designated slop tank. Personnel should read and understand the material safety data sheets accompanying the cleaning agents. It is important to realize that certain cleaning chemicals may pose a toxicity hazard to personnel as well as a possible flammability hazard in the space. Personnel handling cleaning chemicals that pose a health risk should be outfitted with proper respiratory equipment and protective clothing.

TANK COATINGS

Modern tank vessels are usually constructed of mild steel which must be properly protected against corrosion. The surfaces of the cargo and ballast tanks are frequently lined in an effort to prevent corrosion or potential cargo contamination and to reduce the accumulation of scale and facilitate tank cleaning. A number of approaches are utilized to physically protect the steel structure of a vessel, including hard coatings, stainless steel, and sacrificial anodes (Figure 13-12a) Image provided by International Paint Ltd.

Hard Coatings

Hard coatings are the most common method of protecting the cargo and ballast compartments of a tank vessel. The tanks on a product or parcel car-

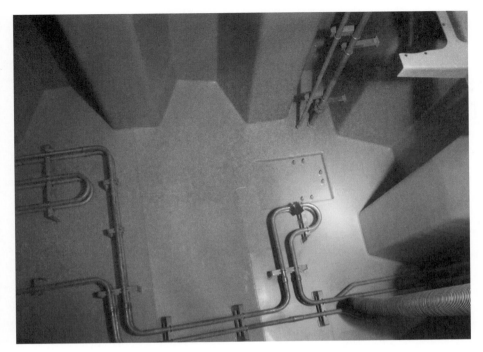

Figure 13-12a. Epoxy-coated cargo tank. Image courtesy of International Paint Ltd.

rier are typically lined with specially formulated paint based on the durability of the coating and its chemical resistance to the cargoes transported. These coatings provide a hard barrier between the steel surfaces of the tank and the cargo or ballast water. Some of the hard coatings commonly used in marine application today include:

1. Epoxy Phenolic (Interline 994), refer again to Figure 13-12a
2. Pure Epoxy (Interline 704)
3. Zinc Silicate (Interline 344)
4. Polymers

Regardless of the type of coating used on a vessel, the need for regular inspection cannot be overstated. Proper protection of the vessel structure requires careful inspection and a maintenance program to deal with any localized failure of the coatings, particularly in high-stress areas.

Major paint manufacturers around the world provide detailed cargo resistance guides based on their unique paint formulations such as the example provided by International Paint Ltd. (in Appendix E on the enclosed disc). Operators should consult the coating manufacturer for guidance if in doubt to prevent the possibility of incurring expensive damage to the coatings on the vessel (Figure 13-12b). Image provided by International Paint Ltd. Experience has shown that some coatings can be adversely affected by the following:

1. Temperature extremes (thermal shock)
2. High stress areas in the structure of the vessel
3. Excessive vibration
4. Aggressive chemical cargoes (acids and alkalis)
5. Abrasive cargoes
6. Impurities in cargoes
7. Tank cleaning
8. Inert gas

Figure 13-12b. Example of a failed coating. Image courtesy of International Paint Ltd.

Stainless Steel

An alloy of iron, stainless steel contains relatively high proportions of chromium, nickel, vanadium, and cadmium. In certain applications, cargo tanks are lined (clad) with stainless steel Figure 13-13) to protect the vessel from attack by aggressive chemicals and corrosives. In this system, the mild-steel plating of a tank is lined with a veneer of stainless steel usually around 2 mm thick. As in the case of hard coatings, manufacturers manuals should be consulted for proper maintenance of stainless steel tanks and piping.

Sacrificial Anodes

A cathodic protection system operates by sacrificing zinc anodes in lieu of the steel of the vessel. Zinc anodes are fastened to the surfaces of the tank and are sacrificed through electrolytic action that occurs when the space is filled with ballast water. The shortcoming associated with the use of anodes is the fact that they only provide protection when fully immersed in the ballast water. The ballast spaces can therefore experience serious wastage when left empty as is the case with segregated-ballast tanks during the loaded passage.

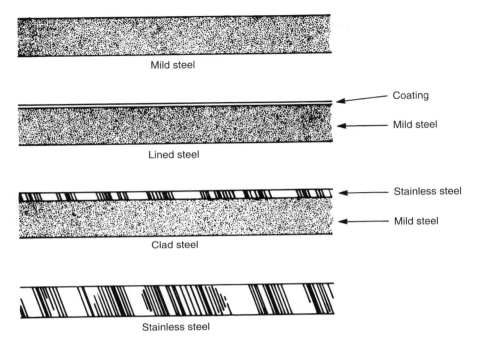

Figure 13-13 Cargo tank materials used in chemical tankers. Copyright International Maritime Organization (IMO), London.

REVIEW

1. List the typical reasons for cleaning tanks.
2. What is the recommended cleaning procedure if the last cargo carried was commercial motor gasoline and the plan calls for loading jet fuel in the same tank?
3. What is meant by the cycle time of a cleaning machine as specified in the manufacturer's manual?
4. Why is the supply pressure to the tank cleaning machines of particular importance to the success of the cleaning operation?
5. When performing the bottom wash of a tank, what is meant by cleaning over a "dry" bottom?
6. List the factors that influence the number of drops and the amount of time spent at each drop when cleaning a cargo tank with portable machines.
7. When water-washing an inerted cargo tank with portable machines, explain the precautions to be followed and the status of the inert gas system.
8. When should a continuity test be performed on a portable tank cleaning hose? What continuity reading is cause for rejection of the hose?

9. Describe the proper sequence of connection of the portable tank cleaning machine and hose when preparing to start the cleaning operation.

10. How are cargo pipelines, heating coils, and other associated piping cleaned for a change of cargo, and when it is necessary to gas-free them?

CHALLENGE QUESTIONS

11. As the number of double hull crude oil tankers in service increases describe the minimum number of cargo tanks to be crude oil washed during a typical discharge. How has the COW cleaning program changed since the advent of the double hull tanker?

12. What restrictions are typically imposed when cleaning a coated cargo tank?

13. Describe some of the typical reasons for an unsuccessful cleaning operation.

14. When is a diesel product wash employed on tank vessels today?

15. When transporting a crude oil considered unsuitable for washing, what options are available to vessel personnel?

CHAPTER 14

Pollution Regulations

During the past several decades, increased public awareness of the deteriorating condition of the marine environment has prompted corrective action both on the national and international level. One of the tasks early on was to identify the primary sources of pollution from the cargo system of tank vessels.

SOURCES OF POLLUTION FROM TANK VESSELS

The following lists, separated according to general category, show a number of potential sources of pollution from tank vessels:

Operational

1. *Deballasting*—the disposal of contaminated ballast water into the marine environment. Present-day concerns over the content of the ballast water go much farther than contamination by the cargo. Historically, the primary focus of authorities was contamination of the ballast water by oil and noxious liquid substances (NLS). Today, vessels are known to transport ballast water with unacceptable quantities of bacteria, plant and marine organisms, and solids (silt/mud) from distant regions of the earth.
2. *Tank washings*—disposal of cargo/water mixtures from the washing and flushing of cargo tanks into the sea.
3. *Pipeline clearing*—flushing the contents (residual cargo) of the cargo piping into the sea.
4. *Cargo transfer*—accidental discharges of cargo during loading and discharging operations. The transfer of cargo is associated with a heightened risk of pollution from the following:
   Overfill of a cargo tank during topping-off
   Overfilling a slop tank
   Ruptured cargo hose or loading arm
   Overfill of a cargo tank resulting from gravitation between tanks
   Vessel equipment or piping failure
   Structural failure of the vessel
   Human error

5. *Mucking*—the disposal of accumulated cargo (solid) residues and scale that has been physically removed from the tanks.
6. *Air pollution*—the uncontrolled release of the cargo tank atmosphere during loading and ballasting operations on the tank vessel.
7. *Bunkering operation*—the accidental discharge of oil while conducting a fuel oil transfer (bunkering) operation on a vessel.

Vessel Casualties
Cargo may be released into the environment as a result of a vessel casualty:

1. Grounding
2. Collision
3. Fire (explosion)
4. Structural failure

FEDERAL POLLUTION LEGISLATION

A number of national regulations have been implemented to reduce pollution of the marine environment. These include the Federal Water Pollution Control Act 1972, the Act to Prevent Pollution from Ships 1980, and the oil Pollution Act of 1990.

Federal Water Pollution Control Act (FWPCA)
Under the Federal Water Pollution Control Act (FWPCA), the discharge of oil within the waters of the United States is strictly prohibited as conveyed in the wording of the placard (Figure 14-1) that is required to be displayed on board vessels (Title 33 CFR Part 155.450).

Act to Prevent Pollution from Ships 1980
The Act to Prevent Pollution from Ships 1980 was implemented by federal regulation and can be found in Title 33 CFR Part 151.01 (subchapter O—Pollution). The thrust of this regulation was to adopt annexes I, II, and V of the international agreement known as MARPOL 73/78 into national rules. A noteworthy part of this regulation was the establishment of fines and penalties for violations of the act. Today, these penalties include fines up to $250,000 for individuals and up to $500,000 for organizations—with possibly half going to the informant—in addition to a jail sentence of up to six years.

Oil Pollution Act of 1990 (OPA 90)
One of the most controversial and widely debated pieces of U.S. legislation dealing with pollution is the Oil Pollution Act of 1990. This act covers many facets of the transportation and handling of oil cargoes. In addition to mandating changes in vessel construction (the double-hull requirement discussed in chapter 1) the act addresses preparedness and the ability to respond to a spill. Under OPA 90, owners must develop and provide a plan of action to be followed by vessel personnel when a pollution incident occurs or is likely to occur. The vessel response plan, as it is known, contains general information and operational instructions that must

DISCHARGE OF OIL PROHIBITED

The Federal Water Pollution Control Act

prohibits the discharge of oil or oily waste into or upon the navigable waters of the United States, or the waters of the contiguous zone, or which may affect natural resources belonging to, appertaining to, or under the exclusive management authority of the United States, if such discharge causes a film or discoloration of the surface of the water or causes a sludge or emulsion beneath the surface of the water. Violators are subject to substantial civil penalties and/or criminal sanctions, including fines and imprisionment.

**Report all discharges to the
National Response Center at 1-800-424-8802
or to your local U.S. Coast Guard office
by phone or VHF radio, Channel 16.**

Figure 14-1. The U. S. Coast Guard requires all vessels to display the placard regarding discharge of oil. Courtesy U.S. Coast Guard.

be approved by the U.S. Coast Guard. The following categories must be addressed in the plan:

1. Introduction and general information (i.e., ship's name, call sign, official number, International Maritime Organization (IMO) international number, and principal characteristics)
2. Notification procedures
3. Shipboard spill-mitigation procedures (i.e., procedures dealing with operational spills and vessel casualties)
4. Shore-based response activities (i.e., qualified individual's responsibilities and authority as well as the organizational structure of the response team)
5. List of contacts (i.e., governmental, corporate, and contractors)
6. Training requirements for vessel personnel and shore response team
7. Drills (frequency and exercises to determine spill-response preparedness)
8. Review and update procedures
9. Geographic-specific appendices
10. Vessel-specific appendix

INTERNATIONAL POLLUTION
LEGISLATION, MARPOL

Legislation aimed at reducing pollution from tankers has evolved on the international level through the efforts of the International Maritime Organization (IMO), which has been instrumental in hammering out the rules limiting the discharge of cargo residues from tankers. The most comprehensive of the IMO conventions dealing with marine pollution is the International Convention for the Prevention of Pollution from Ships, 1973, as modified by the protocol of 1978 commonly referred to as MARPOL 73/78. There are six annexes within MARPOL, each dealing with a different form of pollution as follows:

Annex I—Regulations for the prevention of pollution by oil

Annex II—Regulations for the control of pollution by noxious liquid substances

Annex III—Regulations for the prevention of pollution by harmful substances carried in package form

Annex IV—Regulations for the prevention of pollution by sewage from ships

Annex V—Regulations for the prevention of pollution by garbage from ships

Annex VI—Regulations for the prevention of air pollution from ships

Annex I: Prevention of Pollution by Oil

The following chart, taken from the USCG Marine Safety Manual, summarizes the MARPOL 73/78 requirements concerning the discharge of oil from the cargo system of a tanker (Figure 14-2).

It can be seen that MARPOL prohibits the discharge of oil and oily mixtures from the cargo system of a tanker within designated "special areas" of the world. In addition to imposing strict limits on the location, quantity, and rate at which oil can be legally discharged into the sea, MARPOL also requires detailed record-keeping.

The oil record book contains a comprehensive list of the cargo and ballast operations that must be chronologically recorded on the vessel. Vessel personnel must enter the date, operation code, and item number in the appropriate columns and any particulars in the space provided (Figure 14-3).

OIL RECORD BOOK

Entries shall be made for the following cargo and ballast operations that take place on a tank-by-tank basis:

1. Loading oil cargo
2. Internal transfer of oil cargo
3. Unloading of oil cargo
4. Ballasting of cargo tanks
5. Cleaning of cargo tanks including crude oil washing
6. Discharge of ballast except from segregated ballast tanks
7. Discharge of water from slop tanks

Control of Discharge of Oil from Cargo Tank Areas of Oil Tankers

Sea Areas		Discharge Criteria
Within a SPECIAL AREA		NO DISCHARGE except clean* or segregated ballast
Outside a SPECIAL AREA	Within 50 nautical miles from land	NO DISCHARGE except clean or segregated ballast
	More than 50 nm from land	NO DISCHARGE except either: (a) clean or segregated ballast; (b) or when: (1) the tanker is enroute; and (2) the instantaneous rate of discharge of oil does not exceed 60 litres per nautical mile; and *[Changed to 30 litres per nm]* (3) the total quantity of oil discharged does not exceed 1/15,000 (for existing tankers) *[1/30,000 for new tankers]* of the total quantity of cargo which was carried on the previous voyage; and (4) the tanker has in operation an oil discharge monitoring and control system and slop tank arrangements as required by Regulation 15 of Annex I of MARPOL 73/78

* "Clean ballast" is the ballast in a tank which has been so cleaned that the effluent therefrom does not create a visible sheen or the oil content exceed 15 PPM

Note: Items in brackets [ ] are not in the current MSM but show changes due to current regulation changes

Figure 14-2. Summary of the regulations governing the discharge at sea of cargo residues from oil tankers. Courtesy U.S. Coast Guard.

8. Closing of all valves after slop tank discharge
9. Closing of valves necessary for the isolation of dedicated clean ballast tanks
10. Disposal of oil residue

In the event of an emergency, accidental or other exceptional discharge of oil or oily mixture, a statement shall be made in the oil record book of the circumstances of, and the reasons for, the discharge.

Entries in the oil record book shall be fully recorded, kept up-to-date and signed by the person in charge of the operation and each completed page shall be signed by the Master. The oil record book must be kept onboard the vessel and

---
EXAMPLE

Name of Ship: _____

Official Number or Call Sign: _____

CARGO/BALLAST OPERATIONS (OIL TANKERS)

MACHINERY SPACE OPERATIONS (ALL SHIPS)

(circle one)

DATE	CODE	ITEM	Record of operations/signature of officers in charge.
8/11/82	G	27	No. 5 Port tank
		28	Port Shaw, CA
		29	1 hour
		31.1	8/11/82 J.B. Smith
8/14/82	A	1	Port Shaw, CA
		2	Heavy fuel oil 1-5C, 1-5 SB and 1-5P
		3	1500 barrels 8/14/82 J.B. Smith
8/16/82	C	6	Port Pine, Texas
		7	1 C 3 C and 5 C
		8	yes
8/18/82	B	4.1	2 C
		4.2	5 C
		5	No 8/18/82 D.B. Miller

Signature of Master: _____

Figure 14-3. Sample entries from an oil record book for cargo/ballast opertions. Courtesy U.S. Coast Guard.

readily available for review by port state inspectors. (Figure 14-3). In the United States, ownership of the oil record book resides with the United States Government and it must be retained onboard for a period not less than three years. A word of caution regarding oil record book entries is in order at this point. In recent times, port states have imposed substantial monetary fines and criminal penalties on seafarers and shipping companies alike for failure to maintain and/or making false entries in the oil record book. Therefore, individuals responsible for maintaining the oil record book are cautioned to be meticulous with respect to the timeliness and accuracy of their entries.

OIL DISCHARGE MONITORING AND CONTROL EQUIPMENT

The oil discharge monitoring and control equipment frequently referred to as the ODME is a system required under MARPOL that provides for continuous monitoring, recording and control of the overboard discharge of oily water mixtures to the sea. Today, the oily water mixtures are those resulting from dirty (storm) ballast, tank washings, pipeline flushing, pumproom bilges and any other slops generated within the cargo system of a tanker. The ODME calculates and records the instantaneous rate of discharge of oil in liters per nautical mile and the total quantity of oil discharged to the sea during the voyage.

The ODME recording usually includes a print out of the following parameters:

1. Instantaneous rate of discharged oil (liters/nm)
2. Instantaneous oil content (ppm)
3. Total quantity of oil discharged (cubic meters or liters)
4. Time and date (GMT)
5. Ship's speed in knots
6. Ship's position (Latitude and Longitude)
7. Effluent flow rate
8. Status of the overboard discharge control
9. Oil type selector
10. Alarm condition
11. Failure (no flow, fault etc.)
12. Over ride action (flushing, calibration etc.)

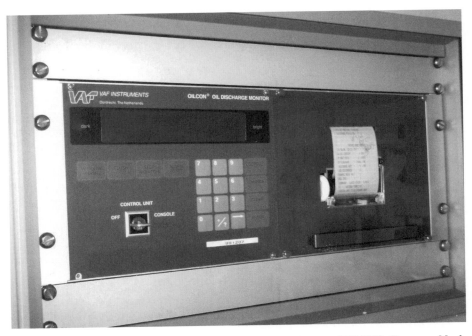

Figure 14-4. Oil Discharge Monitoring and Control Equipment (ODME). Courtesy Mark Huber.

ALARM CONDITIONS

The ODME should be provided with audible and visual alarms that indicate the following:

1. Instantaneous rate of discharge of oil exceeding 30 liters per nautical mile.
2. Total quantity of oil discharged reaches 1/30,000 of the previous cargo carried.
3. Operational fault in the system
   power failure
   loss of sample
   loss of measuring/recording

The system also includes a control feature on the overboard discharge to reduce the possibility of discharging oily water mixtures that violate MARPOL regulations. The printed record of the overboard discharge of oily water mixtures must be retained onboard the vessel for at least three years. The ODME should be tested monthly to ensure continued smooth operation of the system. Some of the more common sources of problems with the ODME are linked to dirt and corrosion (moisture, oil and particulate) in the air supply to valves and controllers. The accuracy of this equipment should be verified as part of the ongoing IOPP survey.

INTERNATIONAL OIL POLLUTION PREVENTION CERTIFICATE (IOPP)

A valid IOPP certificate must be maintained by the vessel. This certificate details the ship's arrangements and equipment that enable it to meet the applicable Annex I requirements. It is normally valid for a period not to exceed five years.

ANNEX II: Control of Pollution by Noxious Liquid Substances (NLS)

This annex contains the regulations for the control of pollution by noxious liquid substances (NLS) transported in bulk. In an effort to minimize the discharge of these substances, Annex II of MARPOL 73/78 requires that the design, construction, equipment, and operation of chemical tankers be in accordance with either the International Code for the Construction and Equipment of Ships Carrying Dangerous Chemicals in Bulk (IBC) or the Code for the Construction and Equipment of Ships Carrying Dangerous Chemicals in Bulk (BCH) as appropriate, based on the date of construction of the vessel.

NLS CATEGORIES

The substances in Annex II have been reclassified (effective 1 January 2007) into new pollution categories based on the Global Harmonized System (GHS) considering the level of hazard they pose to the marine environment:

Category X: noxious liquid substances which present a major hazard to marine resources or human health resulting from deballasting operations and the disposal of tank washings into the sea. These substances require the most stringent pollution control measures due to the possibility of a bioaccumulation hazard, biodegradation, acute and chronic toxicity to marine life and humans.

Category Y: noxious liquid substances which present a hazard to marine resources or human life resulting from deballasting operations and the disposal of tank washings into the sea. These substances require special pollution control measures due to possible bioaccumulation of short duration, possible tainting of seafood, or moderate toxicity to aquatic life.

Category Z: noxious liquid substances which present only a minor hazard to marine resources or human life resulting from deballasting operations or the disposal of tank washings into the sea. These substances only require special operational measures due to the fact that they range from slightly toxic to nontoxic to aquatic life.

OS: noxious liquid substances not considered a hazard to the marine environment.

DISPOSAL OF CARGO RESIDUE

IMO established detailed criteria governing the proper disposal of cargo residues in each NLS category. These include a number of control measures designed to significantly reduce the quantity of NLS cargoes discharged into the sea:

1. *Cargo stripping:* efficient stripping of the cargo tanks to reduce the quantity of residues remaining on board at the completion of discharge. New tankers contracted after 1 January 2007 must now comply with a new more stringent limit on the quantity of cargo remaining in a tank and piping upon completion of stripping to no more than 75 liters for category X, Y and Z substances.
2. *Cleaning and disposal procedures (CDP):* includes prewash of cargo tanks at the discharge port (following procedures outlined in Annex II and the Procedures and Arrangement Manual) including the use of reception facilities
3. *Ventilation procedures:* removal of substances with a high vapor pressure

In addition to these control measures, a number of conditions must be met with respect to the discharge of noxious liquid substances into the sea:

1. Vessel speed while proceeding en route (during discharge)
2. Discharge of effluent below the waterline, taking into account the location of sea suctions
3. Vessel location with respect to any designated "special areas" as well as the minimum distance offshore
4. Maximum quantity of substances per tank which may be discharged to the sea
5. Maximum concentration of substances (effluent) in the ship's wake
6. Minimum depth of water at sea during the discharge

CARGO RECORD BOOK

The cargo and ballast operations performed on a vessel carrying noxious liquid substances in bulk must be recorded in an approved cargo record book. The cargo record book must be completed for each tank in which the following occur:

1. Loading of cargo
2. Internal transfer of cargo
3. Unloading of cargo
4. Cleaning of cargo tanks
5. Mandatory prewash in accordance with the procedures and arrangements (P&A) manual

6. Ballasting of cargo tanks
7. Discharge of ballast from the cargo tanks
8. Disposal of residues to reception facilities
9. Discharge into the sea or removal of residues by ventilation
10. Accidental or other exceptional discharge of cargo

As in the case of the oil record book, entries must be completed by the person in charge of the operation and signed by the master of the vessel. The cargo record book is subject to inspection and review by competent port state authority.

SURVEYS

Vessels engaged in the transport of noxious liquid substances in bulk are subject to an initial survey as well as periodic surveys by the flag state administration. Continuing compliance with the requirements of annex II also rests with the inspectors of the port states where the vessel is trading.

PROCEDURES AND ARRANGEMENTS MANUAL

Vessels transporting noxious liquid substances in bulk must have an approved *Procedures and Arrangements* (P&A) manual on board. The P&A manual must incorporate the revisions to Annex II effective 1 January 2007. This manual contains detailed information of the vessel's physical layout and cargo handling equipment as well as operational procedures that must be followed in an effort to comply with the requirements of annex II of MARPOL 73/78.

The manual should contain operational instructions for vessel personnel conducting cargo handling, tank cleaning, handling of slops, and the ballasting/deballasting of cargo tanks.

The manual should contain the following information and operational instructions:

1. A list of the NLS cargoes the vessel is certified to transport and cargo specific information
2. A list of the cargo tanks and the noxious liquid substance(s) that may be carried
3. A description of the equipment and arrangements in the cargo system including such items as the following:
   Line drawing of the cargo pumping and stripping systems
   Cargo heating and temperature control system
   Identification of tanks to be used for slops
   Description of the discharge arrangements
4. The detailed procedures to meet the standards for the specific vessel, including such items as:
   Stripping methods (cargo tanks)
   Methods of draining cargo pumps and pipelines
   Prewash programs for the cargo tanks:
   Cleaning machine positions to be used
   Slops procedures
   Requirements for hot water washing
   Number of cycles of the tank cleaning machine or time required for wash

Minimum operating pressure
Instructions for the use of cleaning agents
Ballasting and deballasting procedures
Procedures for the discharge of cargo/water mixtures
Procedures to be followed when a problem develops involving a deviation from the approved discharge procedure
A table of the quantity of cargo residue in each tank upon completion of stripping operations
A table indicating the quantity in the tank after performing the water test

The master of the vessel shall ensure the discharge of cargo residues is conducted in accordance with the guidance contained in the P&A manual.

CERTIFICATES

Upon satisfactory completion of a survey, including approval of the P&A manual, the vessel is issued an International Pollution Prevention Certificate for the Carriage of Noxious Liquid Substances in Bulk or a Certificate of Fitness for the Carriage of Dangerous Chemicals in Bulk. Both are usually valid for a period not to exceed five years.

POLLUTION REDUCTION EFFORTS

Tremendous strides continue to be made within the tanker industry to reduce both accidental and intentional pollution of the marine environment. These efforts, shown in the following list, include operational measures as well as improved vessel design and equipment:

1. Load-on-top
2. Segregated-ballast designs
3. Shore reception facilities
4. Crude-oil-washing systems
5. Vapor control systems
6. Double-hull construction
7. Efficient cargo stripping systems
8. Prewash procedures and slops disposal
9. Ballast water exchange and treatment

REVIEW

1. List the operational sources of marine pollution from the cargo system of a tank vessel.
2. Describe the methods used in the tanker industry to reduce sea pollution both operationally and through vessel design.
3. Which of the IMO conventions is considered the most comprehensive with respect to pollution of the environment?
4. List the present pollution regulations governing the discharge of cargo residues from an oil tanker.

5. What is a "special area"?
6. What forms of pollution are addressed in each of the MARPOL annexes?
7. How are noxious liquid substances (NLS) categorized regarding pollution hazard?
8. What information must be contained in a "Cargo Record Book"?
9. What information is contained in a *Procedures and Arrangements* manual?
10. What certificates are required to be maintained on vessels transporting oil and noxious liquid substances in bulk?

CHALLENGE QUESTIONS

11. In recent times, what actions have been taken by Port State authorities against vessel personnel and shipping companies for making false entries in the Oil Record Book?
12. During the overboard discharge of the water from the settled storm ballast tanks at sea an alarm sounds on the ODME. List the possible causes for an alarm condition during this operation.
13. List the items that should be contained in the vessel spill response locker onboard.
14. During a loading operation an oil spill occurs at a facility. Describe your initial actions as the person in charge of the cargo transfer and discuss the notifications and the plan of action that must be carried out in accordance with the vessel response plan.
15. During the ballast leg of a voyage list some of the reasons for not being able to decant the oily water mixtures in the cargo system using the load on top technique.

CHAPTER 15

Enclosed Space Entry

Without a doubt, the entry of personnel into an enclosed space is potentially the most hazardous operation performed on a tank vessel. It is a necessary function in connection with the following operations: inspection, maintenance and repair, and hot work.

RISKS ASSOCIATED WITH ENCLOSED SPACE ENTRY

Entering a compartment that has been used to transport a variety of hazardous cargoes and sealed for an extended period of time poses serious risks and should only be done under the close supervision of experienced personnel. Today, these risks are compounded by the widespread use of inert gas systems with their associated requirement to maintain an oxygen-deficient atmosphere in the cargo tanks. This chapter seeks to warn the reader of the potential hazards and to describe the precautions that should be followed whenever personnel enter an enclosed space. Vessel personnel should consult the various sources of information (company manuals, National Fire Protection Association, *International Safety Guide for Oil Tankers and Terminals,* International Maritime Organization) for guidance as to the preparation, testing, and procedures for entering an enclosed space. The primary hazards associated with entry into enclosed spaces can be classified as physical and atmospheric.

Physical Hazards
The physical size of the compartments on a vessel can vary from extremely confined spaces (i.e., voids) to cavernous cargo tanks. For example, in a double-bottom ballast tank, personnel must crawl sideways through small lightening holes to access the various bays of the space (Figure 15-1 and Figure 15-1a).

Figure 15-1. A view into the double bottom of a tanker showing the lightening holes and restricted access to the adjacent bays. Courtesy Kevin Duschenchuk.

Figure 15-1a. Double sides and double bottom.

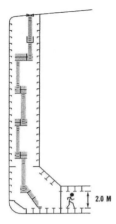

Figure 15-1b. The ballast spaces of modern double hull tankers present a particular challenge for entry by personnel. Courtesy Chevron Shipping Company LLC.

Fugure 15-2. The open expanse of a cargo tank can be seen in this photograph. Courtesy Salen & Wicander.

On the other hand, the cargo tanks on some vessels are vast open areas (Figure 15-2) that have considerable depth and large distances between bulkheads. Each type poses significantly different problems to the individuals who must enter and work in these compartments.

A related concern is the limited access to most spaces—small manholes or tank hatches that make ingress, egress, and the rescue of an individual in distress difficult at best. Entry into such compartments is not only physically challenging but personnel must be alert to the ever-present danger of slips, falls, and entrapment. Prior to certifying a space safe for workers, the access ladders, railings, and platforms in the compartment should be carefully checked for loose connections, deterioration, and slick surfaces. Individuals should avoid physical contact with cargo residue (muck) remaining on the surfaces of a tank; personnel entering tanks should wear proper protective clothing such as coveralls, gloves, boots, and eye protection.

Atmospheric Hazards

A more insidious problem concerns the atmosphere in enclosed spaces. A serious hazard is posed by the possible inhalation of dangerous concentrations of cargo vapors and oxygen-deficient inert gas in a space. Cargo vapors and inert gas are particularly troublesome given the fact that gases can move about freely and accumulate in concentrated pockets. Hydrocarbon vapors, for example, having a greater vapor density than air, tend to settle in the lower regions of a space. This could become an issue when determining the adequacy of the testing performed on the atmosphere of a space. Casualties have been linked to inadequate testing because sampling at one location may miss the "pockets" of gas at another. Another concern is the possible contamination of an atmosphere resulting from the leakage of vapors and inert gas through bulkhead fractures, piping, heating coils, and valves. To protect against possible leakage through piping connected to the space being entered, the lines must be positively isolated. On inerted vessels, the deck pressure should be reduced to a low positive pressure to minimize potential leakage through lines and bulkheads. It is important to realize that in sufficient concentration, these vapors are capable of not only incapacitating an individual, but actually killing the person. Table 15-1 lists the adverse effects experienced by individuals who breathe an atmosphere that is deficient in oxygen.

Table 15-1
Physiological Effects of Oxygen Deficiency

Oxygen Percentage	Effects
19.5–16	No visible effect.
16–12	Increased breathing rate. Accelerated heartbeat. Impaired attention, thinking, and coordination.
14–10	Faulty judgment and poor muscular coordination. Muscular exertion causing rapid fatigue. Intermittent respiration.
10–6	Nausea, vomiting. Inability to perform vigorous movement, or loss of ability to move. Unconsciousness, followed by death.
Below 6	Difficulty breathing. Convulsive movements. Death in minutes.

Courtesy MSA.

In addition to the concerns posed by oxygen-deficient (inerted) spaces, certain cargo vapors, such as hydrogen sulfide (H_2S) gas derived from "sour" crude oils, can quickly render a person unconscious. Characterized by a rotten-egg odor, hydrogen sulfide is a colorless gas which has a nasty property of accumulating in concentrated pockets. Unfortunately, inhalation of hydrogen sulfide deadens an individual's sense of smell, thereby stifling most forewarning of potential overexposure. Table 15-2 gives airborne concentrations of hydrogen sulfide and their physiological effects to personnel.

Table 15-2
Airborne Concentration of Hydrogen Sulfide and Effect on Personnel

PPM Level Hydrogen Sulfide	Physiological Effect
18–25 ppm	Eye irritation
75–150 ppm for several hours	Respiratory irritation
170–300 ppm for one hour	Marked irritation
400–600 ppm for ½ to 1 hour	Unconsciousness, death
1,000 ppm	Death in minutes

Courtesy MSA.

Prior to entering a cargo tank, an individual must be aware of the specific properties of the cargo and the concerns associated with exposure to that cargo. It is advisable to consult the appropriate Material Safety Data Sheets (MSDS) for the recommended occupational exposure limits and to check any additional precautions that may be necessary when working in an enclosed space.

GAS-FREEING PROCESS

The process of gas-freeing usually involves mechanically ventilating a compartment with fresh air to drive out the remaining cargo vapors and inert gas in the atmosphere. For gas-freeing to be successful, the tank must first be properly cleaned (mechanically washed) and purged with inert gas. Purging reduces the hydrocarbon concentration of the space to a point below 2 percent hydrocarbons by volume, where subsequent ventilation with fresh air will *not* result in the cre-

Figure 15-3. Portable high-capacity fan commonly used for gas-freeing cargo tanks. Courtesy Coppus Engineering.

ation of a flammable atmosphere. See chapter 16 for a detailed discussion of inert gas systems. When the cargo tank has been properly purged, the inert gas delivery is stopped and gas-freeing commences. Gas-freeing is accomplished through the use of fixed or portable fans. The method employed is usually dependent on the number of tanks to be gas-freed. For example, when it is necessary to gas-free a small number of tanks for inspection or repair, it is common practice to use portable fans mounted in the openings on deck. Figure 15-3 illustrates a typical portable fan positioned over the tank cleaning opening in the deck.

Fans should be electrically bonded to the hull and positioned in such a way that all areas of the space can be adequately ventilated. The outlet point from the tank should result in the longest possible path through the tank to ensure good cross ventilation. Modern portable fans have a high-volume output and can operate in either a supply or exhaust mode. These fans can be driven pneumatically, hydraulically, or by using steam acting on a turbine. When it is necessary to gas-free the entire vessel in preparation for the shipyard, the inert gas fans can be used in the gas-free mode. This is accomplished by removing a blank in the fresh-air inlet line on the suction side of the IG fans. Figure 15-4 shows the location of the fresh-air inlet line in the IG system.

The fans deliver fresh air to the cargo tanks via the deck distribution piping (IG main and branch lines). Gas-freeing the vessel in this way is not only convenient but also permits ventilation of the cargo piping in preparation for the shipyard. A number of questions frequently arise concerning proper gas-freeing of a space such as the required amount of time and the necessary number of volume changes of the tank atmosphere. It is important to realize that gas-freeing is not a timed operation. Ventilation of a space should continue until safe readings are attained

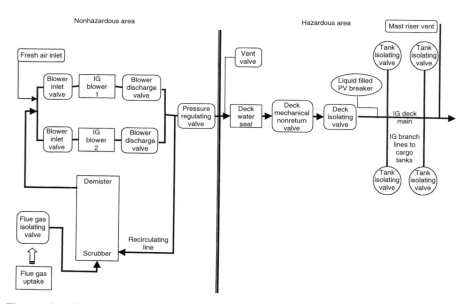

Figure 15-4. The inert gas system can be used in the gas-free mode. Note the location of the fresh air inlet on the suction side of the fans. Courtesy Richard Beadon and Eric Ma.

with suitable atmosphere testing equipment. The time required to accomplish this depends on many variables:

1. Adequacy of the cleaning job
2. Time elapsed since the previous cleaning
3. Quantity of muck in the tank
4. Gas-freeing method employed
5. Size and complexity of the compartment
6. Temperature and humidity conditions
7. Leakage from adjacent compartments and lines
8. Cargoes previously carried

TESTING

To ascertain the success of the gas-freeing operation, thorough testing of the atmosphere in the space is essential. While gas-freeing operations are underway, personnel should first check all atmosphere testing equipment. The accuracy of the testing equipment is critical to the safety of the operation, since the decision to enter a space is based largely on its readings. The testing equipment should be checked using a calibration gas from the manufacturer and following the procedures outlined in the manufacturers manuals. The testing equipment is connected to a cylinder of "measured" gas provided by the manufacturer and the reading is compared to the stated value. At this time, the battery-charge and alarm functions should also be checked. A log of instrument maintenance and testing should be maintained on board the vessel.

Testing for Gases

After the accuracy of the testing equipment is confirmed, the compartment to be entered is tested remotely using clean sample tubes connected to the instrument with the blowers off for a minimum of ten minutes. According to the *International Safety Guide for Oil Tankers and Terminals (ISGOTT)*: "Care should be taken to obtain a representative cross-section of the compartment by sampling at several depths and through as many deck openings as practicable." Figure 15-5 shows a compartment being tested remotely using a sampling tube prior to entry. Cargo tanks with partial bulkheads or complex internal configurations may require further sampling to fully assess the condition of the atmosphere. If the work in a space involves opening or removal of cargo piping, pumps, and heating coils, they should also be thoroughly tested. Testing is performed by drawing a sample through the instrument using an aspirator bulb or battery-operated pump. (The manufacturer's manual will give the time or number of squeezes of the bulb necessary for the sample to reach the instrument based on the length of sample tube employed.)

The performance of the instrument should be carefully monitored while testing the piping components and the compartment space. The space should first be tested for oxygen content with a portable analyzer to determine that the level is safe for personnel and sufficient for the combustible-gas indicator to function correctly. The space is then tested with a combustible-gas indicator. Certain types of these indicators operate by measuring the increase in resistance caused when the sample heats a hot wire filament. Instruments that function on a burning principle will not respond correctly if the atmosphere being tested is oxygen deficient. Personnel should refer to the manufacturer's manual for information regarding instrument

Figure 15-5. Remote sampling of the atmosphere in a cargo tank. Courtesy Kelly Curtin and Mark Huber.

limitations and the use of response curves when interpreting the readings. Figure 15-6 shows a Watchman Multigas Monitor from Mine Safety Appliances (replaces MSA model 260/360/361) used for testing atmospheres on a tank vessel.

Toxicity

After the compartment has been tested for oxygen and hydrocarbon content relative to the lower explosive limit, additional testing for the presence of potentially health-threatening substances may be necessary. The need to test for toxic vapors in the atmosphere is usually based on the makeup of the cargoes previously carried in the space. Figure 15-7 shows the detector tubes typically used to perform this test.

The detector tubes are chemically treated to react to the presence of a specific gas in the atmosphere. For example, a cargo tank previously used to transport commercial gasoline (regulated cargo) would need to be checked for the presence of benzene in the atmosphere. The permissible exposure limit (PEL) for substances such as benzene, a known carcinogen, is so low (1 ppm) that a conventional combustible-gas indicator is not suitable and should not be relied upon to measure the toxicity level in the atmosphere. The following instrument readings derived from *ISGOTT* are considered acceptable for certifying a space safe for workers (cold work):

Oxygen—21 percent
Combustible-gas indicator—less than 1 percent lower explosive limit (LEL)
Toxicity—below recommended/regulatory occupational exposure limits

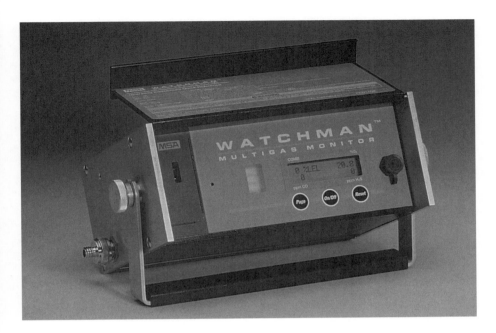

Figure 15-6. The Watchman Multigas Monitor. Courtesy Mine Safety Appliances Company (MSA).

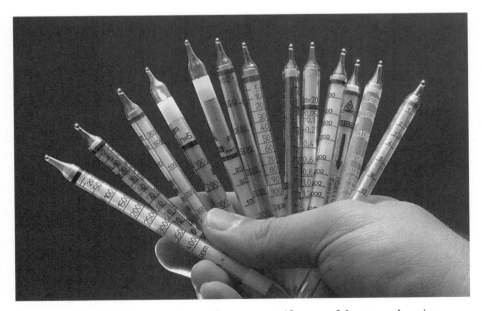

Figure 15-7. Detector tubes used to perform gas-specific tests of the atmosphere in a space. Courtesy Mine Safety Appliances Company (MSA).

It is necessary to attain these readings to permit entry into a compartment without requiring external breathing apparatus. Based on the test results of the atmosphere, personnel may be required to wear a respirator or meet other stipulated conditions for entry. There are two classifications an atmosphere generally receives: "safe for workers" and "safe for hot work." The rating assigned to a compartment generally indicates the type of work that may be conducted in the space. For example, when a space is given a "safe for workers" rating, it is possible for personnel to enter and perform cold work without the need for external breathing apparatus.

Hot Work

The presence of sufficient cargo residues (muck) in the space, which are capable of generating a flammable atmosphere, makes any fire-producing actions unsafe. The use of burning or welding equipment in such a space could create a fire hazard; therefore, additional cleaning and the removal of physical residues is usually necessary for a "safe for hot work" rating. The spaces adjacent to the compartment in which hot work is being conducted and the cargo piping and heating coils must also be checked and properly treated to prevent the spread of fire. In the United States, a "marine chemist" certificated by the National Fire Protection Association (NFPA) is required to perform the testing, inspection, and issuance of certificates when conducting hot work in or on cargo tanks, fuel oil tanks, and any pumps, pipelines, heating coils, or fittings connected to such spaces. Figure 15-8 is a copy of the marine chemist certificate, which is issued after the proper tests and inspections have been conducted. (Figure 15-8a)

Situations requiring the services of a "marine chemist" are described in Title 46 CFR Part 35.01-1. When the services of a marine chemist are not available, such as at sea, the senior officer present frequently determines the condition of a space. Company manuals detail the requirements that must be met prior to permitting personnel to enter a compartment or conduct hot work. Figure 15-9 is an example of a standard checklist used by one company to assist senior personnel with the process of preparing a compartment for entry.

Vessel personnel should also familiarize themselves with the guidance contained in the *International Safety Guide for Oil Tankers and Terminals* and in the National Fire Protection Association manual no. 306, *Standard for the Control of Gas Hazards on Vessels to be Repaired,* concerning entry and work in enclosed spaces.

Figure 15-8a. Photo showing hot work casualty

ENTRY PROCEDURE

Once a compartment has been thoroughly tested and an entry permit issued, the actual entry procedure must be carefully executed. All personnel involved in the operation should be properly trained and must clearly understand their duties when it is necessary to enter an enclosed space. Practical drills and demonstrations should be conducted regularly to prepare individuals in emergency response and procedures regarding rescue from an enclosed space. Preparation is always the key to a safe operation, particularly in the event of an emergency. Past incidents have repeatedly shown that hasty, ill-conceived rescue attempts involving personnel in distress in an enclosed space often result in tragic and unnecessary loss of life. The following notice has been published by the International Chamber of Shipping, and it conveys an important message—"Time and time again circumstances have shown that when well established and proven precautions have not been observed, accidents result. The majority of fatalities could have been prevented by simple supervision, and by following agreed procedures. Ignoring the need for such procedures puts lives at risk." With this message in mind, the following safety equipment should be positioned at the point of entry and remain there for the entire time that personnel are in the space:

1.   Self-contained breathing apparatus (SCBAs) with spare bottles
2.   Lifelines and harnesses with suitable tripod or other fairlead and hoisting equipment
3.   Fall arresting gear
4.   First aid kit, resuscitator, stretcher, AED (automatic external defibrillator)
5.   Protective clothing and vests with reflective tape and hard hats
6.   Approved lighting equipment
7.   Approved communications equipment
8.   Emergency escape breathing device (EEBD)
9.   Fire fighting gear readied
10. Approved atmosphere testing equipment
11. Completed safety checklist, permission of senior personnel, posted entry permit, and appropriate logbook entries
12. All valves in the pipelines connected to the space tagged (secured)

Additional testing of an enclosed space may be necessary to check for pockets of gas. In addition, a physical inspection of the space should be made by the first person entering the compartment. The first entry into a compartment must be made according to the most stringent safety precautions.

A designated individual must serve as a standby for the person or persons in the space. The standby should keep visual contact with the persons in the space and maintain necessary communications capability at all times. This individual should have no other assigned duties that might distract attention or provide a reason to leave the immediate location. In the event of an injury or other problem resulting in a person in distress, the first action of the standby is to raise the alarm. No one should attempt a rescue before the trained response team is on the scene and outfitted with the proper safety equipment. See Chapter 17 for a number of step-by-step guides for rescue operations concerning various emergency situations.

Mechanical ventilation should be operating the entire time personnel are working in an enclosed space. Periodic testing of the atmosphere must be performed to detect any adverse changes in the conditions of the space. Many of the instruments in use today continuously monitor the space through the use of a rechargeable battery-operated pump. These instruments provide a constant readout, and they are also equipped with audible and visual alarms that indicate when predetermined limits are reached.

MARINE CHEMIST'S CERTIFICATE

MARINE CHEMIST CERTIFICATE
SERIAL NO. A 00000

Survey Requested by	Vessel Owner or Agent	Date
Vessel	Type of Vessel	Specific Location of Vessel
Last Three (3) Loadings	Tests Performed	Time Survey Completed

SAMPLE

In the event of physical or atmospheric changes affecting the STANDARD SAFETY DESIGNATIONS assigned to any of the above spaces, this certificate is voided; spaces not listed on the Certificate are not to be entered unless authorized on another Certificate and/or maintained in accordance with OSHA 29 CFR 1915; or if in any doubt, immediately stop all work and contact the undersigned Marine Chemist. Unless otherwise stated on the Certifcate, all spaces and affected adjacent spaces are to be reinspected daily or more often as necessary by the competent person in support of work prior to entry or recommencement of work.

QUALIFICATIONS: Transfer of ballast, cargo, fuel, or manipulation of valves or closure equipment tending to alter conditions in pipelines, tanks, or compartments subject to gas accumulation, unless specifically approved in this Certificate, requires inspection and a new Certificate for spaces so affected. All lines, vents, heating coils, valves, and similar enclosed appurtenances are considered "not safe" unless otherwise specifically designated. Movement of the vessel from its specific location voids the Certificate unless shifting of the vessel within the facility has been specifically authorized on this Certificate.

STANDARD SAFETY DESIGNATIONS (partial list, paraphrased from NFPA 306):

ATMOSPHERE SAFE FOR WORKERS: In the compartment or space so designated (a) the oxygen content of the atmosphere is at least 19.5 percent and not greater than 22 percent by volume; (b) the concentration of flammable materials is below 10 percent of the lower explosive limit; (c) any toxic materials in the atmosphere associated with cargo, fuel, tank coatings, inerting mediums, or fumigants are within permissible concentrations at the time of the inspection.

NOT SAFE FOR WORKERS: In the compartment or space so designated, entry is not permitted.

ENTER WITH RESTRICTIONS: In the compartment or space so designated, entry for work is permitted only if conditions of proper protective equipment, or clothing, or time, or all of the aforementioned, as appropriate, are as specified.

SAFE FOR HOT WORK: In the compartment or space so designated (a) the oxygen content of the atmosphere is not greater than 22 percent by volume; (b) the concentration of flammable materials in the atmosphere is less than 10 percent of the lower explosive limit; (c) the residues, scale, or preservative coatings are cleaned sufficiently to prevent the spread of fire and are not capable of producing a higher concentration than permitted by (a) or (b); (d) all adjacent spaces containing or having contained flammable or combustible materials are sufficiently cleaned of residues, scale, or preservative coatings to prevent the spread of fire, or they are to be inerted. Ship's fuel tanks, lube tanks, or engine room or fire room bilges, or other machinery spaces, are to be treated in accordance with the Marine Chemist's requirements.

NOT SAFE FOR HOT WORK: In the compartment or space so designated, hot work is not permitted.

SAFE FOR LIMITED HOT WORK: In the compartment or space so designated (a) portions of the space are to meet the requirements for SAFE FOR HOT WORK AND PARTIAL CLEANING, as applicable; (b) the space is to be inerted, adjacent spaces are to meet the requirements for SAFE FOR HOT WORK, and hot work is restricted to specific locations; (c) portions of the space are to meet the requirements for Safe for Hot Work, as applicable, and the nature or type of hot work is to be limited or restricted.

CHEMIST'S ENDORSEMENT: This is to certify that I have personally determined that all spaces in the foregoing list are in accordance with NFPA 306, *Standard for the Control of Gas Hazards on Vessels*, and have found the condition of each to be in accordance with its assigned designation.

The undersigned acknowledges receipt of this Certificate under NFPA 306 and understands conditions and limitations under which it was issued, and the requirements for maintaining its validity.	This Certificate is based on conditions existing at the time the inspection herein set forth was completed and is issued subject to compliance with all qualifications and instructions.

Signed _____

Name	Company	Date

Signed _____

Marine Chemist	Certificate No.

VESSEL POSTING

© 2008 National Fire Protection Association   NFPA 306

Figure 15-8. Marine Chemist Certificate. Reprinted with permission from NFPA 306, Control of Gas Hazards on Vessels, copyright 2008, National Fire Protection Association, Quincy, MA 02269. This reprinted material is not the complete and official position of the Natinal Fire Protection Association on the referenced subject, which is represented onlu by the standard in its entirety.

If conditions within the space warrant, personnel should be instructed to evacuate. All those involved in the operation should be observant of anyone exhibiting signs of acute exposure. The telltale signs of acute exposure include eye, nose, or throat irritation; nausea; dizziness; euphoria; and headaches. If necessary the space should be rewashed and ventilation continued until safe readings are restored. The number of persons in the space should be kept to a minimum consistent with the needs of the job.

CONFINED SPACE ENTRY PERMIT

Vessel:

THIS PERMIT TO WORK RELATES TO ENTRY INTO ANY CONFINED SPACE

General

Location of confined space:

Reason for entry:

This permit is valid: FROM: ____ hrs.   Date: ____   TO: ____ hrs.   Date: ____   *(See Notes)*

Section 1 - Pre-Entry Preparations (To be checked by the master or Authorized Officer in Charge)

- Is the span gas check of the combustible gas meter complete?   ☐ Yes   Time: ____
- Has the space been segregated by blanking off or isolating all connecting pipelines?   ☐ Yes   ☐ NA
- Have valves on all pipelines serving the space been secured through LOTO?   ☐ Yes   ☐ NA
  (If yes, ensure required LOTO forms are completed)
- Has the space been cleaned?   ☐ Yes   ☐ NA
- Has the space been thoroughly ventilated?   ☐ Yes
- Pre-entry atmosphere test readings *(See Note 2 & 3)*

Toxic Gases: ____ ppm **CO** (25 ppm)   **Oxygen** ____ % vol (21%)

(PEL) ____ ppm **H²S** (5 ppm)   **Hydrocarbon** ____ % LFL (Less than 1%)

____ ppm **Benzene**(0.5 ppm Action Level)
- Have arrangements been made for frequent atmosphere checks to be made while the space is occupied and after work breaks?   ☐ Yes
- Have arrangements been made for the space to be continuously ventilated throughout the period of occupation and during work breaks?   ☐ Yes
- Is adequate illumination provided?   ☐ Yes
- Is rescue and resuscitation equipment available for immediate use by the entrance to the space?   ☐ Yes
- Has a responsible person been designated to stand by the entrance to the space?   ☐ Yes
- Has the Officer of the Watch (bridge, engine room, cargo control room) been advised of the planned entry?   ☐ Yes
- Has a system of communication between the person at the entrance and those entering the space been agreed and tested?   ☐ Yes
- Are emergency and evacuation procedures established and understood?   ☐ Yes
- Entrants to be accounted for and listed below?   ☐ Yes
- Is all equipment used of an approved type?   ☐ Yes
- Has a JHA been reviewed by all entrants prior to entry?   ☐ Yes

Section 2 - Pre-Entry Checks (To be checked by the Authorized Officer in Charge of the team entering the space)

- Section 1 of this permit has been thoroughly completed.   ☐ Yes
- I am a aware that the space must be vacated immediately in the event of ventilation failure or if atmosphere tests change from safe levels   ☐ Yes
- I have agreed upon a reporting interval of ____ minutes   ☐ Yes
- Emergency and evacuation procedures have been agreed and understood   ☐ Yes

When the above steps are complete, the initial entry can be made for additional atmospheric testing and a visual inspection of the space by the officer in charge and other authorized entrant.

- Has the initial entry confirmed that the tank is safe for entry?   ☐ Yes

Section 3 – Authorized Entrants

1. _____   4. _____   7. _____
2. _____   5. _____   8. _____
3. _____   6. _____   9. _____

Section 4 – Restrictions / Special Instructions

Section 5 – Permit Authorization and Confirmed By:

Master or Chief Engineer: _____   Date: ____   Time: ____
Authorized Officer in Charge: _____   Date: ____   Time: ____
Standby Person: _____   Date: ____   Time: ____

THIS PERMIT IS RENDERED INVALID IF VENTILATION OF THE SPACE STOPS (>10 MINUTES) OR IF ANY OF THE CONDITIONS NOTED IN THE CHECKLIST CHANGE

Section 6 – Cancellation of Permit and Verification of Completion

The work has been completed and all persons under my supervision, materials and equipment have been withdrawn.

Authorized Officer In Charge: _____   Date: ____   Time: ____

Notes:
1. The maximum period of validity may not exceed one work shift (12 hours). If workers have left the space and continuous monitoring was suspended for more than 2 hours, retesting is necessary before re-entry and shall be noted in section 4. and a new permit issued.
2. Samples shall be taken from several depths and through as many openings as possible. Ventilation shall be stopped for approximately 10 minutes before the pre-entry atmosphere tests are taken. (see ISGOTT Section 10.3)
3. Tests for specific toxic contaminants, such as carbon monoxide, benzene and hydrogen sulfide, should be undertaken depending on the nature of the previous contents of the space. Place **N/A** on the line if contaminant not applicable for space being entered.
4. Permit must be cancelled when the all work and/or inspections are complete- see section 5.

Figure 15-9. Confined Space Entry Permit.

INSTRUMENTATION

Most tank vessels carry a variety of portable instruments that are used to assess the atmospheric conditions within a space. The most common instruments used are single-gas monitors (oxygen, combustible gas, carbon monoxide, hydrogen sulfide), multiple gas monitors, and detector tubes. Figure 15-10 shows three single-gas monitors from Mine Safety Appliances.

As mentioned earlier, the reliability and accuracy of the testing equipment is critical to the safety of the operation. Equally important is an individual's understanding of the operating principles and limitations of each instrument. Vessel personnel should consult manufacturers manuals to familiarize themselves with proper operation, maintenance, and interpretation of the performance of instruments. Many types of portable instruments are found on vessels; one of the most popular is known as a multigas monitor. Figure 15-11 illustrates three models of multigas monitors from Mine Safety Appliances.

These instruments have multiple sensors capable of simultaneously detecting and measuring three to five different gases. On an inerted crude-oil carrier, for example, an owner might opt for a multigas monitor capable of measuring oxygen, combustible gas, and hydrogen sulfide. Portable gas monitors use different sensors, which generally fall under one of the following categories:

1. Catalytic
2. Electrochemical
3. Photoionization detector

Figure 15-10. Single-gas monitors. Courtesy Mine Safety Appliances Company (MSA).

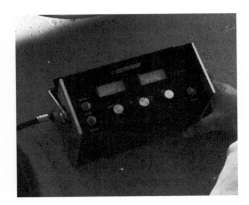

Figure 15-11. Multigas monitors.
Courtesy Mine Safety Appliances
Company (MSA).

Figure 15-11a. Model 260.

Figure 15-11b. Watchman Multigas Monitor.

Figure 15-11c. Passport FiveStar Personal Alarm.

Catalytic and electrochemical sensors are the most common types found in portable instruments used to detect gases.

Catalytic Sensors

A combustible-gas indicator uses a catalytic sensor to detect and measure the level of a wide array of flammable vapors in a space. The sensor consists of a wire filament that is heated from a battery supply. The heated wire readily burns (oxidizes) any flammable gases in the sample being drawn through the instrument. The wire heats up, causing an increase in its electrical resistance which results in an imbalance of a "Wheatstone bridge." This change in resistance is converted to a meter-reading which is expressed as a percentage of the lower explosive limit (LEL) in the case of a combustible-gas indicator. Several points are worth mentioning concerning this type of instrument:

1. Combustible-gas indicators consume a substantial amount of power; therefore the battery charge should be checked before and during use.
2. Combustible-gas indicators that operate on a hot-wire (burning) principle require sufficient oxygen for the instrument to function properly. There-

fore, this type of instrument is *not* suitable when sampling an oxygen-deficient compartment such as an inerted cargo tank.

3. Consult the manufacturer's manual for a listing of substances that could adversely affect the performance of an instrument. Atmospheres containing lead, silicones, and silicates should be avoided as they can cause inaccurate readings with a combustible-gas indicator.

4. In general, a combustible-gas indicator is useful for detecting and measuring the level of vapors given off by a flammable liquid relative to the lower explosive limit (LEL). The LEL represents the minimum concentration of vapor that must be present in air to support combustion. If the liquid has a high flashpoint, such as jet fuel or diesel, the catalytic sensor cannot readily measure the concentration in a space due to the absence of vapors to support combustion.

5. A combustible-gas indicator will not respond correctly when measuring an atmosphere where the concentration of flammable vapors is above the upper explosive limit (UEL), commonly referred to as a "rich" mixture. The catalytic sensor may initially react to the presence of a high concentration of vapors, but it will then fall off as the rich mixture is drawn through the instrument. A rich mixture is incapable of supporting combustion; this affects the performance of the hot-wire filament, causing erroneous readings from the instrument.

6. The combustible-gas indicator is typically calibrated by the manufacturer, using a representative gas such as pentane. The characteristics of pentane are similar to a wide range of hydrocarbons, making it the gas of choice for calibration of the instrument in the factory. In the field, however, when sampling for the presence of a combustible gas with substantially different characteristics from pentane, it is necessary to apply response factors provided by the manufacturer. The meter-reading of the instrument is multiplied by the appropriate response factor to find the actual concentration of a specific gas in the space. Table 15-3 lists the relative responses to various combustible gases for an instrument calibrated using pentane. In the case of styrene for example, note that an instrument (meter) reading of 10 percent LEL equates to an actual reading of 19 percent LEL in the atmosphere of the space when one applies the factor of 1.9.

Electrochemical Sensors

Electrochemical sensors are used to detect the presence of oxygen and a variety of toxic gases. An electrochemical sensor works in a similar way to that of a small battery, but one chemical component, which is necessary to produce electric current, is missing from the sensor cell. When a suspect gas such as hydrogen sulfide is sampled, it diffuses through a membrane at the top of the sensor. The hydrogen sulfide reacts with the chemicals on the sensing electrode, which results in an electrical current that can be measured. These sensors are designed to detect a specific gas; therefore the operator must know the makeup of the atmosphere prior to testing.

Oxygen sensors work in much the same way as other electrochemical sensors. Oxygen from the sample diffuses into the cell and reacts to produce an electrical current. Oxygen sensors typically use the oxidation of lead as the basis of opera-

Table 15-3
Relative Responses to Combustible Gases for
Instrument Calibrated Using Pentane

Combustible Gas	Factor by Which to Multiply LEL Meter Reading
Acetone	1.1
Acrylonitrile (1)	0.8
Benzene	1.1
Carbon disulfide (1)	2.2
Cyclohexane	1.1
Ethane	0.7
Ethylene	0.7
Gasoline (unleaded)	1.3
Heptane	1.1
Isobutyl acetate	1.5
Methanol	0.6
Methyl tertiary butyl ether	1.0
Pentane	1.0
Propane	0.8
Styrene (2)	1.9
Toluene	1.1
Vinyl acetate	0.9
VM & P naptha	1.6

Notes:

1. The compounds may reduce the sensitivity of the combustible-gas sensor by poisoning or inhibiting the catalytic action.

2. The compounds may reduce the sensitivity of the combustible-gas sensor by polymerizing on the catalytic surface.

3. For an instrument calibrated on pentane, multiply the display percent of LEL value by the conversion factor to get the true percent of LEL.

4. These conversion factors should be used only if the combustible gas is known.

5. These conversion factors are typical for a Watchman Multigas Monitor. Individual units may vary ±25 percent from these values.

Courtesy MSA.

tion. The oxygen cell in a portable instrument generally has a service life ranging from one to two years depending on use.

In addition to multigas detectors, many vessels are also equipped with single-gas monitors and detector tubes for measuring the concentration of potentially toxic vapors in a space. Figure 15-12 shows the detector tubes and a hand-operated bellows pump used for drawing the sample from the space being tested.

The detector tubes are made of glass and contain granules that are chemically treated to react to the presence of a specific gas in the sample. When it is necessary to test a space, the appropriate detector tube is selected, and the pointed tips of the glass tube are broken off. The tube is inserted into the pump assembly or holder in the remote sampling line. A measured quantity of the atmosphere from the space in question is drawn through the detector tube using the pump. The concentration of vapor is indicated by the length of discoloration of the granules, which can be read off the scale printed on the tube. The scale on the detector tube may give the reading directly, or a guide must be consulted to determine the level. Detector tubes generally have a stipulated shelf life, and this date should be indicated by the manufacturer on the packaging.

Photoionization Devices

Photoionization devices are useful for measuring extremely low concentrations of various substances that may be present in the atmosphere of a space in a vaporous state. These instruments use ultraviolet light to first ionize the gas being sampled. The gas then passes between two charged plates that separate the gas ions and free electrons. As the gas ions move to the plates, a current is generated that can be measured. Photoionization instruments are currently used to detect low-level concentrations of volatile organic compounds (VOCs).

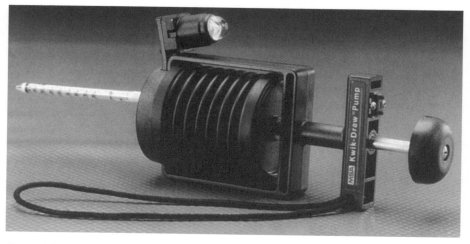

Figure 15-12. Detector tube and hand-operated bellows pump. Courtesy Mine Safety Appliances Company (MSA).

REVIEW

1. What are the primary hazards associated with entry into an enclosed space?
2. List two methods of gas-freeing a cargo tank.
3. When preparing a compartment for entry, how long should the space be ventilated with air?
4. Prior to testing the atmosphere of a compartment, what checks should be performed on the instruments?
5. Prior to entry into an enclosed space, the atmosphere should be tested using what instruments?
6. Referring to the previous question, what instrument readings are considered acceptable according to *ISGOTT* to certify a space "safe for workers"?
7. What additional work is usually necessary to make a space "safe for hot work"?
8. List the equipment that should be available at the point of entry while personnel are in an enclosed space.
9. In the event a person working in an enclosed space is in distress, what is the first action that should be taken by standby personnel?
10. When are the services of a marine chemist required by federal regulation?
11. What publications contain excellent guidance concerning the preparation, testing, and procedure to be followed when entering an enclosed space?
12. What are three types of sensors employed in atmospheric testing equipment?
13. List the factors that can adversely affect the performance of a catalytic type combustible-gas indicator.
14. Detector tubes are typically used to test a space for the presence of what vapors?
15. How does an electrochemical sensor function?

CHALLENGE QUESTIONS

16. As a standby, you suspect that personnel at the bottom of a cargo tank are exhibiting acute effects of exposure to cargo vapors. What action and precautions should be taken?
17. Describe how you would test the atmosphere of a "J" type ballast tank on a double hull tanker?
18. Due to a sudden release of cargo from a failed pipeline in the pumproom a crewmember is overcome and unconscious at the lower level of the cargo pumproom. Give a detailed description of the rescue operation of the individual.
19. Why is it necessary to apply a "response factor" when interpreting the meter reading of a typical combustible gas indicator (catalytic type)?
20. List some of the typical causes of marine casualties and near misses involving the entry of personnel into enclosed spaces on a tanker.

CHAPTER 16

Inert Gas Systems

In December of 1969 the shipping world was rocked by a series of explosions that occurred aboard three VLCCs, the *Marpessa,* the *Mactra,* and the *Kong Haakon VII.* Each was a newly constructed supertanker operated by a reputable company. The photograph showing the aftermath of the explosion aboard the *Kong Haakon VII* (Figure 16-1) is a sobering reminder of the potential hazards associated with cleaning tanks on a crude-oil carrier.

Shipowners launched an extensive investigation to determine the probable cause of the explosions. Although it was difficult to pinpoint the exact cause in each case, the investigation pointed to a static electrical discharge as the probable source of ignition. Factors common to each incident included steaming cargo tanks at sea in a ballasted condition, employing fixed tank washing machines, and tank cleaning (center tanks) at the time of the explosion.

The use of high-capacity (fixed) washing machines (HCWM) delivering water—and, in some cases, recirculated oily-water mixtures—at high velocity and pressure against the cargo tank surfaces was linked to the development of a significant static charge. The benefits of fixed tank washing had been recognized for some time, and rather than abandon this method of cleaning, the tanker industry decided to vigorously pursue the concept of controlling the atmosphere in the cargo tanks while conducting a wash. It was determined that maintaining a low oxygen content in the cargo tank was the key to ensuring the safety of the operation. "Inert gas," as it is commonly called, is defined as a gas or mixture of gases containing insufficient oxygen to support the combustion of hydrocarbons.

The events in 1969 prompted the development of international regulations calling for the installation of inert gas (IG) systems on crude tankers. "Inerting" then became an industry standard in the crude-oil tanker fleet worldwide. The rules governing the design, operation, and maintenance of these systems first appeared as regulation 62 in the Safety of Life at Sea Convention (SOLAS), 1974. The initial inerting requirements were obviously directed at the larger crude carriers that were employing fixed washing machines in their cargo tanks. The success of these systems in the crude-oil trade eventually led to expansion of the inerting requirements to include product carriers and vessels of smaller tonnage.

Figure 16-1. The aftermath of the explosion aboard VLCC *Kong Haakon VII* that occurred during the tank cleaning process on the ballast passage. Courtesy U.S. Salvage.

SOURCES OF INERT GAS

Three types of inert gas systems are in common use today: (1) flue gas systems (steamships); (2) oil-fired inert gas generators (motor ships and barges); and
(3) nitrogen systems (parcel tankers). On steamships, the exhaust from a marine
boiler burning a residual fuel oil results in a mixture of gases with the following
breakdown (approximate values):

Nitrogen	78–80%
Carbon dioxide	12–14%
Oxygen	2–4%
Sulfur oxides	0.3%
Water vapor	5%
Soot	300 mg/m^3
Heat	200–300°C
Carbon monoxide	1,000 ppm
Nitric oxide	400 ppm

The exhaust is drawn off the uptakes, processed in the scrubber, and delivered via heavy-duty fans to the cargo tanks. Figure 16-2 shows the typical layout of an IG system on a steamship.

In new construction, owners are installing oil-fired inert gas generators as a result of the shift toward diesel propulsion and the customers' demand for better "quality" inert gas. On parcel tankers transporting cargoes sensitive to contamination, concerns over quality assurance have led to the installation of nitrogen systems.

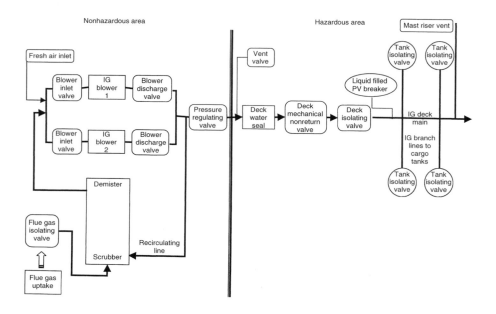

Figure 16-2. A block diagram showing the major components of an inert gas system. Courtesy Richard Beadon and Eric Ma.

GENERAL REQUIREMENTS FOR AN INERT GAS SYSTEM

The inert gas system must be designed and operated as necessary to maintain the atmosphere in the cargo tanks in a nonflammable condition throughout the operating cycle of the vessel unless the tanks are gas free. Regardless of the type of inert gas system installed on the vessel, it should be capable of supplying a gas or a mixture of gases with an oxygen content of 5 percent or less by volume. This enables the operator to maintain an oxygen content of 8 percent or less by volume in the cargo tanks. The onboard IG manual should always be consulted as some companies and terminals have more stringent requirements concerning the oxygen level to be maintained in the cargo tanks. The other general requirement for the IG system relates to the operator's ability to maintain a positive deck pressure at all times. The purpose of this requirement is to prevent the ingress of air that could potentially compromise the inert status of the cargo tank or the entire vessel. By regulation, the operator must maintain a minimum of 4 inches (100 mm) water gauge (wg) during the operating life of the vessel. The United States rules concerning inert gas systems have been harmonized with the SOLAS requirements to eliminate the discrepancies that once existed between the systems found on international and domestic vessels (refer to Title 46 CFR Part 32.53-10). It is important to realize that over the past thirty years, the rules governing the design and operation of IG systems have been amended several times as operators gained more experience with these systems. To stay abreast of any changes that have been approved internationally, operators should consult the appropriate manuals from IMO.

FLUE GAS SYSTEM COMPONENTS

All personnel involved with the operation and maintenance of the inert gas system should be familiar with its components.

Boiler

Starting at the boiler uptake, one or more boiler-uptake valves or flue-gas-isolating valves are fitted to isolate the IG plant when it is secured. These valves are located in an extremely hostile environment—hot, dirty, and corrosive. This can affect their ability to maintain a gas-tight seal. It is recommended these valves be operated on a regular basis and sootblown to ensure proper operation. Some manufacturers have opted for an air-seal arrangement in lieu of attempting to maintain a mechanical seal at the uptake valves. The sootblowers on the boiler should never be operated in conjunction with the inert gas system; therefore, interlocks are installed on the uptake valves to prevent such an occurrence. From the uptake valves, the hot-gas inlet line directs the flue gas to the base of the scrubbing tower.

Uptake Bellows

To accommodate the thermal variances (expansion and contraction) of the piping during operation, an expansion joint generally referred to as the bellows is installed in the hot-gas inlet line. According to surveyors, the bellows represents a weak link in the piping system; it is subject to metal fatigue and holing over time, possibly resulting in air leaks in the system.

Scrubber

The scrubber is the primary processing unit that converts the hot, dirty, corrosive flue gas into a usable inert gas which can be safely delivered to the tanks. The scrubbing process has four functions:

1. Cooling
2. Removing the solids (soot)
3. Removing the corrosives (sulfur oxides)
4. Removing entrained water

Numerous scrubber designs use varying methods to accomplish these functions; therefore, the manufacturer's manual must be consulted for the details of the unit on the vessel. In a typical wet bottom scrubber (Figure 16-3), the flue gas enters the base of the tower through a water seal.

Scrubber–typical layout

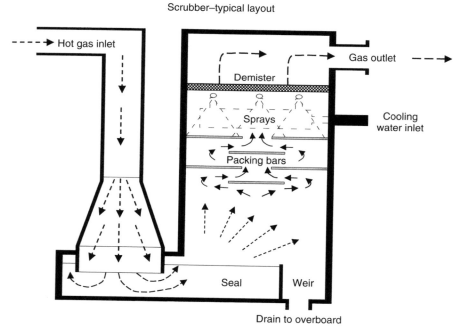

Figure 16-3. Internal view showing the components of a wet bottom scrubber. Courtesy Richard Beadon and Eric Ma.

As the gas bubbles through the water, it is initially cooled and the larger soot particles are removed. This initial cooling is very important as it causes the corrosive sulfur oxides to be flushed out of the gas while it passes through the seal water and seawater sprays. As the gas moves vertically through the tower, it passes through a series of baffles or trays that cleanse the gas of the finer soot particles. The continuous contact of the gas with the seawater sprays results in further cooling action and removal of impurities in the gas. At the top of the tower (outlet) the final process involves removal of physical water droplets from the gas. The first method of removing entrained water is accomplished by passing the gas through a demister which causes the water to condense on the cellular-like material and remain in the tower. In the second method, the gas moves through a cyclone or vortex separator which imparts a spinning action, throwing the water droplets outward into a funnel-shaped catcher where it is drained away. At this point the gas is fully processed and ready for delivery to the deck. It is important to realize that the scrubber does *not* achieve complete removal of the undesirable elements in the gas.

Over time, deterioration of the components in an IG system can be expected based upon the efficiency of the scrubbing process. To protect the internal surfaces of the scrubber, either special coatings or liners are employed against the corrosive elements in the gas. The anticorrosive features must be carefully inspected to ensure proper protection of the scrubber shell. Upon shutdown of the system, it is recommended the scrubber be flushed with seawater for a period of 30 to 60 minutes. Experience with flue-gas systems has shown that routine inspection and preventive maintenance is the key to ensuring safe, uninterrupted operation of the IG plant.

Fresh-Air Inlet

Upon leaving the scrubber, most IG plants have a fresh-air inlet on the suction side of the fans which permits the system to be used for gas-freeing purposes.

Inert Gas Fans

The next major component in the system is the IG fans. Two inert gas fans are required, the combined capacity of which must be rated at 125 percent of the maximum discharge capacity of all the cargo pumps that can be operated simultaneously. The intent of this requirement is to enable the operator to maintain a positive deck pressure throughout the cargo discharge. Installations vary, but most owners opt to exceed this requirement by installing two 125-percent-rated fans. The advantage claimed for this arrangement is that if either fan fails, it would not adversely impact the discharge operation of the ship. The IG fans are typically driven by heavy-duty electric motors and equipped with isolation valves on the suction and discharge sides. The fans should be visually inspected on a regular basis through ports located in the top of the casing for evidence of deterioration such as scoring, acid damage, and carbon buildup. During operation, the fan should be checked for excessive heat and vibration. Upon shutdown of the system, a freshwater rinse of the fan blades is recommended for a period of fifteen to twenty minutes. On the discharge side of the fans, two sensors are required, one for monitoring the temperature of the gas and the other for monitoring oxygen content. The alarm settings are as follows:

High gas temperature—U.S./IMO: 149°F (65°C)
High oxygen alarm—U.S./IMO: Operator-set value (but in no case higher than 8 percent oxygen by volume)

The reliability of these instruments is critical to the operator when determining the suitability of the gas to send to the deck. Off-specification gas is either returned to the scrubber through a recirculation line or, in later model systems, vented off directly to the atmosphere.

Gas Pressure Regulating Valve (GRV)

To control the flow of inert gas to the deck distribution piping, most systems are equipped with a gas pressure regulating valve, also known as the main control valve.

This valve has two modes of operation, manual and automatic. In the automatic mode the gas pressure regulating valve operates as necessary to maintain a desired deck pressure in the cargo tanks. In the manual mode this valve is controlled directly by the operator of the system. At this point the inert gas piping leaves the afterhouse and proceeds forward onto the cargo tank area. Safe operation dictates that any IG system permit flow in only one direction, specifically good quality inert gas flowing toward the cargo tanks. This location constitutes the division between what are commonly called the *nonhazardous* and the *potentially hazardous* areas of the vessel. To prevent the return flow of flammable cargo vapors in the system, a series of nonreturn devices are required, providing a barrier between these two areas.

Vent Line

In the 1981 amendments to SOLAS, a vent line was added to the system requirements. This vent provides an additional safeguard in the event the nonreturn devices fail. According to IMO, the vent line and valve should be installed in the run of piping between the gas pressure regulating and the deck isolation valves. This vent valve should be open when the inert gas plant is off to prevent returning cargo vapors and inert gas from building up pressure in the line against the gas pressure regulating valve. The vent also provides a convenient location to check the integrity of the nonreturn devices by sampling this section of the piping for the presence of flammable vapors using portable instruments.

Deck Seal

Several types of deck seals are commonly found aboard tank vessels. Regardless of the design, the deck seal is the principle barrier against return flow of flammable cargo vapors into the gas-safe areas of the vessel. The wet-type seal shown in Figure 16-4 is designed to function as a nonmechanical, nonreturn device, providing virtually fail-safe protection in operation.

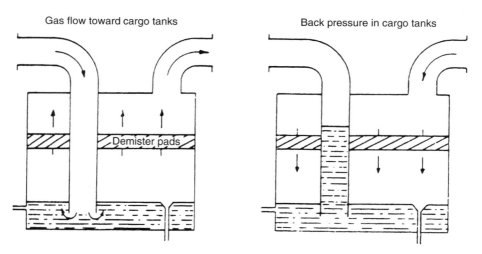

Figure 16-4. Typical wet-type deck water seal. Copyright International Maritime Organization (IMO), London.

The deck seal works on a simple principle whereby the returning IG and cargo vapors exert pressure on the larger water surface in the base of the seal, forcing the water up the inlet pipe to form a water plug. The water plug prevents the gas from entering the nonhazardous areas of the vessel. Other varieties of deck seals are known as the semidry and dry-type seal. These seals operate in much the same way as the wet-type seal; however, they are designed to minimize water carryover into the deck main and cargo tanks during normal operation.

The semidry-type seal (Figure 16-5) employs a venturi to draw the water away from the inlet pipe, permitting the gas to move through the seal without having to physically bubble through the water as in the wet-type seal. In the dry-type seal (Figure 16-6), seal water is drained away at the start-up of the system and the gas moves through the empty compartment without coming into contact with water. Some manufacturers employ a demister similar to that found in the scrubber to physically trap the water and keep it from passing to the deck main. Several checks should be performed on the deck seal including the following:

1. Seal water level (normal water supply)
2. Overboard drain from the seal
3. Level sensors
4. Coatings or liners in the deck seal
5. Heating coils (in cold weather)
6. Demister
7. Venturi line and any valves in a semidry or dry-type seal

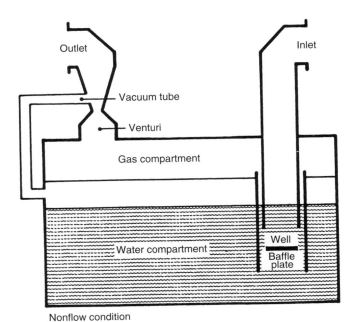

Figure 16-5. Semi-dry-type deck water seal. Courtesy Wilson Walton International.

In addition to the checks mentioned above, the performance of the deck seal should be monitored through periodic testing for the presence of cargo vapors, using a hydrocarbon analyzer just aft of the deck seal. The presence of flammable cargo vapors at this point in the system would indicate a problem with the nonreturn devices which would warrant further investigation.

Nonreturn Valve

On most systems, immediately forward of the deck seal is the mechanical nonreturn valve. A lift or swing check valve is generally employed; it is either weight- or spring-loaded. Given the operating environment in which this valve functions, operators should be aware that these valves are prone to leaking or remaining in the open position, thereby permitting return flow. The photo in Figure 16-7 shows a nonreturn valve which was replaced after it was determined that it was frozen in the open position.

The nonreturn valve is required to be provided with a positive means of closure or, alternatively, a second valve should be installed forward of the nonreturn valve.

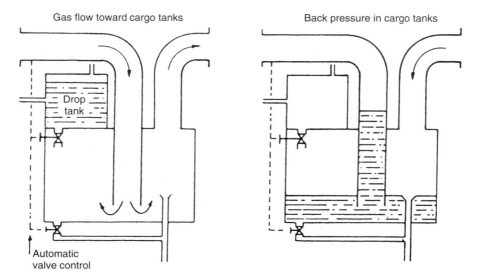

Figure 16-6. Dry-type deck water seal. Copyright International Maritime Organization (IMO), London.

Figure 16-7. A mechanical nonreturn valve that was replaced after it was found to be frozen in the open position.

Deck Isolation Valve

The deck isolation valve is generally a manually operated valve (typically butterfly or gate) that blocks the cargo tank area from the after end of the system.

Deck Distribution System

From the deck isolation valve, the inert gas is directed to the cargo tanks via a branching deck main. This network of piping on deck not only serves to supply the inert gas to the cargo tanks but also serves, in many instances, as the vent and vapor control piping for the vessel (see chapter 5 for details concerning venting and vapor control systems). An alternative piping arrangement less commonly seen for the supply of inert gas to the cargo tanks is a connection between the deck main and the bottom piping of the vessel. This permits the gas to enter the bottom of an empty cargo tank via the bellmouth. The deck main and branch lines should be equipped with low point drains to permit the removal of any liquid that may collect in the piping system on deck. Liquids tend to accumulate in the deck main for a variety of reasons including condensation due to temperature variation, water carryover from the deck seal, and cargo accumulation due to misting and tank overfill. The smaller branch lines off the inert gas deck main enter the cargo tank via the side of the tank hatch (Figure 16-8) or by vertically penetrating the deck (Figure 16-9).

Cargo Tank Isolation
The cargo tanks must be equipped with some form of block valve or blanking arrangement. Systems vary from those that use a single butterfly valve in the branch line to those that employ a spade blank inserted where the piping enters the coaming of the tank hatch. Several methods of isolating a cargo tank from the rest of the vessel are shown in Figure 16-10. Cargo tanks are routinely isolated for a variety of reasons:

1. Manual gauging, water cuts, sampling, or taking temperatures in the tank
2. Gas-freeing and entry by personnel
3. Segregation of dissimilar cargoes and their vapors

Pressure-Vacuum Relief Devices
One of the consequences of any closed operation on a tank vessel is the possibility of the occurrence of structural damage. With this in mind, most inert gas systems are equipped with one or more of the following pressure-vacuum relief devices to protect the system (see Figure 16-11):

Figure 16-8. The branch line off the IG main in this case enters the tank through the side of the tank hatch.

1. Cargo tank pressure-vacuum relief valves (for vessels fitted with branch line valves)
2. A centrally located pressure-vacuum relief valve on the mast riser or a bullet valve for vessels without branch line valves
3. One or more liquid pressure-vacuum breakers (located on the inert gas main)

Figure 16-9. The branch line off the IG main is connected to the cargo tank via a deck penetration. The branch line shown here is equipped wtih a butterfly valve.

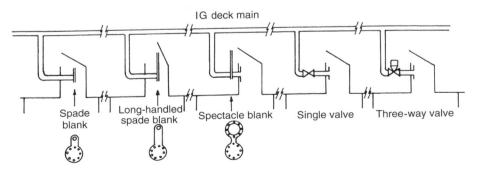

Figure 16-10. Several methods of isolating the cargo tanks from the gas main. Courtesy International Chamber of Shipping and OCIMF.

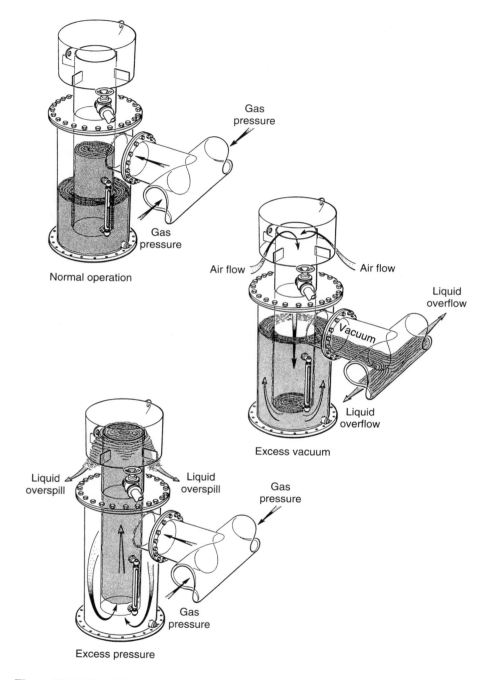

Figure 16-11. Vessels fitted with inert gas systems are typically fitted with a liquid-filled PV breaker to protect the system from extreme pressure or vacuum conditions. Courtesy Permea Maritime Protection.

Figure 16-11a. Liquid filled PV breaker. Courtesy Stacy DeLoach.

Structural protection of the vessel is imperative as expensive damage can result from mechanical malfunctions or human error such as the failure to line up the system correctly. Figure 16-12 shows the damage that occurred when one of the cargo tanks overpressurized during a loading operation. (For further details concerning the method of operation and the typical settings for pressure-vacuum relief devices, consult chapter 4.)

Deck Pressure

One of the general requirements of an IG system concerns maintaining a positive deck pressure on the inerted vessel to prevent the ingress of air. To assist the operator, all installations are required to have a gauge which gives a continuous readout as well as a permanent recording of the deck pressure. Audible and visual alarms are required for both high and low deck pressure conditions. In the bar graph, Figure 16-13, some typical values for the high and low deck pressure alarms are given. An operator can expect variances in the deck pressure at sea as well as during cargo transfer in port.

Owners are moving toward the installation of individual tank pressure readout affording the person in charge the ability to detect any extreme conditions (pressure/vacuum) before the actuation of the P/V relief devices.

Figure 16-12. Over pressurization of a cargo tank during a cargo transfer resulted in structural damage to the vessel.

DECK PRESSURE AT SEA

The deck pressure of a vessel will vary considerably based upon changes in ambient conditions. For example, the deck pressure can be expected to climb in the later afternoon as temperatures approach their maximum; conversely, the deck pressure typically drops off because of the cooler nighttime temperatures. Similar fluctuations in the deck pressure are experienced with changes in seawater temperature and as a result of cargo volatility. It is crucial that vessel personnel monitor the deck pressure for any extreme conditions while underway. In the event the deck pressure is reaching

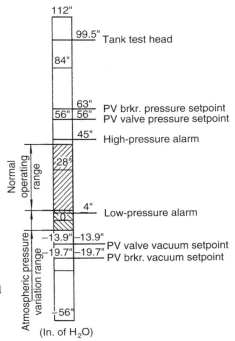

Figure 16-13. Bar graph showing the typical operating pressures and settings of the PV relief devices in an inert gas system.

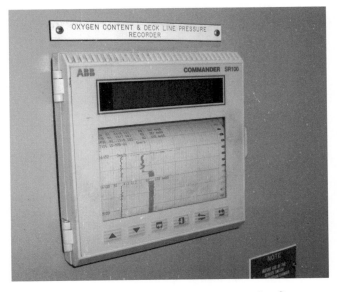

Figure 16-13a. Oxygen and Deck Pressure recorder. Courtesy Stacy DeLoach.

the upper acceptable limit, it is generally necessary to vent off the excess pressure to atmosphere. On the other hand, should the deck pressure be approaching the lower acceptable limit, it may be necessary to start the IG system and top up the deck pressure. "Topping up" is defined as the introduction of inert gas into a tank already in the inert condition with the object of raising the deck pressure to prevent any ingress of air.

DECK PRESSURE IN PORT

During cargo operations, the deck pressure must be carefully monitored by the PIC to prevent the development of any extreme conditions. Following are some typical causes of problems with deck pressure:

1. Improper lineup of the IG/vent system
2. Excessive loading rate
3. Excessive cargo pumping rate
4. Mechanical malfunctions (i.e., faulty PV valves)
5. Constricted line (i.e., liquid plug in the line)
6. Cargo polymerization
7. Shore vapor recovery problems (i.e., restricted line or blower assist)

Alarm and Shutdown Features

The inert gas system must have several alarm and shutdown features to protect the system in the event of a critical fault. The required alarm and shutdown features are as follows:

1. Low cooling water supply to the scrubbing tower
2. High water level in the scrubber
3. High gas temperature—U.S./IMO: 149° F (65° C)
4. High gas temperature shutdown—U.S./IMO: 167°F (75°C)
5. Blower failure alarm
6. Power failure to the autocontrol feature
7. Loss of water supply to the deck seal

Suitable interlocks must be installed such that any fault condition resulting in a shutdown of the blowers also results in the closing of the gas pressure regulating valve. The automatic shutdown features of the IG system should be tested on a regular basis to ensure proper operation.

HIGH OXYGEN READINGS

During the operation of the IG system, the PIC may encounter elevated oxygen readings as sensed at the fixed oxygen analyzer on the discharge side of the fans. Should this occur, the operator must determine the source of the oxygen and take corrective action before the high oxygen alarm sounds. If the efforts to improve gas quality are unsuccessful, it may be necessary to divert the off-specification gas either by placing the system in recirculation or venting to atmosphere. Some typical causes of elevated oxygen readings include the following:

1. Poor combustion control
2. Low boiler load
3. Leaks on the suction side of the fans

4. Prolonged recirculation
5. Failure to secure the air seal arrangement of the uptake valve
6. Faulty oxygen analyzer

EMERGENCY PROCEDURES

In the event of a total failure of the inert gas plant to deliver the required quantity or quality of inert gas to the deck, the operator of the system must take certain actions. Under domestic rules, the operator is prohibited from allowing the creation of a flammable atmosphere in the cargo tanks; therefore, it may be necessary to suspend all cargo operations and hold the deck pressure in the vessel. In other words, the operator must not compromise the inert status of the vessel by resuming the cargo discharge without the benefit of the IG system. Several safety concerns, particularly in the case of crude tankers, warrant this action. Also, local and terminal requirements may prohibit the resumption of cargo operations. The following list represents some of the potential ignition sources that may pose a threat to the safety of the operation:

1. Pyrophoric iron sulfide ignition
2. Static generation from the IG system
3. Static generation from the use of high-capacity washing machines

The risk from an internal source of ignition is judged to be significant enough to warrant the suspension of cargo operations. Should an operator wish to resume cargo operations, the only viable options are (1) to fix the inert gas system or (2) to connect to an external supply of IG.

Port state inspectors will randomly check tank vessels to ensure they are in compliance with the inerting requirements.

GAS REPLACEMENT

During the operating life of an inerted vessel, it is necessary to perform gas replacement. Types of gas replacement include the following:

1. Primary inerting
2. Purging
3. Gas-freeing
4. Reinerting

A complete understanding of gas replacement is necessary in order to avoid the creation of a flammable atmosphere in the cargo tanks. Two gas replacement methods are routinely performed on inerted vessels: dilution and displacement. Which method a particular vessel employs is dictated by the supply (entry) and venting (exit) arrangements of the cargo tanks.

Dilution

The dilution (mixing) method is accomplished through the introduction of gas at high velocity, achieving maximum penetration into the cargo tank and considerable turbulence within the tank atmosphere. The aim is to create a homogeneous atmosphere in the tank by thoroughly mixing the existing atmosphere with the incoming gas. Experience with this method has shown that it generally takes 3 to 5 volume changes of the cargo tank atmosphere to achieve full gas replacement. To achieve the best results using the dilution method, the supply of gas is generally directed to a limited number of tanks, preferably one at a time, thereby achieving the maximum velocity to that tank.

Displacement

In the displacement (layering) method, the gas enters the cargo tank at low velocity, forming a stable horizontal interface between the incoming and exiting gases. The gas enters the top of the tank and acts like a piston pushing the exiting gas out via a purge pipe or other suitable arrangement. To minimize turbulence in the tank atmosphere, the gas is generally directed to a number of tanks simultaneously, thereby slowing the velocity. The displacement method usually takes 1½ to 2 volume changes of the tank atmosphere to accomplish full gas replacement. Figure 16-14 illustrates the typical supply and venting arrangements and the methods

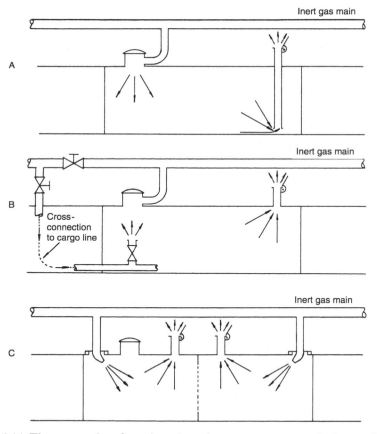

Figure 16-14. Three examples of supply and venting arrangements used when perfoming gas replacement. (A) Displacement method. (B) Dilution method via bottom piping. (C) Dilution method via deck piping. Courtesy Howden Engineering.

that should be employed when performing gas replacement. For detailed guidance concerning the methods and time necessary to perform gas replacement consult the vessel's inert gas manual.

There are generally two ways to determine the success of a gas replacement operation. One is by an experience factor with the particular vessel, the other is by testing and monitoring the tank atmosphere with portable instruments. Figure 16-15 illustrates a typical hydrocarbon analyzer (model 100T Gascope or Tankscope from MSA) commonly used to verify that a space has been properly purged prior to gas-freeing. A reading on the analyzer of 2 percent hydrocarbons or less by volume is required before ventilating with air. Reducing the hydrocarbon concentration of the space to this level prevents the tank atmosphere from becoming flammable when it is subsequently ventilated with air (gas-freed). An operator's understanding of gas replacement methods is the key to successfully controlling the cargo tank atmosphere and ensuring the vessel is maintained in a nonflammable condition throughout its operating life.

Figure 16-15. The hydrocarbon analyzer seen here is used to verify the success of the purging operation. Courtesy Mine Safety Appliances Company (MSA).

INERT GAS GENERATORS

Inert gas generators (IGG) are typically installed on vessels that do not possess a ready supply of oxygen-deficient gas for use in the cargo tanks. In recent times more inert gas generators have been installed because of their application to tank barges and the shift from steam to motor tankers. Owners may also opt for an inert gas generator in situations where the quality of the inert gas may pose a risk of contamination to the cargo. A properly operating inert gas generator is capable of producing a superior quality gas based on the type of fuel employed. For example, a diesel-fired inert gas generator does not produce the soot or corrosives typically encountered with a flue gas system. The higher quality fuel produces a cleaner burn which in turn requires a much less sophisticated scrubbing process. Inert gas generators are very compact consisting of a combined combustion and scrubbing unit as seen in Figure 16-16.

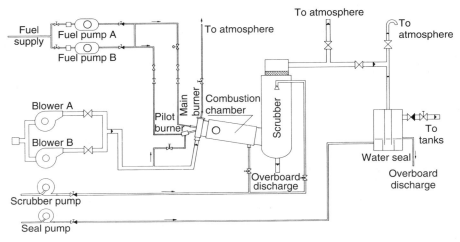

Figure 16-16. Typical inert gas generator. Courtesy Permea Maritime Protection.

Figure 16-16a. IG branch valve and dehumidified air supply Courtesy Christopher Adams.

MAJOR COMPONENTS

Air Blower—generally two electric motor driven fans are provided that supply combustion air to the burner unit at the required capacity (Figure 16-17). On the control screen (Figure 16-18) the operator selects blower A or B based on the number of operating hours indicated. At start up the selected blower operates for approximately 30-40 seconds to purge the combustion chamber prior to lighting off the pilot burner. In an effort to reduce fuel consumption, combustion air bypasses the burner allowing for a minimum flow condition to keep the combustion process going when there is little or no demand for inert gas on deck.

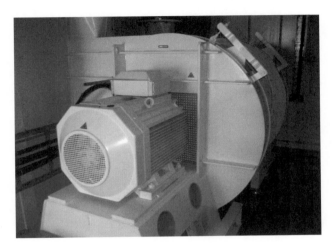

Figure 16-17. IGG air blower.

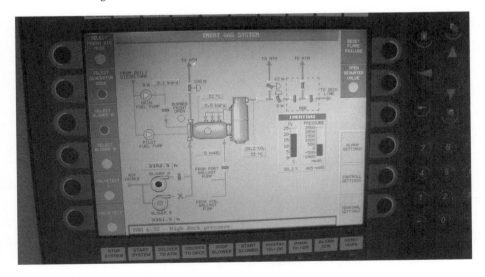

Figure 16-18. IGS control panel.

Fuel pumps—the system is generally equipped with a pilot and main fuel pump. The fuel pumps are gravity feed from a day tank via a duplex strainer shown in (Figure 16-19). The pump feeds the pilot burner at start up where electrodes ignite the fuel/air mixture. The pilot flame subsequently lights off the main burner once the flame scanner detects a normal firing condition.

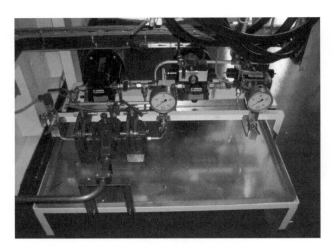

Figure 16-19. Duplex strainer and fuel pump.

Combustion Unit—the combustion unit has a hinged burner door (Figure 16-20) that permits access to the burner for cleaning and maintenance. The combustion chamber is constructed of stainless steel and surrounded by a seawater cooling jacket that is supplied by the scrubber pump. The pilot and main fuel nozzle as well as the combustion air diffuser is shown in Figure 16-21 and Figure 16-21a.

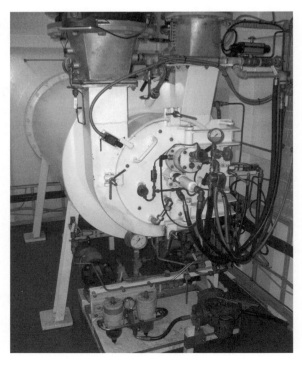

Figure 16-20. IGG combustion unit.

Figure 16-21. Hinged Burner Door.

Figure 16-21a. IGG pilot and main fuel nozzle

Scrubber—the scrubbing tower (Figure 16-22) is the primary processing unit equipped with sea water spray nozzles (Figure 16-23) to cool the gas and a demister (Figure 16-24) to minimize water carryover downstream in the system. The cooling water is supplied by a sea water service pump often referred to as the scrubber pump. The cooling water in the scrubber drains overboard and to protect the system alarms are fitted to warn the operator of fault conditions. Two critical parameters that must be monitored are loss of cooling water supply to the scrubber and high water level in scrubber.

Figure 16-22. Scrubber external view.

Figure 16-23. Scrubber-internal view of the upper half of the tower showing the spray nozzles. Courtesy Knut Kaupang.

Figure 16-24. Demister. Courtesy Knut Kaupang.

Access covers (Figure 16-25)—are provided at the top and side of the scrubber to enable inspection and maintenance of the internal components. From the scrubber the processed gas passes a fixed oxygen analyzer (Figure 16-26) and temperature sensor that permit the operator to monitor the quality of the gas. Based on these readings the gas is either sent to the deck or the surplus gas is vented off. The oxygen set point for the system is generally set at 4% and the temperature of the gas at this point in the system should be close to ambient sea water temperature. The high oxygen alarm is generally set at 6% and the low oxygen alarm sounds at 1%.

Figure 16-25. Scrubbing tower with access covers on the side and top

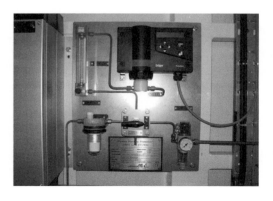

Figure 16-26. Fixed Oxygen Analyzer.

Delivery to Deck—when the IGS ready light illuminates the flow of inert gas to the cargo tanks is controlled via a "deck delivery" valve or main control valve. This valve controls the flow of inert gas to the deck via a pressure controller based on the deck pressure requirement in the tanks. The deck pressure set point for the system is generally 400mmwg, however this value can be changed based on the operation being conducted. The inert gas next passes through a deck water seal (discussed earlier) or a double block and bleed (DBB) arrangement shown in Figure 16-27.

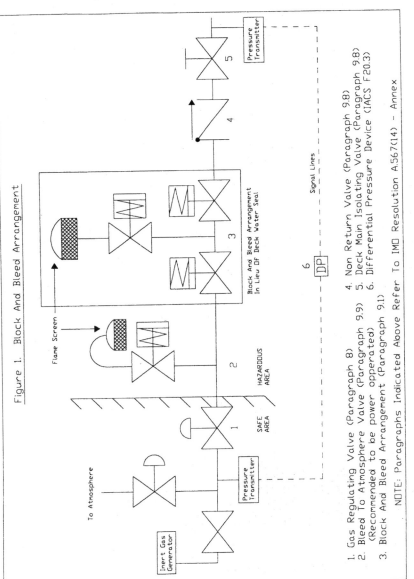

Figure 1. Block And Bleed Arrangement

1. Gas Regulating Valve (Paragraph 8)
2. Bleed To Atmosphere Valve (Paragraph 9.9) (Recommended to be power opperated)
3. Block And Bleed Arrangement (Paragraph 9.1)
4. Non Return Valve (Paragraph 9.8)
5. Deck Main Isolating Valve (Paragraph 9.8)
6. Differential Pressure Device (IACS F20.3)

NOTE: Paragraphs Indicated Above Refer To IMO Resolution A.567(14) – Annex

Figure 16-27. Double block and bleed arrangement. Courtesy OCIMF.

Double Block and Bleed (DBB)—the double block and bleed (Figure 16-27) is an alternative to the deck seal that offers an equivalent level of protection against return of cargo vapors in the system. A DBB normally consists of two butterfly valves in series with a vent line and valve (bleed) to atmosphere between them. It is an approach more commonly seen today on tank barges outfitted with inert gas systems.

At this point, most systems are equipped with a mechanical non return valve and deck isolating valve before entering the inert gas deck main.

Inert gas generators differ from flue gas systems in one important way: the system is under positive pressure while in operation from the combustion chamber to the cargo tanks. As a result they are less prone to air leaks in the system, and they do not recirculate the surplus or off-specification gas to the scrubber as frequently seen in a flue gas system. From the deck isolating valve to the cargo tanks the piping arrangements seen are dictated by the intended trade of the vessel.

NITROGEN SYSTEMS

Nitrogen systems are found on vessels transporting sensitive parcels which require some form of atmosphere control. Inerting of certain parcels on a tank vessel may be required due to concerns about flammability or the contact of sensitive cargoes with oxygen or moisture. These systems are designed to deliver nitrogen of high purity (95 to 99 percent N_2) and dryness to properly protect the vessel and cargo.

One of the more cost-effective methods used to generate nitrogen is via a membrane separator (Figure 16-28), which consists of a cylindrical shell filled with hollow fibers. One or more compressors supply air to the separator, where the oxygen, carbon dioxide, water vapor, and other gases permeate through the walls of the hollow fibers faster than nitrogen. The waste oxygen-rich stream is bled off to atmosphere while the nitrogen stream at the outlet is directed through piping to the tank or tanks protected by the system.

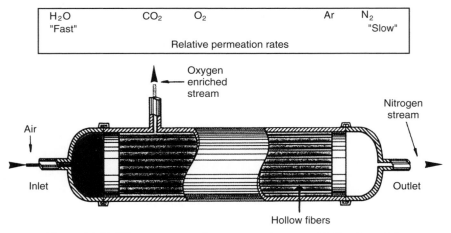

Figure 16-28. Nitrogen generation process. Courtesy Air Products AS

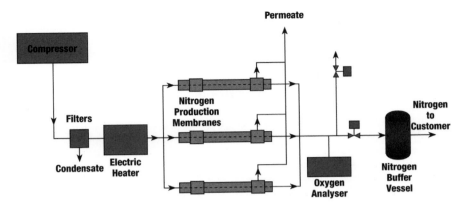

Fig. 16-29 Nitrogen Generator System Components. Courtesy Air Products AS

MAJOR COMPONENTS
(Figure 16-29)

Air Compressor—two or more screw type air compressors are typically fitted which are capable of delivering the required capacity of gas to the protected cargo tanks.

Filter Package—the feed air passes through a series of filters designed to eliminate potentially damaging particles, water and oil.

Heater—the filtered air then passes through a heater that raises the temperature of the air to approximately 50° C thereby reducing the possibility of condensate forming in the membrane bank.

Membrane Bank—one or more banks of hollow fiber membranes separate the nitrogen in the feed air through selective permeation. The oxygen and trace gases in air pass through the walls of the hollow fiber membranes faster than nitrogen resulting in an oxygen rich waste stream.

Flow Control Valve—the purity of the nitrogen stream can be changed by adjusting the flow, temperature and pressure through the membrane system. Under optimum conditions a membrane system is capable of producing nitrogen of 99.9% purity. In practice however, a system designed for inerting of vessel cargo tanks typically has a set point of 5% oxygen which equates to 95% nitrogen purity.

PRECAUTIONS IN THE USE OF IG SYSTEMS

Several decades of operational experience with inert gas systems have identified a number of issues worth reviewing.

Opening a "Closed" System

A properly operated inert gas system requires the maintenance of a closed system to prevent the ingress of air which could alter the atmosphere in the tanks. In the past, it was common practice to open the cargo tanks for cargo surveyors and terminal representatives when performing manual gauging, temperature checks, water cuts, and sampling. It was necessary to vent off the deck pressure prior to opening the cargo tanks to enable personnel to safely conduct these operations. According to IMO it is acceptable to open the tanks for this purpose provided a positive deck pressure is maintained. Reference to the U.S. rules reveals that a minimum deck pressure of 4 inches (100 mm) wg must be maintained in the cargo tanks. Personnel are reminded to exercise extreme caution when opening pressurized cargo tanks. Crewmembers should be trained in the proper procedures to avoid physical injury when it is necessary to open purge pipes, ullage caps, vent covers, tank cleaning covers, and so forth. The need to open inerted cargo tanks was created in some instances by the inaccuracy of the gauging system and installed equipment on the vessel. The reliability and accuracy of the equipment on modern inerted vessels has improved to the point that it is now unnecessary to "open" the tanks. Once an experience factor is established with a particular system and all parties agree to accept the readings of the installed gauging system, the cargo tanks can remain closed. One of the more popular ways to take readings today is to use portable sonic gauging tapes which operate through a standpipe and vapor lock on deck.

Another concern involving an "open" deck on an inerted vessel occurs when water-washing a cargo tank with portable machines. Ideally, a vessel fitted with an IG system should also be equipped with fixed tank cleaning machines. However, many older vessels continue to use portable tank cleaning machines, making it necessary to open the covers on deck. While conducting the wash, the stripping system must be continuously operated; this could possibly create a negative pressure in the cargo tank, resulting in air being drawn in through the deck openings. To avoid the ingress of air during the wash, the inert gas system should be operated and gas flow to the cargo tank should be maintained. This ensures that a positive pressure is maintained in the tank atmosphere; however, it also creates occupational exposure issues for the personnel on deck. Any personnel directly involved in the cleaning operation should be provided with suitable respirators to protect against inhalation of the exiting vapors and inert gas. Owners are addressing this problem in new construction through the installation of fixed washing systems, virtually eliminating the need to open tanks to the atmosphere during the washing operation.

Cargo Segregation

Certain cargoes may be adversely affected by the introduction of water or other impurities such as soot, carbon, and acids carried in the inert gas. These concerns have prompted cargo owners to look toward vessels equipped with inert gas systems that do not pose a risk of contamination to their cargo. In many instances the cargo owner will opt for a vessel fitted with an oil-fired inert gas generator or a nitrogen system. Another issue with multigrade carriers is the possibility of cross contamination via the inert gas piping on deck. Operators must carefully segregate potentially contaminating cargoes by preventing the movement of liquids or vapors via the IG main.

Segregation between cargoes is frequently maintained through the use of valves, blanks, or separate IG piping to the tanks in question. The PIC should be aware of these segregations and carefully verify the correctness of the lineup of the IG system prior to commencing a cargo transfer.

Quality and Quantity of the Gas

The efficient operation of an IG system depends on the quality of the gas from the source. A common operational problem with flue gas systems that can affect the quality and quantity of gas relates to the boiler load. For example, a low boiler load is frequently experienced upon docking, before the cargo pumps are operating. The PIC of the cargo operation may wish to start the inert gas plant and allow it to stabilize before commencing the cargo operation, but the quality and volume of the gas coming from the boiler is poor. To rectify this situation, operators often create an artificial load on the boiler by starting a segregated-ballast pump or by recirculating cargo with a cargo pump. The boiler load can also be a problem toward the end of the discharge during the final stages of stripping the vessel. An insufficient load not only affects the quality of the gas but can result in air being drawn down the stack.

Pyrophoric Oxidation

In vessels carrying crude oils with a significant hydrogen sulfide level (sour), the formation of pyrophoric deposits is a concern. This occurs when iron oxide (rust) in the tanks combines with hydrogen sulfide from the cargo in an oxygen-deficient atmosphere to form iron sulfide deposits. In the normal operation of an inerted vessel, these deposits do not pose a threat unless oxygen is introduced into the tank. If the oxygen level increases, these deposits rapidly oxidize to form iron oxide, sulfur dioxide, and considerable heat. This rapid oxidation can result in individual particles reaching incandescence, creating a potential in-tank source of ignition. With this in mind, operators of crude carriers are cautioned never to compromise the inert status of the vessel. It is imperative that the cargo tanks remain in a nonflammable condition during the operating life of the vessel unless they are gas free.

REVIEW

1. The rules governing the design and operation of inert gas systems were originally published in which IMO convention?
2. What three types of inert gas systems are in common use today?
3. In new construction, why are owners shifting from traditional flue gas systems to oil-fired inert gas generators?
4. Any inert gas system must be capable of supplying a gas with an oxygen content of what value?
5. Operators of an inerted vessel must ensure that the oxygen content of the cargo tanks does not exceed what value?
6. By regulation, what is the minimum deck pressure considered acceptable in an inerted vessel?

7. What is the function of the uptake bellows?
8. List four functions of the scrubber in an inert gas system (flue gas).
9. List two methods employed to remove entrained water from the gas before it reaches the IG fans (flue gas).
10. In a flue gas system, the blowers must be rated to what capacity?
11. In a flue gas system, where is the off-specification or surplus gas detected?
12. What is the purpose of the gas pressure regulating valve (GRV)?
13. What is the purpose of the deck water seal?
14. When the inert gas system is off, must the vent required between the gas pressure regulating valve and deck isolating valve be open or closed?
15. What valves are typically found immediately forward of the deck seal?
16. Define topping up.
17. List the critical fault conditions in an inert gas system that result in an automatic shutdown of the plant.
18. List the reasons for high oxygen readings as sensed on the discharge side of the IG fans.
19. During a crude-oil cargo discharge and crude-oil-washing operation, the inert gas system fails. What action must be taken by the PIC? If the inert gas system cannot be repaired, what options does the PIC have with respect to resuming cargo operations?
20. During an apparently normal discharge operation (inert gas system operating), the deck pressure begins to fall and is approaching the lower acceptable limit. What action should be taken by the PIC?
21. List the conditions that could adversely affect the performance of a liquid-filled pressure-vacuum breaker.
22. Define purging. What reading by tankscope (hydrocarbon analyzer) indicates a successfully purged tank?
23. What are two methods of gas replacement? Describe them in detail.
24. How does a membrane-type nitrogen generator work? List the major components of a nitrogen inerting system.
25. How do pyrophoric iron sulfide deposits form in the cargo tanks on a crude-oil carrier? Why should the operator be concerned about the presence of these deposits?

CHALLENGE QUESTIONS

26. List the items that should be checked when performing an internal inspection of the scrubbing tower.
27. What are the typical settings for the high and low deck pressure alarms on an inerted vessel?
28. What is the typical set point for deck pressure in an IG system?
29. How is it possible to admit air into the cargo tanks of an inerted vessel
30. How does one perform a "bump" test of the fixed oxygen analyzer in the inert gas system?
31. How does one test the alarm functions in an inert gas system during an annual inspection?
32. How does one check the integrity of the non-return devices in the IG system?

CHAPTER 17

Emergency Procedures

This chapter addresses the actions to be taken in various emergency situations that are specific to the cargo area of the vessel. The reader is cautioned that the information contained herein is generic in nature and not necessarily applicable to any particular vessel. Personnel are advised to develop emergency action plans dealing with various casualties that are specific to the vessel and the peculiarities of the operation.

Practical demonstrations and drills following the developed plans should be regularly conducted to familiarize all hands with emergency response. The first priority in any casualty is the safety of the personnel and vessel. The sections that follow address some of the situations that may arise on a tank vessel and list the typical actions to be taken in response.

PUMPROOM RESCUE

If the person in the pumproom is unconscious, the crew should respond in the following way:

1. Sound emergency alarm. If alongside, notify dock to secure all cargo operations and call for medical assistance.
2. Emergency team under the direction of a senior officer responds to the staging area with the following equipment:
   Self-contained breathing apparatus with spare cylinders
   Communications equipment
   Atmosphere-testing equipment
   First aid kit/resuscitator/stretcher/automatic external defibrillator (AED)
   Emergency escape breathing apparatus (EEBD)
   Fire-fighting gear
3. Check pumproom ventilation.
4. Prepare harness and lifeline permanently rigged at the top of the pumproom.
5. Test the atmosphere using a sample tube permanently rigged in the pumproom.
   If the atmosphere is immediately dangerous to life and health (IDLH),
go to step 6.

If there is a physical problem (i.e., heart attack or fall), go to step 12.

6. Rescue personnel properly outfitted with breathing apparatus descend into the pumproom with the harness (lifeline tended from the top of the pumproom).
7. Rescue personnel place unconscious person in harness securely and activate the emergency escape breathing apparatus.
8. On signal from rescuers, personnel outside the pumproom hoist the person out. Use taglines to guide person clear of obstructions such as platforms, valves, and ladders.
9. Once clear of the pumproom, administer first aid to the victim. Check pulse and respiration. Administer CPR if necessary. Consult the MSDS for medical guidance concerning exposure to the specific cargo.
10. Get medical assistance (MedEvac/ambulance) as soon as possible.
11. Determine the cause of the incident and take corrective action.

For physical problem:

12. If atmosphere is safe, rescuers proceed to the victim with the necessary medical equipment and assess the problem. Administer first aid and stabilize the patient for transport.
13. Lower stretcher and board to the patient using the lifeline rigged at the top of the pumproom.
14. Place the patient securely in the stretcher.
15. On signal from the rescuers, personnel outside the pumproom hoist the patient out, using taglines to clear obstructions such as platforms, valves, and ladders.
16. Get medical assistance (MedEvac/ambulance) as soon as possible.

OIL SPILL

In the event that a discharge of oil into the water has occurred during a cargo transfer operation, the PIC should follow this procedure:

1. Stop the transfer operation (emergency shutdown).
2. Secure all valves involved in the operation to limit the extent of the spill.
    If *tank overfill*, take steps to shift the excess cargo to other tanks.
    If *piping failure*, isolate the affected section by immediately closing the appropriate valves.
    For *hull failure (weep)*, identify the location of the leak. Take action to reduce the head pressure in the tank(s) by shifting the cargo to an intact compartment(s) and reducing the inert gas pressure in the space above the cargo. Drop the cargo level in the affected tank below the waterline of the vessel. Verify that the leakage has stopped.
3. Notify vessel personnel, shore facility, and engine room. Restrict access to the area of the spill to essential personnel. Consult the vessel response plan and commence notification to the proper authorities in accordance with the instructions.

4. Eliminate potential sources of ignition and have fire-fighting gear readied.
5. Vessel personnel with proper protective clothing and respirators should commence cleanup of the oil contained on deck. Use the equipment in the spill response locker including portable pumps, shovels, absorbent pads, squeegees, rags, sawdust, brooms, plastic bags, and so forth.
6. Upon notification, shore facility personnel should begin deploying the containment boom (spill barricade) around the vessel (if not previously rigged) to limit the movement of the spill.
7. Consult the company representatives concerning coordination with the cleanup contractor, qualified individual of the company, media, and state and federal authorities as outlined in the vessel response plan.

GROUNDING

In the event of a vessel grounding, the following actions should be taken:

1. Assess the condition of the vessel. Carefully inspect the water around the vessel to determine if any compartments are damaged, allowing cargo to leak or water to flood.
   Check the cargo level (ullages) in each tank. Record the ullages.
   Take soundings of ballast tanks, bunker tanks and compartments that are normally empty (voids/cofferdams).
   Monitor the vessel for any changes in trim or list.
   Maintain positive stability of the vessel.
   Consult the vessel response plan and make the necessary notifications to the appropriate authorities in accordance with the instructions.
   Display the appropriate signal (day/night) for a vessel aground and notify any vessel traffic of the situation.
2. If the vessel is leaking cargo, identify the damaged tanks that are losing cargo.
   Check for changing cargo levels (ullages) in the tanks and dropping deck pressure readings. On deck, personnel should be alert for any signs of a vacuum being created in the tanks (air leaks or PV valves lifting) as well as distortion in the deck plating.
   Eliminate all potential sources of ignition on or near the vessel and have fire-fighting gear readied.
3. If the vessel has a high deck pressure, reduce the inert gas pressure (deck pressure) to a minimum positive pressure in the vessel. Doing so will reduce the outflow of cargo.
4. Isolate the tanks that are leaking cargo by closing *all* valves in the cargo piping system.
5. If it has been determined that it is safe to do so, transfer cargo from the damaged compartment(s) to other intact compartments until the leakage has stopped.
6. Take soundings around the vessel (forward, amidships, aft) to determine where the hull is touching bottom.
7. Determine the range and the present stage of the tide in the locality.
8. Determine the direction and velocity of the tidal currents for the locality.

9. Obtain a weather report that includes wind speed and direction, sea state and swell.
10. Determine the type of bottom around the vessel.
11. Consult with company/salvage experts concerning the most prudent action to take to maintain vessel stability and minimize hull stresses while aground. If necessary, take appropriate action to prevent the vessel from going further aground and sustaining greater damage. Do not attempt to refloat the vessel or move the vessel until the extent of the damage has been determined.
12. The actions taken on the vessel should be appropriate to limit the environmental damage resulting from the grounding while at the same time ensuring the safety of the crew and the vessel.
13. Assist the spill response contractors and salvage people by providing the necessary information concerning the vessel and its condition.

COLLISION

In the event of a collision in which one or more cargo tanks have been breached the following instructions apply:

1. Sound the emergency signal, muster all hands, and account for the entire crew.
2. Fire-fighting gear should be readied, given the heightened risk of a fire on the vessel or on the water alongside.
3. In a collision between two vessels, personnel should immediately assess the extent of the damage to each vessel.
   Check cargo tank levels (ullages) and sound ballast tanks and voids (cofferdams).
   Maintain positive stability.
   Isolate the damaged cargo tanks to minimize the outflow of cargo.
4. Immediately notify the USCG or local national authority as well as the company. Consult the vessel response plan for specific guidance in this situation.
5. Do not attempt to separate the vessels until the condition and stability of each vessel has been assessed. Consult company/salvage experts concerning the most prudent action to be taken to maintain vessel stability, movement of the vessel(s), or anchoring.
6. Monitor the vessel for uncontrolled flooding and any change in trim or list.
7. If there is loss of stability and a threat of sinking, the best action may be to intentionally ground the vessel.
8. Assist fire-fighting and spill response contractors by providing the necessary information concerning the vessel and its condition.
9. Obtain weather forecast, tide, and tidal current information.

FIRE—GENERAL GUIDELINES

By the very nature of the cargo being transported, a tank vessel poses a significant fire risk. Vessel personnel must be ever-vigilant to minimize the possibility of a fire during cargo operations and at sea. Should a fire occur, early detection and prompt action by vessel personnel is essential for the safety of all hands and survival of the vessel. For this reason many companies require the fire main and other fixed extinguishing systems on the vessel to be in a readied status. Regular demonstrations and drills should be conducted to ensure that all personnel are aware of their responsibilities in a fire and are familiar with the location and operation of fire-fighting equipment. In general, the following steps should be addressed when dealing with any fire situation:

1. Sound the alarm and muster.
2. Evaluate the fire.
3. Establish the method of attack (direct or indirect).
4. Get the fire under control.
5. Extinguish the fire.
6. Guard against reignition.
7. Overhaul the fire and investigate the cause.

FIRE—CARGO MANIFOLD

In the event of a fire at the cargo manifold and containment area (trough) on the vessel during connection or removal of the hoses or mechanical arms, personnel should follow these suggestions:

1. Sound the alarm and pass the word "fire." Muster the crew and account for all personnel. Shut down all cargo and ballast operations. Secure all cargo system and vent valves.
2. Notify the engine room and dock. Request the assistance of the shoreside fire department. Fire-fighting systems should be readied.
3. Based on the type of substance (cargo) that is burning, determine the appropriate extinguishing agent(s) to be employed and the method of attack.
4. Fire-fighting teams in full gear should approach the manifold area from upwind using high-velocity fog (water) and low-velocity fog applicators if appropriate.
5. Verify that all cargo valves to the manifold are secured.
6. Cool the manifold piping, containment area, and deck. If water spray is incapable of extinguishing the fire, use the water shield to protect the firefighters approaching the area with suitable (i.e., dry-chemical) portable extinguishers.
7. If the spilled cargo (from the pipelines) has spread the fire over the deck, employ the deck monitors to blanket the area with foam.
8. When the fire is out, continue to cool down the piping and deck in the vicinity of the fire. Maintain a fire watch.

FIRE—MAST OR KING POST VENT

In the event of a vent fire at the top of a mast or king post during loading operations (i.e., due to a lightning strike) the PIC should direct the following actions:

1. Sound the alarm and pass the word "fire." Secure all cargo and ballast operations. Secure all cargo system and vent valves.

2. Notify the engine room and dock. Request the assistance of the shoreside fire department.
3. Muster vessel personnel and account for all crewmembers.
4. Evaluate the situation and decide on the best course of action to attack the fire. These actions may be included:

> Cut off the exiting cargo vapors (fuel) by closing the valve(s) at the base of the vent (mast/king post). Should the fire continue to burn (i.e., because of vent valve leaks), use portable dry-chemical extinguishers from several positions around the vent stack.
>
> Use deck monitors (if they have adequate reach to extend to the top of the vent stack).
>
> Use high-velocity fog from the deck to cool the vent piping in the vicinity of the fire.
>
> Use the inert gas system to prevent the fire from spreading and to aid in extinguishing the fire.

5. When the fire is out, continue to cool down the vent piping with water sprays. Maintain a fire watch in the area.

FIRE—CARGO PUMPROOM

The cargo pumproom is potentially the most hazardous compartment on a tank vessel. The pumproom contains all the necessary elements to start and sustain a fire. Given the confined nature and complexity of this space, fighting such a fire is particularly challenging. Personnel must decide if a direct or indirect attack is an appropriate course of action to deal with such a fire.

1. Sound the alarm and pass the word "fire." Notify the engine room and dock of the situation. Request the assistance of the shoreside fire department.
2. Shut down all cargo and ballast operations. Secure all cargo system and vent valves.
3. Muster the crew and account for all hands. Verify that no personnel are in the pumproom.
4. Fire-fighting teams cool the entrance to the pumproom and secure the pumproom ventilation.
5. In an indirect attack, continue to cool the exposures, seal all doors/vent covers and activate the fixed extinguishing system (carbon dioxide, water spray, or foam) from outside the pumproom. Maintain the pumproom in a sealed condition and monitor the effect on the fire by taking temperature readings over time. Ensure the surrounding compartments are cooled or inerted to eliminate the spread of the fire.
6. In a direct attack, cool the entrance to the pumproom and use low-velocity fog applicators to shield the firefighters. Most likely the fire is located in the bilge of the pumproom; therefore continue to apply water spray and foam from portable extinguishers to knock the fire down through a combination of cooling and smothering.
7. Continue cooling surrounding areas until the fire is out.
8. Maintain a fire watch.

APPENDICES

Conversion Factors

Also on enclosed disc

To convert from:	To:	Multiply by: (Numbers in **boldface type** are exact values.)
barrels	cubic feet	5.614583
barrels	cubic meters	0.15898729
barrels	gallons (U.S.)	**42**
Barrels	Liters	158.98284
bars	psi	14.5
centimeters	inches	0.39370079
centimeters	meters	**0.01**
centimeters	millimeters	**10**
cubic feet	barrels	0.1781076
cubic feet	cubic inches	**1,728**
cubic feet	cubic meters	0.028316847
cubic feet	gallons (U.S.)	7.4805195
cubic inches	cubic centimeters	**16.387064**
cubic meters	barrels	6.289811
cubic meters	cubic centimeters	**1,000,000**
cubic meters	cubic feet	35.314667
cubic meters	cubic inches	61,023.74
cubic meters	gallons (U.S.)	264.17205
cubic meters	liters	999.972
fathoms	feet	**6**
fathoms	meters	**1.8288**
feet	centimeters	**30.48**
feet	fathoms	0.166667
feet	inches	**12**
feet	meters	**0.3048**
gallons (U.S.)	cubic inches	**231**
gallons (U.S.)	liters	3.785306
inches	centimeters	**2.54**

To convert from:	To:	Multiply by: (Numbers in **boldface type** are exact values.)
inches	millimeters	**25.4**
inches of water (4°C)	kg/cm²	0.002539927
inches of water	millimeters of water	**25.4**
inches of water (4°C)	psi	0.03612625
kilograms	pounds (avdp)	2.2046226
kilograms	tons (metric)	**0.001**
kg/cm²	inches of water (4°C)	393.7122
kg/cm²	millimeters of water (4°C)	10,000.28
kg/cm²	psi	14.223343
liters	cubic centimeters	1,000.028
liters	cubic inches	61.02545
liters	cubic meters	0.001000028
liters	gallons (U.S.)	0.2641794
meters	centimeters	**100**
meters	fathoms	0.54680665
meters	feet	3.2808399
meters	inches	39.370079
meters	millimeters	**1,000**
millimeters	inches	0.039370079
millimeters of water (4°C)	kg/cm²	0.00009999709
millimeters of water	inches of water	0.039370079
millimeters of water (4°C)	psi	0.001422293
pounds (avdp)	kilograms	**0.45359237**
psi	inches of water (4°C)	27.6807
psi	kg/cm²	0.070306958
psi	millimeters of water (4°C)	703.089
tons (long)	kilograms	1,016.0469
tons (long)	pounds (avdp)	**2,240**
tons (long)	tons (metric)	1.1060469
tons (metric)	kilograms	**1,000**
tons (metric)	pounds (avdp)	2,204.6226
tons (metric)	tons (long)	0.98420653

APPENDIX G
CargoMax VLCC Software
On enclosed disc

APPENDIX H
USCG Incompatibility Chart
On enclosed disc and also in text Chapter 2

Glossary

ACGIH. American Conference of Governmental and Industrial Hygienists.

acute exposure. Exposure to a toxic substance which causes immediate effects such as breathlessness; irritability; euphoria; irritation to eyes, nose and throat; headaches; dizziness; nausea; and appearance of drunkenness. In the worst case it can lead to convulsions, coma, or death.

Aframax. A system of sizing and freight rate assessment used by the London Tanker Brokers Panel Ltd. known as AFRA (average freight rate assessment). It is used to determine the average cost per ton for various size vessels on a monthly basis. An Aframax-size vessel is in the range of 75,000 to 120,000 dwt, either crude or product carrier.

ANSI. American National Standards Institute.

API. American Petroleum Institute.

API gravity. An arbitrary expression of the weight of a product created by the American Petroleum Institute. API gravity is equal to $141.5/(\text{SG} @ 60° \text{F}) - 131.5°$

API tables. Eleven volumes of tables developed by the American Petroleum Institute to derive the necessary information to perform a cargo calculation.

arrival ballast. Also known as "clean" ballast. The term is traditionally used on crude carriers to describe seawater introduced into cargo tanks that have been crude-oil-washed and water-rinsed. When discharged to the harbor through an oil content monitor, this ballast should not produce a visible sheen. The oil content of such ballast must not exceed 15 ppm.

ASTM. American Society for Testing Materials.

ATB. Articulated Tug and Barge unit

ballast. Seawater introduced into compartments on a tank vessel for the return leg of a voyage to the loading port. It is the additional weight necessary to bring the vessel to a suitable draft and trim and to reduce stresses and improve stability.

Ballast water exchange. An operational technique in which the water in the vessel's ballast tanks is replaced with water from the open ocean where the salinity is above 30 parts per thousand.

barrel (bbl). Standard unit of volume commonly used on U.S. tank vessels. Equivalent to 42 gallons (U.S.).

barrel, gross. The volume of cargo at the observed temperature in the tank.

barrel, net. The volume of cargo corrected to a standard temperature of 60°F.

BCH. Code for the Construction and Equipment of Ships Carrying Dangerous Chemicals in Bulk. These rules apply to chemical tankers constructed before 1 July 1986.

bellmouth. The terminal end of the bottom piping in each tank, through which the tank is loaded and discharged. Its shape is usually flared to approximately 1½ times the original pipe diameter.

bending stress. The stresses resulting from a concentration of weight at a certain location in the hull (uneven load) as well as from the motion of the vessel in a heavy seaway. On a tank vessel bending stress is commonly referred to as a hogging or sagging condition.

benzene. An aromatic hydrocarbon with the composition C_6H_6. It is a regulated cargo (refer to Title 46 CFR Part 197) that requires special handling and safety measures designed to minimize the risk of exposure to vessel personnel.

blank (blind flange). A solid steel disk or plate used to cover and seal the end of a pipeline such as the cargo manifold.

blind flange. See blank.

boiling point. The temperature at which the vapor pressure of a substance is equal to atmospheric pressure.

bonding. The connecting of metal parts to provide electrical continuity.

bonnet. The top housing of a valve typically bolted to the body. This section of a valve usually contains the packing gland, bonnet bushing, and opening for the stem.

bph. Barrels per hour.

bpt. Barrels per ton.

BS&W. Bottom sediment and water.

bullet valve. A high-velocity venting device which is designed to achieve a specified minimum efflux velocity of the atmosphere exiting from the cargo tanks during loading and ballasting operations.

bursting pressure. A pressure rating typically assigned to cargo and vapor hoses.

Butterworth. A brand name of tank cleaning equipment. It is a term commonly used to describe portable tank cleaning equipment and the process of water-washing ("Butterworthing") tanks.

Cam-Locks. One style of quick-connect coupling used to make a secure connection between the vessel and the facility. It consists of several rotating cams and a locking mechanism that holds the two flanges together at the manifold.

cathodic protection. An electrochemical method of preventing corrosion on vessels. This system usually employs sacrificial zinc anodes attached to the surfaces of a tank.

cavitation. A condition in the operation of a pump that occurs when the pressure in the suction line falls below the vapor pressure of the cargo. Vapor pockets are formed in the liquid stream flowing to the pump. The vapor pockets collapse when they reach the high-energy region of the pump, resulting in undue noise and vibration. This condition is frequently caused when cargo vapors, air, or inert gas enter the casing of an operating pump.

CCR. Cargo control room. A space on a tank vessel that contains the necessary equipment to monitor and control a cargo/ballast operation.

centrifugal pump. Kinetic-type pump usually referred to as a main cargo pump. Centrifugal pumps have a high-volume output and require a continuous prime for smooth and efficient operation.

CFR. Code of Federal Regulations (U.S.).

CGI. Combustible-gas indicator. A portable instrument used to detect the presence of explosive gas/air mixtures. It usually measures the concentration of hydrocarbon vapors as a percentage of the lower explosive limit (LEL) or percentage by volume in the space.

chicksan. A mechanical arm constructed of steel used at modern shoreside facilities to connect the shore manifold with a tank vessel's manifold.

CHRIS manual, Chemical Hazards Response Information System from the USCG.

chronic exposure. Long-term exposure to a substance that may cause latent defects to an individual's health, such as liver disorders, chronic skin rashes, blood disorders ranging from anemia to leukemia, and so forth.

clean ballast. Refers to water ballast in a tank which has been so cleaned that effluent from the tank does not produce a visible sheen on the surface of the harbor and the oil content does not exceed 15 ppm.

cleaning chart. A publication that contains written guidance for vessel personnel regarding the proper preparation of a cargo tank and associated equipment prior to loading the next cargo.

clingage. Cargo adhering to the internal surfaces of a tank upon completion of discharge.

cloud point. The temperature at which wax and other solid substances begin to separate from the liquid when the temperature of oil drops under specified conditions.

coatings. Protective paints or linings applied to the surfaces of a tank to extend the service life of the steel. The use of coatings has been found to facilitate cleaning and reduce the quantity of scale in cargo tanks.

cofferdam. A void or empty space used to physically separate the cargo and noncargo areas of a tank vessel.

COI. Certificate of inspection.

commingling. The blending of two or more petroleum products in a cargo tank or pipeline either intentionally or by accident.

compatibility chart. A chart used to determine potential reactivity between different chemical cargoes or materials (See Figure 2-4 and Appendix H on the enclosed disc).

containment boom. A floating barrier that surrounds a vessel during a cargo transfer or bunkering operation. It is designed to limit the movement of a spill should cargo/bunkers be accidentally discharged into the water.

contamination. The presence of unacceptable quantities of solid residues, rust, water, previous cargo, or other contaminant, resulting in a cargo that is off-specification.

continuity test. A test typically performed on portable tank cleaning hoses prior to their use to determine the integrity of the ground wire in the hose. The test measures the resistance (ohms) through the bonding wire in the hose from coupling to coupling.

controlled venting. A method of venting the atmosphere of a cargo tank with the object of minimizing the accumulation of cargo vapors on deck and around the superstructure. Typical methods of venting include mast venting and the use of standpipes equipped with high-velocity vent valves.

COW. Crude-oil-washing. A fixed tank washing system employing the crude-oil cargo as the cleaning medium. This method of cleaning takes advantage of the solvent properties of crude oil to assist in the removal of oil clingage and deposits from the tank.

crossover. Piping used to interconnect different cargo systems or groups.

cubic meter (when used to indicate cargo quantity). Metric unit of volume measurement equal to 6.2898 bbls.

cushion. The process of controlling the initial loading rate of a tank when handling a known static-accumulating cargo. This reduces the splashing and agitation of the cargo, thereby minimizing the development of a significant static charge. This reduced flow to the tank is usually maintained until the bottom framing in the tank is immersed in the liquid.

cycle time. The length of time it takes a tank washing machine to move through all the angles within the tank during a wash. Reference to the manufacturer's manual will give the time necessary for a particular tank cleaning machine.

deck girder. Primary (fore-and-aft) structural framing member of a vessel designed specifically to support the deck plating.

deck pressure alarm. An audible and visual warning system for the vessel operator, designed to activate at predetermined settings (high/low) of pressure and vacuum in the cargo tanks.

deck water seal. A nonmechanical, nonreturn device in the IG system designed to prevent the return flow of cargo vapors into the nonhazardous areas of the vessel.

deep web frame. Primary transverse structural member used to support the shell plating and provide support against side impact.

deepwell (vertical-turbine) pump. Cargo pump designed specifically for installation in the tanks. The pump is located at the bottom of the tank and is connected to a drive unit on deck via a long shaft. Particularly suited for multigrade vessels that require enhanced cargo segregation.

demister. A component of the scrubber in an IG system; designed to remove water droplets from the gas.

demurrage. Compensation from the charterer for time exceeding laytime.

departure ballast. A term used on crude carriers to describe seawater ballast introduced into cargo tanks that have been crude-oil-washed and stripped during the discharge operation.

dilution method. A method of atmosphere replacement utilized in the cargo tanks of an inerted vessel. It is a mixing method requiring high gas velocity to the tank in question.

dip. The term that designates the depth of a liquid in a tank. See innage.

displacement. The weight of water displaced by the hull; it is exactly equal to the weight of the vessel and its contents.

displacement method. A method of atmosphere replacement utilized in the cargo tanks of an inerted tank vessel. Low-velocity entry of gas and minimal turbulence create a layered effect between the incoming and outgoing gases.

DOI. Declaration of inspection. A pretransfer checklist that must be completed by vessel and terminal personnel.

double-hull construction. A method of vessel construction in which the cargo carrying compartments are separated from the sea by an inner and outer hull (two pieces of steel).

Dresser coupling. A slip-on collar used to connect two nonflanged ends of pipe. This type of connection allows for movement of the piping due to thermal variances and vessel stress.

drip pan. A portable container commonly placed under a manifold to collect any leakage from the connections.

DWT. Deadweight tonnage. The amount of cargo, fuel, water, and stores a vessel can carry when fully loaded, expressed in either long tons or metric tons.

earthing. The electrical connection of equipment (i.e., portable gauging equipment) to the hull of the vessel which is at earth potential due to its contact with the sea.

eductor. A jet-type pump commonly used to strip the cargo tanks. An eductor requires a driving (power) fluid that is delivered at high pressure to a small orifice creating a vacuum (venturi) which enables its use as a stripping device.

EEZ. Exclusive economic zone. The area extending 200 miles offshore of the United States.

entry permit. A document issued by a responsible person prior to permitting the entry of personnel into an enclosed space.

epoxy. A special two-part resin or paint (hard coating) used to protect tank surfaces from salt-water corrosion and attack by certain aggressive chemical cargoes.

explosimeter. See combustible-gas indicator.

explosion proof. Electrical equipment is defined as "explosion proof" when it is enclosed in a case that is capable of withstanding an internal explosion of a hydrocarbon vapor/air mixture. The device must also prevent the ignition of a flammable mixture outside the case either from a spark or flame resulting from the internal explosion or from the temperature rise of the case following such an explosion. The equipment must normally operate at such a temperature that a surrounding flammable atmosphere will not be ignited.

explosive range. See flammable range.

fire point. The lowest temperature at which a liquid gives off sufficient vapors to support sustained combustion in the presence of an external source of ignition.

fire wire. Emergency towing wires affixed to the vessel with the eye hanging above the water on the offshore side. They are generally located near the forecastle head and at the quarter to enable towboats to move the vessel quickly in the event of an emergency while docked at a facility.

fixed containment. A permanent trough or coaming under the manifold connections designed to collect any cargo spillage during the connecting and disconnecting of hoses or arms.

flame arrestor. Any device of cellular, tubular, or other construction designed to prevent the passage of flames into an enclosed space.

flame screen. A portable device consisting of fine corrosion-resistant wire mesh designed to prevent sparks or the passage of a flame into a tank. According to regulation, a single screen must be constructed of 30×30 (squares-per-inch) mesh; two screens must have at least 20×20 mesh spaced not less than $\frac{1}{2}$ inch or more than $1\frac{1}{2}$ inches apart.

flammable range. The area between the minimum and maximum concentrations of vapor in air which form a flammable or explosive mixture. Usually abbreviated LEL (lower explosive limit) and UEL (upper explosive limit).

flange. Raised flat end of piping used to connect successive lengths of piping. The manifold is usually flanged to permit connection of the cargo hoses or loading arms.

flash point. The lowest temperature at which a liquid gives off sufficient vapors to form a flammable mixture with air in the presence of an external source of ignition. The vapors will ignite momentarily but are not capable of sustaining combustion.

flue gas system. An inert gas system in which the oxygen-deficient exhaust gas is derived from a marine boiler.

free surface. Refers to the free movement of liquid (such as cargo or ballast water) in a tank which has a negative effect on the stability of a vessel.

FWA. Fresh water allowance. The change in draft (sinkage) that occurs when a vessel moves from a salt to fresh body of water.

FWPCA. Federal Water Pollution Control Act.

gascope. See tankscope.

gas free. A tank or other enclosed space is considered gas-free when sufficient fresh air has been introduced into the space to lower the level of flammable and toxic vapors and increase the oxygen level for a specific purpose.

gas-freeing. The process of ventilating a space with air to prepare the space for various operations such as entry, drydock, hot work, and so on. Before gas-freeing an inerted cargo tank, the operator must ensure the space has been properly purged of hydrocarbon vapors to prevent the creation of a flammable atmosphere.

gasket. A fiber, neoprene, or teflon ring inserted between two flanges to prevent leakage.

gauging, closed. A method of measuring the liquid level in a tank by means of a device which penetrates the tank and does not result in the release of the atmosphere in the space. This device enables the operator to maintain a vapor-tight deck. Examples include float-type, electrical resistance, radar, pressure sensing, and magnetic probe.

gauging, open. A method of measuring the liquid level in a tank which does nothing to minimize or prevent the escape of vapor from the tank. Gauging of the tank is performed through an open hatch (ullage opening).

gauging, restricted. A method of measuring the liquid level in a tank by means of a device that results in a small release of the tank atmosphere. An example of such a device is a manual sonic tape inserted through a standpipe and vapor valve on deck.

GRV. Gas pressure regulating valve. The valve in an IG system that controls the flow of inert gas to the deck.

guide ribs. Grooves or channels in the body of a valve that keep the operating disk in alignment while opening and closing.

HCWM. High-capacity washing machine. Any fixed tank washing machine with a throughput of 60 cubic meters per hour and above.

high jet. A device used to vent the cargo tanks in a controlled fashion on modern tank vessels. It is classified as a high-velocity venting device, which is one that achieves a specified minimum efflux velocity of the atmosphere exiting from a cargo tank during loading and ballasting operations.

hog. Bending stress caused by the uneven distribution of weight on a vessel. A hog condition is created when there is a concentration of weight at the ends of the vessel; the deck is subjected to tension and the keel is under compression. See also sag.

hot work. Any fire-producing action or activity capable of increasing temperature to the point of causing the ignition of flammable vapors. This typically includes work such as welding, burning, soldering, grinding, drilling, blasting and so on.

HVV. High-velocity vent valve. A venting device that results in a high exit velocity of the tank atmosphere, usually exceeding 30 meters per second. Examples include bullet valves and hi-jets.

hydrocarbon. Any compound made up of hydrogen and carbon exclusively.

hydrometer. A device used to measure the specific gravity of a liquid.

hydrostatic tables. A tabular form of the curves derived from the hydrostatic data of the immersed portion of a vessel. Entering the hydrostatic tables with the displacement of the vessel, it is possible to determine a number of key values that are needed to perform a cargo calculation, for example, mean draft, MT1, LCB, and so on.

IBC. International Code for the Construction and Equipment of Ships Carrying Dangerous Chemicals in Bulk. These rules apply to chemical tankers constructed on or after 1 July 1986.

ICS. International Chamber of Shipping.

IDLH. Immediately dangerous to life or health—the concentration of a toxic substance that poses an immediate threat to an individual's life or health.

IMO. International Maritime Organization.

inert condition. An atmosphere is said to be in an inert condition when it is incapable of supporting combustion through oxygen deficiency. It consists of gas or a mixture of gases with an oxygen content of 8 percent or less by volume.

inert gas. A gas or mixture of gases containing insufficient oxygen to support the combustion of hydrocarbons.

inerting, primary. The process of gas replacement in which inert gas is introduced into a space with the object of establishing an inert condition. Primary inerting is typically performed leaving the shipyard.

inhibitor. A substance, generally an additive, that has the net effect of slowing or stopping a chemical change in a self-reactive cargo.

innage. A measurement of the depth of liquid in a tank (the distance from the surface of the cargo to the bottom of the tank).

insulating flange. A flanged joint usually installed between the vessel and the shore facility to prevent electrical continuity through hose strings and loading arms. It consists of a special insulating gasket, bolt sleeves, and washers designed to prevent any metal-to-metal contact across the flange faces.

interface detector. An electrical instrument capable of detecting the boundary between oil and water layers in a tank.

INTERTANKO. The International Association of Independent Tanker Owners.

intrinsically safe. An electrical circuit is considered intrinsically safe if any spark or thermal effect produced in normal operation (i.e., by opening or closing the circuit) or accidentally (i.e., by short circuit or fault) is incapable of igniting a prescribed flammable mixture.

ISGOTT. International Safety Guide for Oil Tankers and Terminals.

ITB. Integrated tug-barge unit. According to the USCG, ITB refers to a tug and tank barge with a mechanical system that allows the connection of the propulsion unit (tug) to the stern of the cargo carrying unit (barge) so that the two vessels function as a single self-propelled vessel.

jumper. Flexible hose used to interconnect (cross over) two cargo systems or groups at the manifold.

laytime. The amount of time allowed for cargo loading and discharge as specified by the charter party.

LCB. Longitudinal center of buoyancy.

LCG. Longitudinal center of gravity.

LEL. Lower explosive limit. The minimum concentration of hydrocarbon vapor in air that forms an ignitable mixture in the presence of an external source of ignition. Below the LEL there is insufficient hydrocarbon vapor (lean mixture) for combustion to occur. Also referred to as the lower flammable limit (LFL).

lightening holes. Holes cut in the framing of a tanker to save weight without sacrificing structural strength.

lightering. The transfer of bulk liquid cargo from the vessel to be lightered (VTBL) to a service vessel. Lightering is usually necessary in areas where draft limitations or local regulations prevent the VTBL from proceeding directly to a dock.

light ship. The weight of the empty ship. (Displacement of a vessel with no cargo, crew, stores, fuel, water, and ballast; usually expressed in long tons or metric tons.)

limber holes. Small openings cut into the framing members of the vessel to allow the free flow of liquids (cargo/ballast) through a space. These openings permit a tank to be thoroughly drained (stripped) at the end of a discharge.

liquid wedge formula. A formula used to determine the quantity of cargo remaining in a tank upon completion of discharge on a vessel with a considerable trim.

loading arm. See chicksan.

longitudinal bulkhead. Fore-and-aft bulkhead running the entire length of the cargo area of the vessel. On a typical single-hull vessel, two of them are used to separate the cargo area into three distinct tanks athwartships: a center tank and a set of wing tanks.

longitudinals. Numerous fore-and-aft framing members in the structure of a tanker.

LOT. Load-on-top. An operational technique devised by the tanker industry to retain oily/water mixtures on board in order to reduce sea pollution.

LOOP. Louisiana Offshore Oil Port–a deepwater offshore port in the Gulf of Mexico

manifold. The vessel/shore connection point on a tanker, usually located amidships. The manifold piping extends athwartships, thereby permitting the vessel to tie up with either side to the dock. Each manifold is equipped with its own valve frequently called a "header."

marine chemist. An individual certificated by the National Fire Protection Association (NFPA) in the United States to conduct the necessary tests and inspections to determine the condition of compartments on a vessel prior to entry (inspection) or repair (hot work).

Marine Loading Arm. (see chicksan)

MARPOL. International Convention for the Prevention of Pollution from Ships, 1973 as amended in the protocol of 1978.

MARPOL line. The special small diameter piping that permits the PIC to bypass the large discharge mains when performing the final stripping of the of the vessel's pumps and pipelines to the shore terminal. This line typically runs from the discharge side of the stripping pump to the outboard side of the manifold valve on deck.

mast riser. A method of controlled venting of cargo tanks in which the atmosphere is piped to one or more masts or king posts. The use of a tall stack directs the cargo vapors aloft, thereby minimizing potential accumulations on deck and around the superstructure during loading and ballasting operations.

MAWP. Maximum allowable working pressure. A pressure rating used for cargo and vapor hoses.

mechanical loading arm. See chicksan.

mechanical seal. A method of preventing the leakage of cargo from the openings in the casing of a pump. Mechanical seals are used where the driveshaft penetrates the casing of the pump.

mixmaster. Fixed piping installed at the vessel's manifold that serves as a crossover between two or more cargo systems or groups.

MPM. Multiple-point mooring.

MT1. Moment to change the trim of a vessel one inch.

mucking. The physical removal of cargo residues, scale, sludge, mud, etc., from a cargo or ballast tank using shovels and buckets. It is a necessary operation to prevent the robbing of cargo space, clogging of limber holes, and contamination of cargo. It is also carried out to prepare the vessel for repair work in a shipyard.

naked lights. Open flames or any other potential source of ignition confined or unconfined.

NFPA. National Fire Protection Association.

NIS. Non indigenous invasive species

NLS. Noxious liquid substance.

NOR. Notice of readiness.

NPSH. Net positive suction head. The minimum energy the liquid must possess at the inlet of a pump for the pump to operate.

OBO. Oil/bulk/ore carrier. Also called a combination carrier. A vessel specially constructed to carry various liquid and solid cargoes in bulk.

OBQ. Onboard quantity of cargo.

OCIMF. Oil Companies International Marine Forum.

odor threshold. The minimum concentration of a gas that can be detected by an average person's sense of smell. Usually expressed in parts per million by volume.

oil. Defined by the USCG as petroleum whether in solid, semisolid, emulsified, or liquid form, including but not limited to crude oil, fuel oil, sludge, oil refuse, oil residue, and refined products and, without limiting the generality of the foregoing, includes the substances listed in Appendix I of Annex I of MARPOL 73/78.

oil cargo residue. Defined by the USCG as any residue of oil cargo whether in solid, semisolid, emulsified, or liquid form from cargo tanks and cargo pumproom bilges, including but not limited to drainages, leakages, exhausted oil, muck, clingage, sludge, bottoms, paraffin (wax), and any constituent component of oil. (The term "oil cargo residue" is also known as "cargo oil residue.")

oily mixture. Defined by the USCG as a mixture, in any form with any oil content, including but not limited to slops from bilges, slops from oil cargoes (such as cargo tank washings, oily waste, and oily refuse), oil residue, and oily ballast water from cargo or fuel oil tanks.

OPA '90. Oil Pollution Act of 1990-the landmark piece of legislation in the United States that had far reaching impact on the tanker industry worldwide.

outage. See ullage.

overfill alarm. An audible and visual warning system that activates when the liquid level in a cargo tank reaches a predetermined point. According to the USCG, the alarm must be set to give the PIC ample warning to permit the shutdown of loading before the tank overflows.

oxygen analyzer/meter. A portable instrument used to determine the percentage of oxygen by volume in the atmosphere of a space.

P&A Manual. Procedures and Arrangement Manual-a manual required by MARPOL Annex II containing detailed information on cargo handling equipment, installed systems and operational procedures concerning the NLS cargoes the vessel is authorized to transport.

packing gland. The area around the stem of a valve in which packing material is wrapped to provide a leak-free opening in the top of the bonnet. The gland is tightened down to squeeze the packing material, thereby preventing leakage from the valve.

Panamax. A tanker between 55,000 and 80,000 dwt (the largest vessel capable of transiting the Panama Canal).

PEL. Permissible exposure limit. The maximum level of exposure to a toxic substance that is allowed by appropriate regulatory authority. The PEL is usually expressed as a time weighted average (TWA)—the airborne concentration of a toxic substance averaged over an 8-hour period, usually expressed in parts per million (ppm). PEL may also be expressed according to a short-term exposure limit (STEL)—the airborne concentration of a toxic substance averaged over any 15-minute period, usually expressed in parts per million (ppm).

petrochemicals. Organic chemicals manufactured from petroleum.

petroleum. A compound consisting of a mixture of hydrocarbons. Crude oil is a naturally occurring petroleum from which other products are derived through the refining process.

petroleum gas. A gas evolved from petroleum. The main constituents of petroleum gas are hydrocarbons, but it may also contain other substances, such as hydrogen sulfide or lead alkyls, as minor constituents.

PIC. Person-in-charge. An individual possessing the proper USCG endorsement to control a cargo transfer involving dangerous liquids or liquefied gases on a tankship or barge.

polymerization. A process whereby the molecules of a particular compound link together to form an extended chain referred to as a polymer. A compound is capable of changing from a free-flowing liquid to a viscous one or even a solid with the attendant release of heat. Polymerization may occur if the compound is exposed to heat or a catalyst, or if an impurity is added, which in some cases can be dangerous.

positive-displacement pump. Commonly referred to as a stripping pump, it is self-priming, making it particularly useful when drying up a tank at the end of a discharge. Examples of positive-displacement pumps include reciprocating pumps and rotary pumps.

pour point. The lowest temperature at which a liquid will remain a fluid.

PPE. Personal protective equipment

pressure surge. A sudden increase in the liquid pressure in a pipeline caused by a change in the velocity of the liquid. This can be brought about by closing a valve against the liquid flow in the pipeline. The velocity of the liquid goes to zero, and the energy within the liquid is converted to pressure.

pump relief valve. An automatic valve (spring-loaded) that protects the system from overpressurization; commonly found on the discharge side of a cargo pump. In the event of overpressurization the valve opens, permitting the cargo to recirculate to the suction side of the pump, thereby preventing any further pressure buildup.

purging. A form of gas replacement in which inert gas is introduced into a tank that is in an inert condition. The object of purging is either to reduce the existing oxygen content and/or to reduce the existing hydrocarbon vapor content to a level where subsequent ventilation with fresh air will not result in the creation of a flammable atmosphere within the space. A tank is considered properly purged when the measured hydrocarbon level is found to be 2 percent or less by volume in the space.

PV breaker. Liquid filled pressure-vacuum breaker. It is a nonmechanical pressure-vacuum relief device commonly found on the IG/vent main designed to back up the mechanical PV valves installed on the tanks. It is a container filled to a prescribed level with a liquid of a certain density (usually an antifreeze mixture) that is set to relieve excess pressure or vacuum in the tanks.

PV valve. A mechanical pressure-vacuum relief device that provides structural protection of the cargo tanks on a tank vessel. It consists of a dual disk valve (pressure-vacuum) utilizing springs or weights which are designed to open at a set pressure or vacuum in the tank. In addition to providing structural protection, these valves assist in maintaining a sealed cargo tank thereby minimizing loss of cargo vapors and inert gas deck pressure.

pyrophoric oxidation. Certain sour crudes give off hydrogen sulfide gas which combines with rust (iron oxide) in an oxygen-deficient atmosphere to form iron sulfide. When oxygen is introduced into the tank the process is reversed. Iron sulfide combines with oxygen to form iron oxide, sulfur oxide, and heat. This heat may be enough to cause an explosion if a flammable atmosphere is present.

reach rod. A steel rod or series of rods used to connect a valve situated some distance from the operator with an operating stand (wheel) typically on deck.

reception facility. Usually refers to a shore tank(s) capable of receiving cargo residues (oil/noxious liquid substances) or slops from a tank vessel. In the absence of suitable tanks or capacity ashore, it can also include the use of tank barges, railroad tank cars, tank trucks, or other mobile facilities.

reciprocating pump. A positive-displacement piston-type pump. Types include simplex and duplex reciprocating pumps. They are particularly suited as stripping pumps due to their self-priming ability.

reducer. A fitting commonly installed on the vessel manifold when the flanges on the vessel differ in dimension from those on the cargo hoses or loading arms.

response curves. Charts provided by the manufacturer of atmosphere testing equipment to enable the user to convert the meter reading of an instrument to the actual concentration of a particular gas in a space.

ROB. Cargo "remaining on board" at the completion of the discharge operation.

rotary pump. A positive-displacement pump which utilizes intermeshing lobes, gears, vanes, or screws to draw suction.

runaround. A temporary (U-shaped) piece of piping used to cross over cargo systems or groups at the manifold.

RVP. Reid vapor pressure. The vapor pressure of a liquid as determined by using a standard Reid apparatus. The test involves measuring the resultant vapor pressure in a closed container when a sample of the liquid is heated to a standard temperature of 100°F (38.8°C). It provides a measure of the volatility of a liquid.

saddle. A fairlead or support used to prevent a hose from kinking and chafing.

sag. Bending stress caused by the uneven distribution of weight on a vessel. A sag condition is created when there is a concentration of weight in the midsection of the vessel: the deck is subjected to compression and the keel is under tension. See also hog.

SBT. Segregated Ballast Tanks

SCBA. Self-contained breathing apparatus.

schedule. Refers to the thickness of the wall of a pipe. Different schedules of pipe are used depending on the intended use of the line (cargo, vapor, hydraulic, water, bunker).

scrubber. The primary processing unit in an inert gas system. Its functions include cooling the gas, removing solids (soot), removing corrosives and physical water droplets from the gas.

scupper/weatherdeck drain. Drain lines leading from the main deck over the vessel's side. These openings must be plugged before cargo operations commence.

sea chest. The name given to the reinforced opening in the hull of a tank vessel through which seawater ballast can be loaded or discharged.

segregated-ballast system. A system on a tanker consisting of tanks, piping, and pumps used exclusively for ballast service. A segregated-ballast system should not have any interconnection to the cargo system of the vessel.

shadow areas. The internal surfaces of a cargo tank shielded from the impinging jets of the tank cleaning machines.

shear force. The result of opposing forces (such as buoyant versus gravitational) acting at a particular location upon the hull of a vessel.

SIRE. Ship Inspection Reporting Exchange-a database operated by the Oil Companies International Marine Forum (OCIMF) that permits member companies to share information concerning the inspection performed on tankers.

slop tank. A tank designated on a vessel for the receipt of tank washings, pipeline flushings, and slops.

SOLAS. International Convention for the Safety of Life at Sea.

sour crude oil. Any crude oil containing a significant amount of sulfur and sulfur compounds (hydrogen sulfide gas) which is characterized by the objectionable odor of rotten eggs.

specific gravity. The ratio of the weight of a given volume of a substance at a standard temperature of 60° Fahrenheit to the weight of an equal volume of freshwater.

SPM. Single-point mooring.

spontaneous combustion. The ignition of a material that occurs by the generation of heat within the material through an internal chemical reaction. The material ignites without the need for an external source of ignition.

spool piece. A short section of flanged piping.

spud wrench. An open-ended wrench with a tapered spike handle used for aligning the bolt holes in two flanges when making a connection.

static accumulator oil. A poor conducting cargo that is capable of developing and retaining a significant electrostatic charge.

static electricity. The electricity produced by dissimilar materials through physical contact and separation such as the flow of cargo through a pipeline.

static nonaccumulator oil. A good conducting cargo that readily gives up any electrostatic charge through the shell plating of the vessel to earth.

STCW. Standards of Training, Certification and Watchkeeping for Seafarers.

STEL. Short-term exposure limit. The maximum concentration of a substance to which workers can be exposed continuously for a short period of time, provided the daily TLV is not also exceeded.

strainer. A fitting installed on the suction side of a cargo pump to prevent any foreign objects or debris from being drawn into the pump and possibly damaging the internals.

stripping. The final draining of the contents of a cargo tank or piping system.

submerged pump. A pump specifically designed for installation in a tank. It is particularly suited for multigrade vessels that require enhanced segregation. The pump and drive unit are located at the bottom of the tank.

Suezmax. A tanker in the range of 120,000 to 200,000 dwt.

swash plate. Framing member installed in a tank to dampen the movement of liquids.

tanker. Any vessel designed to carry liquid cargo in bulk.

tank hatch. The raised coaming of a hatch on deck that serves as the opening for access to a cargo tank.

tankscope. A combustible-gas indicator that is capable of measuring the concentration of hydrocarbon vapors by volume in an inerted space. It is particularly useful when determining the success of the purging operation prior to gas-freeing a tank.

tank top. See tank hatch.

TAPS. Trans-Alaska Pipeline System.

thievage (water cuts). The measurement of free water in a cargo tank through the use of an instrument or water-finding paste on a bob.

TLV-TWA. Threshold limit value–time weighted average. The maximum airborne concentration of a substance to which it is believed that nearly all workers may be repeatedly exposed for a normal 8-hour workday or 40-hour workweek, without adverse effect. See also permissible exposure limits.

ton, long. A unit of weight: 2,240 pounds (1,016 kg).

ton, metric. A unit of weight: 2,204 pounds (1,000 kgs).

ton, short. A unit of weight: 2,000 pounds (907 kg).

topping off. The process of completing the loading of a cargo tank to a specified ullage. It is considered a critical operation in the transfer of cargo given the heightened potential for a spill. Under U.S. regulations, any critical operation requires the direct supervision of the person-in-charge (PIC).

topping up. The introduction of inert gas into a tank already in the inert condition with the object of raising the tank pressure to prevent the ingress of air.

toxic. Poisonous to human life.

toxicity test. A measurement of the concentration of a known or suspected health-threatening substance in the atmosphere of a space. It is usually a gas-specific test that is measured in parts per million (ppm).

TPI. Tons per inch immersion.

transverse bulkhead. A solid bulkhead running in the athwartship direction usually separating cargo tanks.

trim arm. The numerical difference between LCB and LCG.

trough. A fixed containment area (coaming equipped with gratings) permanently installed under the cargo manifolds of the vessel, designed to collect any spillage that occurs when connecting/disconnecting hoses or arms.

TWIC. Transportation Worker Identification Card

UEL. Upper explosive limit. The maximum concentration of hydrocarbon vapor in air that forms an ignitable mixture in the presence of an external source of ignition. Above the UEL the concentration of hydrocarbon vapors is too great (rich mixture) to support combustion. It is also known as the upper flammable limit (UFL).

ULCC. Ultra large crude carrier.

ullage. The measurement of free space above the liquid in a tank. It is the distance from the surface of the liquid in the tank to a reference datum on deck, normally the rim of the ullage opening in the hatch. Reference to the ullage/calibration tables for the vessel will give the volume of liquid in the tank (gross volume).

ullage opening. Small opening in the tank hatch on deck that serves as the reference point for measurement of liquid (cargo) in the tank.

vapor. A gas below its critical temperature.

vapor control system. Piping and equipment on a tank vessel that is necessary to control certain cargo vapor emissions during loading and ballasting operations. A typical installation includes vessel/shore vapor collection piping, monitoring equipment, control devices, and vapor processing units ashore.

VCF. Volume correction factor. A multiplier derived from the API tables to convert from the gross volume in the tanks to a standard (net) volume.

VEF. Vessel experience factor. A historical compilation of ship-to-shore cargo volume differences used as a loss control tool to rectify quantity discrepancies with shore tank measurements.

vetting. A term used to describe the inspection of a tank vessel under consideration for hire by a charterer.

viscosity. A measure of a liquid's internal resistance to flow. It is useful when determining the pumpability of the cargo as well as the need to heat it.

VLCC. Very large crude carrier.

VOC. Volatile organic compound.

volatility. The tendency of a liquid to vaporize.

volute. The area within the casing of a centrifugal pump where energy in the liquid is converted from high velocity to a combination of velocity and discharge pressure.

vortexing. The creation of an eddy or whirlpool around the suction point (bellmouth) at the stripping stage in the discharge of a tank. Vortexing action can result in the admission of atmosphere into the suction line reaching the pump.

VRP. Vessel response plan.

wedge, liquid. The cargo remaining in the tank after stripping. Due to the trim of the vessel, the remaining cargo appears as a wedge of liquid against the after bulkhead of the cargo tank.

well/sump. A recessed area below the bottom of a cargo tank (inner bottom) which houses the bellmouth or pump suction.

Yokohama fender. Large fenders commonly placed between two vessels during lightering operations to prevent the hulls from making contact.

zinc anode. A sacrificial metal used in ballast tanks to minimize corrosion of the steel of the vessel.

zinc silicate. Paint loaded with a high percentage of zinc, used to protect the steel plating of a tank from corrosive attack.

Bibliography

"A Guide to the Vetting Process," *Intertanko*, 7ᵗʰ Edition, October 2007.

Andersen, Steinar. "Inexpensive Bulk Nitrogen Production." *Shipbuilding Technology International*, 1989: 157–158.

Angelo, Joseph J. "A Status Report of Coast Guard Implementation of OPA Double Hull Related Requirements." Presented at *Marine Log's* Tanker Legislation 1991 Conference. Washington, D.C., September 24, 1991.

Baptist, C. *Tanker Handbook for Deck Officers*. Glasgow, Scotland: Brown, Son, & Ferguson Ltd., 1991.

Beaver, Earl R. "Permea Gas Separation Membranes Developed into a Commercial Reality." Paper delivered to Seventh Annual Membrane Technology Planning Conference. Cambridge, Md., October 1989.

Berry, M. G. *Operation and Maintenance of Inert Gas and Crude Washing Systems*. Maidstone, England: Interlink Inert Gas Ltd., 1981.

Blenkey, Nicholas. "Living with the Law." *Marine Log*, February 1981: 34–37.

———. "Making Existing Tankers Safer." *Marine Log*, April 1991: 3.

Cameron, Douglas B. "Human Error Times Two." *Proceedings of the* Marine Safety Council, October-November-December 1989: 132–135.

CG-174, A Manual for the Safe Handling of Inflammable and Combustible Liquids and Other Hazardous Products. Washington, D.C.: U.S. Coast Guard, 1975.

CG-446, A Condensed Guide to Chemical Hazards. Washington, D.C.: U.S. Coast Guard, 1985.

Chemical Data Guide for Bulk Shipment by Water (former *CG-388*). Washington, D.C.: U.S. Coast Guard, 1994.

Chemical Hazards Response Information System (CHRIS). Washington, D.C.: U.S. Coast Guard, 1999.

Chemical Tanker Familiarization Model Course 1.03. London: International Maritime Organization, 1991.

Chevron Safety Bulletin. San Francisco, Calif.: Chevron Shipping Company, 1989–1999.

Code for the Construction and Equipment of Ships Carrying Dangerous Chemicals in Bulk (BCH Code). London: International Maritime Organization, 1993.

"Congress Dictates Double Hulls." *Marine Log*, October 1990: 39.

Controlling Hydrocarbon Emissions from Tank Vessel Loading (prepared by the Marine Board of the National Research Council). Washington, D.C.: National Academy Press, 1987.

Crude Oil Washing Systems. London: International Maritime Organization, 1983.

"Crude Tanker Pollution Abatement" (Exxon position paper). Houston: Exxon Corporation, 1976.

Crude Washing of Tankers. Gothenburg, Sweden: Salen & Wicander AB, 1976.

Double Hull Tank Vessels: A Review of Current Regulatory Efforts, Design Considerations, and Related Topics of Interest. Paramus, N.J.: American Bureau of Shipping, 1991.

"Federal Oil Pollution Act of 1990." Summary of provisions prepared by Chevron Shipping Company, San Francisco, 1990.

Fighting Pollution: Preventing Pollution at Sea. London: Witherby & Co., 1991.

Fitch, Robert, and Gordon Marsh. "Coast Guard Requirements for Marine Vapor Control Systems." *Marine Technology,* September 1991: 270–275.

Flynn, Robert. "The Impact of OPA 90 and State Legislation on the Tanker Markets: What's Happened So Far?" Presented at *Marine Log's* Tanker Legislation 1991 Conference. Washington, D.C., September 24, 1991.

Fundamentals of Petroleum. Austin: Petroleum Extension Service, University of Texas, 1982.

Gardner, A. Ward, and R. C. Page. *Petroleum Tankship Safety.* Luton, England: Lorne & Maclean Marine Publishers, 1971.

Guide for Cargo Vapor Emission Control Systems on Board Tank Vessels. Paramus, N.J.: American Bureau of Shipping, 1990.

"Guidelines for Marine Cargo Inspection." *Manual of Petroleum Measurement Standards,* chapter 17, section 1. Washington, D.C.: American Petroleum Institute, 1986.

Hodgson, Brian. "Alaska's Big Spill—Can the Wilderness Heal?" *National Geographic,* January 1990: 4–43.

Horrocks, J. C. S. "U.S. Oil Pollution Act of 1990." Letter to International Chamber of Shipping members. London: International Chamber of Shipping, 1991.

Howden Inert Gas System Operating and Maintenance Manual, volume 1. Hounslow, England: Howden Engineering Ltd., 1983.

Index of Dangerous Chemicals Carried in Bulk. London: International Maritime Organization, 1990.

Inert Flue Gas Safety Guide (prepared by the International Chamber of Shipping and the Oil Companies International Marine Forum). London: Witherby & Co., 1978.

Inert Gas/Crude Oil Washing Syllabus. Linthicum Heights, Md.: Maritime Institute of Technology and Graduate Studies, 1991.

Inert Gas Systems. London: International Maritime Organization, 1990.

"Inspection of Inert Gas Systems." *Marine Safety Manual,* chapter 15. Washington, D.C.: U.S. Coast Guard, 1990.

International Code for the Construction and Equipment of Ships Carrying Dangerous Chemicals in Bulk (IBC Code). London: International Maritime Organization, 1994.

International Safety Guide for Oil Tankers and Terminals (ISGOTT), 5th Edition. Prepared by the International Chamber of Shipping, Oil Companies International Marine Forum, and the International Association of Ports and Harbors. London: Witherby & Co.

Jimenez, Richard. "The Evolution of the Load Line." *Surveyor,* May 1976: 7–11.

Ketchum, Donald E. *A Failure Modes and Effects Analysis of Vapor Collection Systems.* San Antonio, Tex.: Southwest Research Institute, 1988.

King, G. A. B. *Tanker Practice: The Construction, Operation, and Maintenance of Tankers.* London: Stanford Maritime Ltd., 1971.

"Marine Vapor Control Systems: Final Rule." *Federal Register,* June 21, 1990.

MARPOL 73/78 (Consolidated Edition): London: International Maritime Organization, 1991.

"Measurement of Cargoes on Board Tank Vessels." *Manual of Petroleum Measurement Standards,* chapter 17, section 2. Washington, D.C.: American Petroleum Institute, 1990.

"Mid-deck Tanker Can't Spill." *Marine Log,* January 1991: 40.

OCIMF Tanker Management and Self Assessment, A Best-Practice Guide for Vessel Operators, Second Edition, 2008.

Our Petroleum Industry. London: British Petroleum, 1977.

Pendexter, L. A., and W. G. Coulter. *Classification Society Overview for Construction, Inspection, and Repair of OPA 90 Double Hull Tankers.* Paramus, N.J.: American Bureau of Shipping, 1991.

Pocket Guide to Chemical Hazards. Washington, D.C.: U.S. Department of Health and Human Services, June 1994.

"Rebuilding the Exxon Valdez." *Marine Log,* October 1990: 36–38.

Rutherford, D. *Tanker Cargo Handling.* London: Charles Griffin & Company Ltd., 1980.

Saab MaC/501 Tanker Monitoring and Control System, Technical Description. Gothenburg, Sweden: Saab Marine Electronics, 1991.

Standard for the Control of Gas Hazards on Vessels (NFPA No. 306). Boston, Ma: National Fire Protection Association, 1997.

Tanker Cleaning Manual. San Francisco: Gamlen Chemical Company, 1976.

Tanker Safety Course Notes. Kings Point, New York: Global Maritime and Transportation School at United States Merchant Marine Academy, 1999.

Tanker Spills: Prevention by Design (prepared by the Marine Board of the National Research Council). Washington, D.C.: National Academy Press, 1991.

Technical Description, Saab TankRadar. Gothenburg, Sweden: Saab Marine Electronics, 1991.

The Unseen Menace: A Pocket Guide to the Atmospheric Hazards of Confined Spaces. Pittsburgh: Bacharach, Inc., 1989.

Weller, G. Alex. "Learning to Live with OPA 1990." *Marine Log,* March 1991: 36–41.

Index

About the Author and Contributors

Mark Huber is a graduate of the State University of New York Maritime College at Fort Schuyler, Bronx, New York. He received a master's degree in environmental studies from Long Island University. He has worked extensively in the tanker industry in a seagoing capacity with Gulf Oil Corporation, Military Sealift Command, and Keystone Shipping Company, and as a consultant to tanker companies and legal firms. He holds an unlimited master's license and is endorsed as a Tankerman PIC (DL) Dangerous Liquids. A member of the Nautical Institute, Huber is currently a professor in the Department of Marine Transportation at the United States Merchant Marine Academy in Kings Point, New York.

Richard Beadon, master mariner, is currently the director of the Seamen's Church Institute Center for Maritime Studies in New York, which offers courses in marine operations, including those with a focus on tankships. While at the United States Merchant Marine Academy at Kings Point, he served as senior ship simulation consultant at the Computer-Aided Operations and Research Facility and also as deputy director of the department of continuing education.

Scott R. Bergeron, a graduate of the U.S. Merchant Marine Academy, is currently the chief operating officer for the Liberian International Ship and Corporate Registry, the world's second largest ship registry. He is a member of the Chemical Transportation Advisory Committee, which provides advice and consultation to the U.S. Coast Guard on water transportation of hazardous materials in bulk.

Kelly Curtin serves as Division Manager for Nautical Science Programs at the Global Maritime and Transportation School (GMATS) located at the United States Merchant Marine Academy. Prior to GMATS he taught at the State University of New York Maritime College as an Assistant Professor of Marine Transportation and Senior Deck Training Officer aboard the Training Ship Empire State: He has sailed as a deck officer and person-in-charge on oil, product, and chemical tankers. Kelly holds a bachelor degree in Business Administration from the University of Southern California, a Masters in Transportation Management from the State University of New York Maritime College, and an Unlimited Master Mariner license.

Margaret Kaigh Doyle has over twenty-five years of experience in maritime consulting specializing in the tanker sector. She is currently employed as the General Manager of the Marine Response Alliance, Marine Response Alliance, LLC (MRA), a US based emergency response partnership comprised of Crowley Marine Services, Marine Pollution Control, Titan Salvage, and Marine Hazard Response. Ms. Doyle is best known for her accomplishments as Chemicals Manager for the International Association of Independent Tanker Owners (INTERTANKO) as well as serving as Executive Director of the Chemical Carriers' Association (CCA) for over a decade. Throughout her tenure with INTERTANKO and CCA, Ms. Doyle represented 85 percent of the worldwide chemical tanker fleet at the international, federal and state level. Margaret Doyle is considered an industry expert on the chemical tanker industry, she is an appointed member of the US Coast Guard's Chemical Transportation Advisory Committee, a position she has held since 1995. She has been a guest lecturer on the industry in many forums including the continuing education programs at USMMA, Mass Maritime Academy and Maine Maritime Academy. Ms Doyle has a Bachelor of Science from the US Merchant Marine Academy at Kings Point, a Masters in Engineering Management from the

George Washington University and a Master of Engineering in Environmental Pollution Control from the Pennsylvania State University.

Kevin Duschenchuk, a graduate of the State University of New York Maritime College at Fort Schuyler, Bronx, New York, is currently sailing as Chief Mate with Polar Tankers Inc., a subsidiary of ConocoPhillips. He is also an instructor for GMATS at the United States Merchant Marine Academy. He holds an unlimited master's license, and has served aboard product, chemical, and crude oil tankers, as well as working in the tug industry in New York harbor.

John O'Connor is a graduate of the State University of New York Maritime College at Fort Schuyler, Bronx, New York. He holds a U.S. Coast Guard unlimited tonnage license as chief officer and has served on numerous vessels including product, chemical, and crude-oil tankers. A member of API's Committee on Measurement Accountability, he has been active in the field of petroleum measurement for over fifteen years. He is currently president of International Marine Consultants.

Figure 17-00. Ocean Reliance Dusk. Courtesy Crowley Maritime Corporation.

US $50.00

9 780870 336201   55000

ISBN: 978-0-87033-620-1